D0643626

Methodology of Longitudinal Surveys

WILEY SERIES IN SURVEY METHODOLOGY

Established in Part by Walter A. Shewhart and Samuel S. Wilks

Editors: *Robert M. Groves, Graham Kalton, J. N. K. Rao, Norbert Schwarz, Christopher Skinner*

A complete list of the titles in this series appears at the end of this volume.

Methodology of Longitudinal Surveys

Peter Lynn
Dept of Survey Methodology,
Institute for Social and Economic Research,
University of Essex, UK.

WILEY
A John Wiley and Sons, Ltd, Publication

This edition first published 2009

Registered office
John Wiley & Sons Ltd, The Atrium, Southern Gate, Chichester, West Sussex,
PO19 8SQ, United Kingdom

For details of our global editorial offices, for customer services and for information about how to apply for permission to reuse the copyright material in this book please see our website at www.wiley.com.

A catalogue record for this book is available from the British Library.

ISBN: 978-0-470-01871-2 (H/B)

Set in 10/12pt Times by Integra Software Services Pvt. Ltd, Pondicherry, India
Printed in the UK by CPI Antony Rowe, Chippenham, Wilts.

Contents

Preface

Longitudinal surveys have become highly valued by researchers and policy makers for their ability to provide insights that cannot be obtained by any other means. The contribution of longitudinal surveys to social justice agendas, for example, is now well established. In this era when the realities of anthropogenic climate change have finally achieved broad acknowledgement, study of the determinants and impacts of individual-level behavioural change has become of crucial significance to our future survival. Longitudinal surveys should come into their own in this and other arenas.

While the successful implementation of a longitudinal survey can be extremely rewarding, there are considerable complexities involved in designing and carrying out a longitudinal survey over and above those that apply to other surveys. Survey methodologists have been studying these complexities and developing better methods for years, but the results are scattered about throughout the general survey methods, statistics and indeed social science literature. A notable exception – a book devoted to methods for longitudinal surveys – is the Wiley book *Panel Surveys*, which was edited by Daniel Kasprzyk, Greg Duncan, Graham Kalton and the late M. P. Singh and published in 1989. That book contained monograph papers presented at the International Symposium on Panel Surveys held in November 1986 in Washington, DC. While the volume has remained helpful to survey researchers, there have been important changes and developments in the field in the past two decades.

In 2004 I proposed the organisation of a conference that would in some sense be a successor to the 1986 Symposium. The idea was that the conference would bring together researchers interested in the methodology of longitudinal surveys from around the world and that a series of invited monograph papers would form the edited volume that you now have in front of you. The proposal was considered by the Steering Group of the UK Longitudinal Studies Centre (ULSC), a centre located at the Institute for Social and Economic Research (ISER), University of Essex. The remit of the ULSC includes the advancement of methodology and the promotion of best practice for longitudinal surveys, so the conference and book seemed a good fit. The Steering group agreed and the decision was taken to adopt the conference and the book as a project under the auspices of the ULSC.

The next steps were to assemble both a scientific committee to pull together the programme for the conference and a local organising committee to make all the arrangements necessary for a successful international conference. The scientific committee consisted of the following people: Roeland Beerten (Office for National Statistics, UK), Paul Biemer (Research Triangle Institute and University of North Carolina at Chapel Hill, USA),

Nick Buck (ISER, University of Essex, UK), Marco Francesconi (Dept of Economics, University of Essex, UK), Stephen Jenkins (ISER, University of Essex, UK), Graham Kalton (Westat, USA), Femke De Keulenaer (University of Antwerp, Belgium – now Gallup Europe, Belgium), Peter Lynn (Chair: ISER, University of Essex, UK), Gad Nathan (Hebrew University of Jerusalem, Israel), Cheti Nicoletti (ISER, University of Essex, UK), Randy Olsen (Ohio State University, USA), Ian Plewis (Institute of Education, University of London, UK – now University of Manchester, UK), Ulrich Rendtel (Freie Universität Berlin, Germany), Joerg-Peter Schraepler (DIW Berlin, Germany) and Mark Wooden (University of Melbourne, Australia). I am grateful to all of them for their role in screening submitted abstracts and shaping the programmes of both monograph and contributed papers for what was to become known as the *International Conference on the Methodology of Longitudinal Surveys*, or *MOLS 2006* for short.

The conference took place on the campus of the University of Essex in July 2006 and attracted over 200 delegates from 26 countries, with 85 papers presented. The local organising committee consisted of five members of the ISER: Randy Banks, Ann Farncombe, Peter Lynn (Chair), Emanuela Sala and Janice Webb. Randy was responsible for creating and maintaining the conference website, which was well-received by participants and continues to provide a valuable archive of presented papers (www.iser.essex.ac.uk/ulsc/mols2006). Janice took charge of bookings, accommodation arrangements, meals and a host more besides. She will be well known to many of the conference participants, having helped many of them with tasks as diverse as travel arrangements, payments, dietary requirements and finding their way across the University of Essex campus. Janice's professionalism and dedication to the job was instrumental to the success of the conference.

The role of these ISER staff in organising and administering the conference was made possible by the support of the ULSC. The ULSC is funded by the Economic and Social Research Council with support from the University of Essex and I am deeply grateful to both organisations for ultimately making this whole thing possible.

Gratitude is also due to many others: the University of Essex Conferences Office for their efficient role in the local arrangements on campus; the then Vice-chancellor of the University, Professor Sir Ivor Crewe, who generously hosted a welcome reception in a marquee in the campus grounds; the deputy mayor of Colchester, councillor Ray Gamble, who hosted a civic reception in Colchester Town Hall and attended the conference dinner in Colchester's historic Moot Hall; and Colchester Borough Council Museums Service who organised guided tours of Colchester Castle Museum for conference participants.

On the day before the conference commenced, two associated short courses were held, each of which was attended by more than 40 participants. The courses were 'Handling incomplete data in longitudinal surveys', presented by Joop Hox and Edith de Leeuw, and 'Multilevel modelling for longitudinal survey data', presented by Sophia Rabe-Hesketh and Anders Skrondal. The courses benefited from funding by the ESRC National Centre for Research Methods (www.ncrm.ac.uk).

With the conference over, attention turned to producing this monograph volume. The strength of the book lies in the quality of the chapters, which in turn reflects the experience, knowledge and energy of the chapter authors. I am grateful to the thirty or so researchers around the world who reviewed the draft chapters. All the chapters were subsequently presented in monograph sessions at *MOLS 2006* and further benefited from the insightful comments of knowledgeable discussants. The Editor's role has been to mould the chapters into a coherent whole, a task made easier by the responsiveness and cooperation of the authors and, above all, by the professionalism of the staff at Wiley, particularly Kathryn Sharples, the commissioning editor who showed faith in the project, Susan Barclay, project editor, who showed commendable patience and restraint in her regular reminders regarding my tardy delivery, and Beth Dufour, content editor, who skillfully guided the book through its final stages. Any weaknesses of the book remain solely the responsibility of the Editor. In addition to supporting the production of the book, Wiley sponsored a prize for best student paper at *MOLS 2006*. This was awarded to Mario Callegaro of the University of Nebraska-Lincoln (now at Knowledge Networks Inc.), with runner-up Hari Lohano of the University of Bath, UK.

The task of editing this book inevitably impinged considerably on family life and I wish to acknowledge the forbearance of Elisabeth, Henry and Adrian who had too often to put up with me being inseparable from my laptop.

The result of this effort I hope is worthwhile. The book covers the range of issues involved in designing, carrying out and analysing a longitudinal survey. In discussing ways to improve longitudinal surveys, chapters draw upon theories and ideas from psychology, sociology, statistics and econometrics. As well as this multidisciplinarity, an international flavour can be detected in the volume. Though the book's UK roots may be apparent, the branches reach out to provide glimpses of survey practice in Canada, USA, Australia, Germany and The Netherlands. The community of survey methodology researchers is truly global and the community specialising in methods for longitudinal surveys is pretty small in most countries, making international communication and collaboration particularly valuable. Such collaboration is now continuing through the biennial *International Workshop on Panel Survey Methods*, which was in some sense spawned by *MOLS 2006*. These workshops are relatively small, informal gatherings of like-minded methods researchers willing to share and discuss research. The first took place at the University of Essex in July 2008 (www.iser.essex.ac.uk/seminars/occasional/psmw2008) and the second is planned for Spring 2010 in Mannheim, Germany. I hope that the friendly international collaboration and spirit of common purpose that I have experienced in putting together this book and in the conference and workshops will continue in future research and publication ventures.

This volume aims to be of particular value to researchers involved in commissioning, designing or carrying out longitudinal surveys, as well as to those planning the analysis of longitudinal survey data. It may also serve as a useful secondary text for students of research methods. I hope that it might additionally help to chart out the known unknowns and hence to stimulate further methodological research.

Peter Lynn
Colchester, October 2008

CHAPTER 1

Methods for Longitudinal Surveys

Peter Lynn
University of Essex, UK

1.1 INTRODUCTION

A *longitudinal survey* is one that collects data from the same sample elements on multiple occasions over time. Such surveys are carried out in a wide variety of contexts and for a wide variety of purposes, but in many situations they have considerable analytical advantages over one-time, or *cross-sectional*, surveys. These advantages have been increasingly recognised and appreciated in recent years, with the result that the number of longitudinal surveys carried out has multiplied. This growth of interest has occurred in the government, academic and private sectors.

For example, in the UK the Economic and Social Research Council (ESRC) has invested heavily in enhancing the country's already impressive portfolio of academic longitudinal surveys. The three existing long-term birth cohort studies that began life in 1946, 1958 and 1970 (Ferri *et al.*, 2003) have been joined by a fourth, the *Millenium Cohort Study* (Dex and Joshi, 2005), which started collecting data in 2001 – and may soon be joined by a fifth. And an ambitious new household panel survey, *Understanding Society: the UK Household Longitudinal Study*, was commissioned in 2007 and begins data collection in January 2009 (www.understandingsociety.org.uk). With a target sample size of 40 000 households and a large boost of ethnic minority households this survey represents the largest single investment in academic social research resources ever launched in the UK. Meanwhile, the UK government has also initiated major new longitudinal surveys. The *Longitudinal Study of Young People in England (LSYPE)* started in 2004, involving annual interviews with a sample of over 15 000 young people born in 1989–1990 and their parents (http://lsype.notlong.com). The *Wealth and Assets Survey* entered the field in 2006 and plans to have biennial waves with a sample of over 30 000 persons. The *Offending, Crime and Justice Survey* (OCJS) began in 2003, with the main component being a longitudinal survey of young people aged 10–25, with annual interviews (Roe and Ashe, 2008).

Methodology of Longitudinal Surveys P. Lynn

This expansion in longitudinal surveys has also been visible in the International Arena. The *Survey of Health, Ageing and Retirement in Europe* (SHARE) began in 2004 with a first wave of data collection with a sample of over 30 000 individuals aged 50 or over in 11 countries (Börsch-Supan and Jürges, 2005). Further waves took place in 2006 and 2008. The *European Union Statistics on Income and Living Conditions* (EU-SILC) is a panel survey of households in all countries of the EU that began in 2004 and features a four-wave sample rotation pattern with annual interviews (http://eu-silc.notlong.com). The international suite of broadly comparable household panel surveys, which for years comprised only the UK, USA, Germany and Canada, has more recently been joined by Switzerland (1999), Australia (2001), New Zealand (2002) and South Africa (2008).

Reflecting this burgeoning interest in longitudinal surveys, and recognising that such surveys have distinct methodological features, this book aims to review the current state of the art in the design and implementation of longitudinal surveys. All aspects of the survey process are considered, from sample design through to data analysis and including overarching issues such as ethics and data linkage. This chapter aims to provide an overview of the features of longitudinal surveys that are distinct from other surveys and a nomenclature and framework within which to consider the other chapters.

1.2 TYPES OF LONGITUDINAL SURVEYS

The definition of a longitudinal survey presented in the first sentence of this chapter encompasses a multitude of different survey designs that are used in practice. Longitudinal surveys can vary on a number of key dimensions that tend to have implications for various aspects of methodology. These include features of the survey objectives, such as the study topics and the study population of interest (Binder, 1998), and features of the design such as the interval between waves, the mode of data collection, the treatment of new entrants to the population, and so on (Kalton and Citro, 1993). Some of the different types of longitudinal surveys are outlined below. This is not intended as a comprehensive typology, but merely to give a flavour of the variety that exists:

- *Surveys of businesses carried out by national or regional statistical offices.* These surveys tend to collect a limited range of information, typically restricted to key economic indicators. Data may be collected at frequent intervals, such as monthly or quarterly, and the main objectives are usually to publish regular series of statistics on totals, means and net change between periods, often for cross-cutting domains such as regions and industries. The rationale for a panel design is often to improve the precision of estimates of net change rather than an interest in estimates of gross change or any other micro-level longitudinal measures;
- *Surveys of school-leavers, graduates or trainees.* Institutions offering education or training, such as universities, or government agencies responsible for related policy, often wish to assess the outcomes of such education or training at a micro (student) level. These outcomes are often medium term or long term and consequently it is necessary to keep in touch with students/trainees for some time after they have completed their study. Longitudinal surveys are often used for this purpose, collecting

data from students/trainees on several occasions, perhaps beginning while they are still students/trainees and for up to several years after they have completed the course. The information collected is often quite complex, including perhaps full employment and activity histories between each survey wave and maybe also reasons for changes and decisions made. Sometimes, one or more 'control' groups may be included in the survey in an attempt to assess the impact of the education/training;

- *Household panel surveys*. In several countries, long-term panel surveys of the general household population are carried out. The oldest, the Panel Survey of Income Dynamics (PSID) in the USA, has been interviewing the same people since 1968 (Duncan *et al.*, 2004). These surveys are multitopic and general purpose, collecting behavioural, attitudinal and circumstantial data on a range of social and economic issues. The main objective is to provide a rich data resource to be used by a wide range of users for a broad set of purposes. The data structure is complex, involving interviews with each person in the household of each sample member at each wave, in addition to household-level information and, often, additional survey instruments such as self-completion questionnaires or health measurements;
- *Birth cohort studies*. As already referred to in Section 1.1 above, the UK has a suite of four long-term birth cohort studies. Similar studies are carried out in many other countries. For some, the sample consists of all the births in a particular location (often a country) over a particular short time period such as a week. For others, such as the Millenium Cohort Study, a sample is taken of all births over a longer period, such as a year. Many of these studies began with a focus on either maternal health or child development (or both), but were later continued into the sample members' adult life when the value of doing so became apparent. Data about the sample members are typically collected from the mother in the first years of the study and later on from the sample members themselves;
- *Epidemiological studies*. Longitudinal medical studies often follow up samples of people over periods of time in order to study the progression of a disease or other changes in health, wellness and illness. The samples often consist of patients who have presented with a particular condition, either at a particular hospital or across a broader area such as the area of jurisdiction of a health authority. Some of these studies collect data through interviews or self-completion questionnaires while others collect only medical data via tests and diagnoses;
- *Repeated-measures studies*. This is a term found in the literature referring to studies that repeatedly measure the same variables in the same way on each sample unit. In practice, most longitudinal surveys of the types outlined above include some repeated measures, though they may vary in the proportion of the questionnaire or interview items that are repeated measures. For example, most items collected on household panel surveys are repeated (though not necessarily at every wave) in order to build up histories of economic activity, health status and so on (though an interesting development over the last decade or so is the recognition that simply repeating an item in identical fashion at every wave may not be the best way to obtain accurate measures of change or stability – see Chapters 5 and 6). Similarly, epidemiological studies take the same measurements repeatedly in order to chart disease progress. On the other hand, birth cohort studies often have very few repeated items, at least during the sample members' childhood years, as the questions that are appropriate to ask and the ways of testing development are somewhat age-specific. Surveys of school leavers or graduates tend to contain a mix of repeated items and wave-specific items;

- *Panels and cohorts*. The description above makes a distinction between household panel surveys and birth cohort studies. In fact the distinction is somewhat artificial and in practice consists only of differences of emphasis both in design and in substantive focus. A household panel contains samples of every birth cohort – but of course the sample of each cohort tends to be much smaller than would be found in a birth cohort study that concentrates on just a single cohort. And samples from a series of birth cohort studies can be combined to provide a picture of the situation of people of different ages at any one point in time. At the extreme, if a new birth cohort study were started every year, say, with a sample of births over a one-year period, then the combination of the studies would have the same population coverage as a household panel survey. The design focus of birth cohort studies, with a sample who are all very close to one another in age, tends to be associated with substantive research topics that are age-specific, such as growth and development. But there is no reason in principle why such topics could not be addressed with a household panel survey design, with age-specific survey instruments. The advantage would be that every age cohort is covered, providing some kind of population representation. The disadvantage is that in practice it would be impossible for any narrow age cohort to be afforded the kinds of sample sizes that are typically extant in birth cohort studies.

The connections between the two kinds of surveys may be even more obvious if we consider the case of surveys of graduates or school leavers. These typically involve samples from a one-year age group or stage group and therefore constitute another kind of cohort survey. Some of these surveys are carried out only once every n years, where n may take a value like 5 or 10. But some do indeed take place every year and thus, over time, they in principle cover the entire population of persons who have experienced the particular school system or educational establishment that is being sampled.

1.3 STRENGTHS OF LONGITUDINAL SURVEYS

Longitudinal surveys present a number of options, both for data collection and for analysis, that are either simply impossible with cross-sectional surveys or cannot be achieved in a satisfactorily accurate or reliable way. Often these features are key elements of the rationale for carrying out a longitudinal survey. However, it is important to distinguish between longitudinal *surveys* and longitudinal *data*. Longitudinal surveys are a source of longitudinal data, as the resultant data include items that refer to different points in time. But there are also other ways of obtaining longitudinal data, including diary methods and the use of retrospective recall within a single survey instrument. We briefly discuss here the strengths of longitudinal data, but our particular focus is on the strengths of longitudinal surveys.

1.3.1 Analysis Advantages

It is of course artificial to separate analysis advantages of longitudinal surveys from data collection advantages, as the reason for collecting a certain type of data is in order to be able to carry out certain types of analyses. The key advantages of longitudinal data (which in most cases can only be accurately or adequately collected by longitudinal surveys) are analytical:

- *Analysis of gross change.* Analysis of gross change is perhaps one of the most common objectives of longitudinal surveys. Repeated cross-sectional surveys can be used to estimate net change, for example change in the proportion of employees who regularly cycle to work. But only a longitudinal survey can identify the extent to which this is composed of different elements of gross change. For example, suppose that the proportion of employees who regularly cycle to work is estimated to be the same at two points in time. It would be of interest to know whether it is exactly the same set of employees who cycle on both occasions or whether there are equal and opposite flows into and out of regular cycling (and, if so, how large they are and what the individual characteristics associated with each flow are, etc). These are the kinds of questions that longitudinal surveys can address.
- *Analysis of unit-level change.* Individual-level change can be of interest independently of interest in population-level net change. For example, analysts may wish to understand the characteristics and circumstances surrounding divorce regardless of whether there is net change in the proportion of people who are married.
- *Aggregate measures.* There may be a desire to derive measures that combine observations from multiple time points. An example might be combining 12 measurements of monthly expenditure to obtain an estimate of annual expenditure.
- *Measures of stability or instability.* Another reason for combining observations from multiple time points is to provide measures of stability or instability. Often individual-level change can only be well interpreted in the context of changes over a considerable period of time. With a multiwave panel that collects measures of income at each wave, it should be possible to identify people with different patterns of change, such as steady growth, fluctuation around a low level, sudden decline followed by stability, etc. The characteristics associated with such patterns are likely to be of considerable interest to policy makers. Patterns such as these can only be identified using accurate data referring to lengthy time periods. For example, poverty analysts have used household panel data to demonstrate that there is considerable instability over time in the poverty status of many individuals and households in Westernised countries (Jenkins and Rigg, 2001). While the proportion of households in poverty may remain relatively stable over time, there may be many entrants to and exits from poverty. A large proportion of households may experience at least one spell of poverty over a long period of time, while very few households may remain continuously in poverty throughout the period. This insight provided by longitudinal surveys may have shifted the policy focus from (stable) characteristics associated with poverty propensity at a point in time to better understanding poverty dynamics and the factors associated with falling into poverty or persistently failing to exit from poverty.
- *Time-related characteristics of events or circumstances.* Often characteristics such as frequency, timing or duration are of central interest to researchers. For example, understanding the duration of spells in a particular state and the factors associated with exiting from the state (sometimes referred to as 'persistence') are important for poverty, unemployment, marital and partnership status, participation in education and training, company profitability and many other topics. Hazard modelling and survival analysis are techniques used to better understand the propensity for change

(in any status of substantive interest) and the factors associated with such change. These techniques require longitudinal data.

- *Identifying causality*. Policy makers are ultimately interested in understanding what causes certain changes to occur. Most policies are designed to bring about change of some kind or other, but often informed by only limited knowledge of how a changed policy might have an impact on the desired outcomes. Thus, a key objective of researchers is to shed light on causality rather than mere association. Analysis of the ordinal nature of events, which requires longitudinal data, often sheds light on issues of causality. For example, a cross-sectional survey can establish an association between factors A and B. But a longitudinal survey might establish that, for most population units that have experienced both A and B, A happened before B, making it rather more likely that A caused B than *vice versa* (though of course a third factor, C, may have caused both A and B and this and other possibilities must always be considered). Chapter 17 discusses ways in which longitudinal surveys can help to understand causality.

A number of recent advances in methods for analysing longitudinal survey data are the topics of Chapters 16, 18, 19, 20, 21 and 22.

1.3.2 Data Collection Advantages

There are several ways in which longitudinal surveys provide advantages in terms of data collection. These are mostly connected with either the quantity or quality of data that can be collected compared with alternatives such as the use of retrospective recall. In many situations, the advantages are so clear that most researchers would conclude that a longitudinal survey is the only worthwhile option. In other situations, the advantages may be marginal. Additionally, there are certain specialised situations in which a longitudinal survey may have cost or logistical advantages over repeated cross-sectional surveys, even though the latter may in principle also be able to provide adequate data to address the research objectives.

Length of Histories

It is possible to collect much *longer* continuous histories of events and transitions with a longitudinal survey than could be collected retrospectively in a single interview, simply due to the volume of data involved (and hence the length of the interview or questionnaire).

Accuracy of Recall

Longitudinal surveys are often able to obtain more accurate data than could be collected in a single interview with retrospective recall if the data might be subject to recall error. This advantage accrues primarily because of the ability to limit the reference period over which respondents must recall information to a relatively short recent period – such as the past month, quarter or year – while ultimately obtaining data that relate to much longer periods. The reference period often equals the interval between waves, if the aim is to build up continuous histories. Recall accuracy tends to be associated with events and circumstances that are rare, salient and recent (Tourangeau *et al.*, 2000). For events and circumstances that do not have these characteristics, data recalled over long time periods may be subject to severe recall bias. Recall error can take a number of forms. Events can be forgotten completely, they can be misplaced in time or misclassified, or associated characteristics may be misremembered. For example, most survey respondents are likely to be able to recall with reasonable accuracy the date(s) when they got married or had children, even if those events took place decades ago, but they are rather less likely to remember accurately when or how many times they visited a doctor, even if only asked about the past year.

Bounding

For many purposes, accurate dating of events is at least as important as accurate recall of the details of the event. But respondents may not be able to recall accurately the date of a specific event, even if they can recall the event itself. Consequently, retrospective recall questions asked in a single interview may produce biased estimates of frequencies and associated measures. A commonly reported phenomenon is 'telescoping', whereby survey respondents report events as having taken place within a reference period when in fact they took place longer ago. This is the result of errors in dating increasing the longer ago the recalled event took place (Rubin and Baddeley, 1989). Panel surveys have an extra advantage when collecting dates of events. Each interview after the first is *bounded* by the previous interview, so any events reported previously can be discounted from the reports in the current interview in order to avoid telescoping. This of course assumes that it can be unambiguously concluded whether or not reports in two consecutive interviews refer to the *same* event. Sometimes this is difficult, particularly when a respondent has a tendency to experience frequent events of a similar nature, but for many types of survey data it can usually be achieved.

Between-respondent Variation

The difficulty of any recall task will vary between respondents depending on the number, nature and timing of events they have experienced. This variation can be very considerable in the case of economic activity histories for example, causing a dilemma for survey designers. If a survey aims to collect complete activity histories over several years for a sample of people who are likely to vary greatly in their experiences, such as a cross-section of the general population, the ideal interval between survey waves will be very different for different sample members. But it is rarely possible to predict this in advance, nor is it often practical to have different between-wave intervals for different individuals. Instead, a standard interval is chosen. Interviews at annual intervals may be inefficient for persons whose circumstances change little (e.g. retired persons or those who remain in the same job for many years). The marginal amount of information collected in each interview, relative to the considerable cost, will be small. But annual interviews might present a considerable recall challenge to persons who experience large numbers of short spells of employment, perhaps interspersed with spells of unemployment or other activities. The advantage of longitudinal surveys in dealing with this challenge is that a lot of relevant information is known about sample members in advance of each wave (apart from wave 1). This information can be used to predict the likely complexity of recall tasks in various domains and tailored questioning strategies can be adopted. For example, those respondents who experience the highest frequency of changes in a particular domain may be administered a short telephone interview between waves in order to update information and thereby shorten the reference period. Or a group for which particularly low levels of change are observed might be administered an extra module of questions on a different topic in order to maximise the value of the data collection exercise. Such tailored approaches can have disadvantages and will not always be advisable, but they are options that are not available to cross-sectional surveys. Thus, to gain maximum benefit from the ability of longitudinal surveys to collect longer and/or more accurate histories of events, the survey designer needs to understand the recall and reporting task being asked of respondents, how it relates to between-wave interval and how this relationship might vary over sample members.

Expectations and Intentions

Subjective measures such as expectations, intentions, attitudes and reasons for making particular choices may be particularly prone to recall error. Though some researchers have retrospectively collected attitudes (Jaspers *et al.*, 2008) or reasons for making decisions (De Graaf and Kalmijn, 2006), others believe that attempting such measurement is simply not worthwhile, in particular because respondents' memory of such things is likely to be tainted by subsequent experiences and post hoc rationalisation may lead respondents to reassess their earlier motivations or expectations (e.g. Smith, 1984). Longitudinal surveys are able to collect information about expectations and reasons for choices untainted by subsequent events and outcomes. Later waves can of course also collect information about the relevant subsequent events and outcomes. This enables analysis of associations between, say, expectations and outcomes, that would simply not be possible with any other data source.

Maintenance of Samples of Rare Populations

For certain rare study populations, the cost of assembling a representative sample may be very high relative to the cost of collecting survey data from the assembled sample. This may be the case for populations that are not identified on any available sampling frame and for which it is therefore necessary to carry out a large-scale population screening exercise (Kalton and Anderson, 1986). In such cases, it may be cost-efficient to carry out repeated surveys with the same sample rather than select a new sample for each new study of the same population. The motivation in this case, therefore, is not a need for longitudinal data but the result is nevertheless the use of longitudinal data collection methods.

1.4 WEAKNESSES OF LONGITUDINAL SURVEYS

1.4.1 Analysis Disadvantages

Longitudinal surveys are often not as good as cross-sectional surveys at providing cross-sectional estimates. This may be perceived as a weakness, but it is simply not something that longitudinal surveys are designed for. Compared with estimates from a cross-sectional survey, cross-sectional estimates from a longitudinal survey (from wave 2 onwards) may be more likely to suffer from coverage error if the sample does not include, or under-represents, additions to the population since the sample was selected. This coverage error may increase over waves as the time since sample selection, and hence the undercoverage rate, increases. The design of a longitudinal survey can often be adjusted to improve the quality of cross-sectional estimates that can be made, though this may be resource-intensive and may detract from the central longitudinal aims of the survey. Such design options are discussed in Chapter 2. Also, a longitudinal survey sample may suffer from a lower net response rate than a cross-sectional survey on top of the lower coverage rate (though lower coverage and response rates do not necessarily imply greater error: this depends on the association between the coverage and response propensities and key survey estimates).

A difficulty with longitudinal analysis may be the analysis-specific nature of the population to which estimates refer. This can create problems in understanding and communicating effectively the nature of the reference population and consequent difficulties of inference. Definitions of study populations are discussed further in Section 1.5 below.

1.4.2 Data Collection Disadvantages

There are some aspects of survey data collection that are unique to longitudinal surveys and potentially detrimental or problematic.

Panel Conditioning

Panel conditioning refers to the possibility that survey responses given by a person who has already taken part in the survey previously may differ from the responses that the person would have given if they were taking part for the first time. In other words, the response may be *conditioned* by the previous experience of taking part in the survey. This therefore relates to all data collected by longitudinal surveys other than those collected at the first wave. There are two ways in which conditioning can take place. The way in which respondents *report* events, behaviour or characteristics might change; or the actual *behaviour* might change.

For example, a two-wave survey of unemployed persons might find that more people report a particular type of job search activity at the second wave than at the first wave. This might reflect a genuine increase in the extent to which that activity takes place (independent of taking part in the survey) but it could also be affected by panel conditioning. This could be because the first wave interview made some sample members aware of possible job search activities in which they were not currently engaged, so they subsequently started doing those things. Thus, there was a genuine increase in the extent of the activity, but only amongst sample members – not amongst the population as a whole. The *behaviour* of the sample members has been conditioned by the experience of the first interview. Alternatively, the experience of the first interviews may have affected the way in which some sample members respond to the questions in the second interview, even though their actual job search behaviour may not have changed. Perhaps in the first interview they discovered that reporting no activity of a particular type led to them being asked a series of questions about why they did not participate in that activity. So, to make the second interview shorter, or to avoid embarassing questions, they now report that they have participated in this particular activity. In this case, the *reporting* of the sample members has been conditioned by the experience of the first interview.

In the above example the conditioning was caused by having been asked the same question previously, but in other cases the conditioning could be caused merely by having been interviewed previously. Questions that were not asked previously can therefore also be affected. One process by which this might happen is that respondents could build up a greater sense of trust of the interviewer and the survey organisation as time passes. If asked some questions on a sensitive topic at wave 3, say, they might then be more willing to reveal truthful answers than they would have been at wave 1. This example also illustrates that some forms of conditioning can result in improvements to data quality. It is not the case that conditioning always damages quality. Another way in which quality might improve over waves of a panel is if respondents learn that it is useful to prepare for the interview by assembling certain documents. Examples of such documents might be pay slips, utility bills, health records or examination results, depending on the survey topics.

Panel conditioning is more likely to occur for certain topics and certain types of questions than for others, meaning that the extent to which the researcher needs to be concerned about panel conditioning – and the measures that might be introduced to minimise any negative consequences of it – should depend on the survey content. Panel conditioning in the context of attitudinal measures is discussed in Chapter 7.

Sample Attrition

Sample attrition (also referred to as *panel attrition*) refers to the continued loss of respondents from the sample due to nonresponse at each wave of a longitudinal survey. The term attrition is usually used to refer to a monotone process whereby sample members can change from being respondents to nonrespondents but not vice versa. In fact, many longitudinal surveys continue to attempt to collect data from sample members after they have been nonrespondents at one or more waves, and are often successful in doing so. Generally, then, the issue is that many sample members may have been nonrespondents at one or more wave, whether or not their response pattern is one of monotone attrition. The response rate at any one wave of a longitudinal survey may be just as good as that for any other survey but after, say, five waves the proportion of sample units that have responded at every wave may be quite low. Thus, the effective response rate for longitudinal analysis – for which data from every wave is required – may be lower than the response rates that we are used to having on cross-sectional surveys. After several waves there is also a risk that responding sample sizes can become unacceptably small. Minimising sample attrition is consequently a major preoccupation of researchers responsible for longitudinal surveys. The factors affecting sample attrition are discussed in Chapter 10. Chapters 11, 12 and 13 discuss various strategies for minimising attrition while Chapters 14, 15 and 16 address issues of how to deal appropriately with attrition at the analysis stage.

Initial and Ongoing Costs

The time dimension of a longitudinal survey brings with it a set of extra tasks for the survey organisation. In particular, the design and planning work that must take place prior to wave 1 is considerably greater than that required prior to a cross-sectional survey. The wave 1 instrument(s) cannot be designed independently of the instruments for later waves as it is the combination of data from each wave that will provide the necessary survey estimates. Sample design and fieldwork planning must take into account the need to follow sample members throughout the life of the survey. An integrated survey administration system is needed so that both field outcomes and survey data from each wave can determine the actions to be taken at later dates (these may include things like keeping-in-touch exercises – see Chapter 11 – and dependent interviewing – see Chapters 5 and 6). The sample management system needs to be able to cope with updated contact information arriving at any time and needs to be able to log the history of changes in such information. If the survey involves face-to-face interviewing, it becomes impossible with a longitudinal survey to maintain the cost advantages of a clustered sample, as the sample will disperse to some extent over time even if it is clustered initially. These considerations tend to result in the costs of an n-wave longitudinal survey exceeding the costs of n cross-sectional surveys, holding sample size and mode constant, though this need not always be the case.

1.5 DESIGN FEATURES SPECIFIC TO LONGITUDINAL SURVEYS

There are some aspects of survey design that are unique to longitudinal surveys, or are substantially different in nature to cross-sectional surveys. Standard survey methods textbooks provide little or no guidance on how to make design decisions regarding these aspects. Yet these aspects warrant careful consideration as design decisions can have weighty implications for data quality and for analysis possibilities.

1.5.1 Population, Sampling and Weighting

Population Definition

As data collection takes place at multiple time points for each sample unit, the data themselves are defined by time. Consequently, the time dimension must also enter the definition of the population being studied. To estimate a quantity such as mean change between time point 1 (t_1) and time point 2 (t_2), say $\hat{\bar{\Delta}}_i$, where $\Delta_i = Y_{it_2} - Y_{it_1}$, the analyst requires a sample from the population of units that existed at both t_1 and t_2, say $P_{t_1} \cap P_{t_2}$. Quantity Δ_i is not defined for any units that existed at only one of the two time points. In consequence, for any longitudinal survey with more than two waves it will almost certainly be the case that the study population will depend on which waves provide data for an estimate. Therefore different subsets of the sample units will represent different populations of interest and will thus be relevant for different estimation purposes. For example, suppose the time points refer to survey waves. An estimate that requires data from waves 1, 2 and 3 can only refer to the population $P_{t_1} \cap P_{t_2} \cap P_{t_3}$, which is a subset of the population $P_{t_1} \cap P_{t_2}$ to which our estimate of $\bar{\Delta}_i$ refers. Clarity is needed about the populations to which inference is required. This has implications for sample design (see Chapter 2), for following rules and for weighting (see below, and Chapter 15 for a discussion in the context of household panel surveys).

Sample Design

Longitudinal surveys face a number of unique sample design issues. Decisions must be made as to whether and how new entrants to the study population (births) should be included, whether a rotating design should be used, whether a repeated design should be used and, if so, whether the repeats should overlap or whether a split design is desirable. These fundamental design options are discussed in Chapter 2.

Additionally, some other features of sample design take on a rather different complex in the context of longitudinal surveys. If face-to-face interviewing is to be used at any or all waves, geographical clustering of the sample might be desirable on cost efficiency grounds. However, the cost parameters are more complex and harder to estimate as they depend on the extent and nature of sample mobility between face-to-face waves. Predictions of design effects should also influence decisions about sample clustering, but there is much less knowledge about design effects for clustered longitudinal surveys – especially for longitudinal estimates – than there is for cross-sectional surveys. To the extent that intracluster correlations are caused by shared personal characteristics, these should persist even as sample clusters disperse geographically over time. But to the extent that geography per se influences survey measures, dispersal might tend to reduce intracluster correlations. This issue warrants further investigation.

Surveys often oversample population subgroups that are of particular interest. For a cross-sectional survey, this is a useful device and is unproblematic provided that relative selection probabilities are known. For a longitudinal survey, the considerations are a little more complex. If subgroup membership is defined by characteristics that are variable over time, such as income level or geographical location, then oversampling the subgroup at wave 1 can become an inefficient strategy at later waves when substantial numbers may have moved into and out of the subgroup. A particular case of this occurs on business surveys, where businesses may, during the course of their inclusion in a panel survey sample, move between strata that were sampled at very different rates (for example, if the strata are defined by number of employees). Even if subgroup membership is defined by time-invariant characteristics, there is a risk that the focus of interest could shift over the duration of the survey, such that the sample becomes inefficient for current research interests.

Weighting

Many longitudinal surveys provide data users with two types of weights – *cross-sectional weights* and *longitudinal weights*. The distinction reflects a difference in the population to be represented, which in turn is related to different estimation objectives, and in the sample cases that can contribute to estimates. Given the focus of longitudinal surveys on longitudinal estimation, the longitudinal weights are perhaps the more important of the two. However, it is not always obvious for which sets of respondents these should be produced.

There are $2^t - 1$ possible populations that can be represented by a t-wave longitudinal survey, of which t are cross-sectional populations and $2^t - (t + 1)$ are longitudinal populations. Potentially, a set of weights could be created for each population. However, for surveys with more than two or three waves it would not be feasible to create all these sets of weights. It may be confusing to users to have so many sets of weights available and it is probably not necessary anyway, as many of the sets of weights would be so similar that the choice between them would make no practical difference to any estimates.

A solution would be to provide users with the data necessary to calculate weights for any combination of waves – and some guidance or even a program that will calculate the weights. Then, each user could produce weights tailor-made to his or her analysis. However, this is rarely done, either because some of the necessary data cannot be released at the unit level or because users who want to use weights prefer to be provided with ready-to-use weights.

An alternative is for the data provider to produce weights for a limited subset of the possible combinations of waves. This should be accompanied by guidance to users on what to do if the combination in which they are interested is not one of those for which weights are provided. The choice of wave combinations should be guided by the (likely) main uses of the data. For example, if the main objective of the survey is to permit analysis of change relative to baseline data that were collected at wave 1, then there is very little point in producing weights for combinations of waves that do not include wave 1. If a module of questions on a particular topic is included only at waves 1, 4, 7 and 10, then that particular combination should be a strong candidate for weighting. For almost all longitudinal surveys, the complete set of waves should be one of the combinations for which weights are produced. The only exception would be if, by design, there are no units eligible for data collection at every wave.

The choice of sets of weights to produce can have a real impact, as analysis based on respondents to a particular set of waves using weights designed for a different set of waves is suboptimal. Consider a three-wave survey for which one set of longitudinal weights is provided, designed to make the set of persons who responded to all three waves representative of $P_{t_1} \cap P_{t_2} \cap P_{t_3}$. Suppose we want to estimate a parameter of change between wave 1 and wave 3, for which we only need to use data collected at waves 1 and 3. We could use sample members who responded only at waves 1 and 3, in addition to those who responded at all three waves. However, the longitudinal weights will be set to zero for the former group, so these cases will be dropped from the analysis.

Another important consideration is that weights are typically produced after each new wave of data becomes available. Thus, as a minimum, at each wave a set of weights will be produced representing the longitudinal population at all waves to date. This means that ultimately weights will be available for every *attrition sample*. If the survey policy is to attempt to collect data only from previous wave respondents this will be all the weights that are needed. Otherwise, the task is to identify which other combinations of waves are sufficiently important to warrant the calculation of weights.

1.5.2 Other Design Issues

Intervals between Waves

The frequency of data collection waves has several implications. For recall data, higher frequency waves are likely to produce higher quality data, as discussed in Section 1.3.2 above. However, the optimal frequency will depend on the nature of the information to be recalled. For indicators of current circumstances, where the intention is to identify changes in circumstances, the optimum frequency will depend on the rate of true change. If waves are too infrequent then important changes might be missed, while if they are too frequent then expensive data collection efforts will yield little useful information.

Nonresponse should also be a consideration in the choice of between-wave interval. For a given overall survey duration (see below), a shorter between-wave interval (and therefore more waves) is likely to lead to higher nonresponse, in the sense that a higher proportion of sample members are likely to be nonrespondents at one or more wave. But for a given number of waves, a shorter interval is likely to be associated with lower nonresponse, not least because smaller proportions of sample members will move home or otherwise change their contact details, rendering it easier to locate and contact them over multiple waves. For a given overall survey duration, a shorter interval also has obvious cost implications as there is a cost associated with carrying out each extra wave.

Duration of the Survey

The longer the period of time over which a longitudinal survey collects data, the richer and more valuable the data are likely to be. However, for some surveys the focus is limited to a particular phase in the life of sample members, for example defined by age, education or employment, so this might naturally imply a limited duration for the survey. A longer duration will have extra costs, independent of the number of waves carried out during that time. These arise due to the need to maintain staffing and to perform tasks such as panel maintenance and tracking, which are independent of data collection tasks.

Respondents and Study Units

In most of the examples mentioned in Section 1.1 above, data are collected from, and about, individuals. In some cases, and for some analysis purposes, the units of interest may be groupings of individuals such as couples, households or parent–child groups, but even then many of the group-level variables will be constructed from individual-level data items, though some may have been collected from a single individual in the group but referring to the group as a whole. Indeed, on all surveys the respondents are individuals, but in some cases they are supplying data that refer to some other entity, such as a business, farm, educational establishment, community or other entity. Longitudinal surveys where the study unit is an establishment of some kind are common (see Cox *et al.*, 1995; American Statistical Association, 2000). The relationship between respondent and study entity has important implications for such surveys: there may be multiple respondents at each wave; the survey organisation may have to take extra measures if they want to know who the respondents were; and respondents might change from wave to wave, for example through staff turnover. These respondent issues may have implications for measurement error as well as implications for contact, motivation and cooperation. On establishment surveys, the changing nature of establishments over time may also create issues for population and sample definition. Establishments can change name, size, nature and even location. A longitudinal survey needs clear definitions of what constitutes a continuing unit over time as opposed to a unit that has ceased to be part of the study population (a *death*) or one that has entered the population (a *birth*).

Tracking and Tracing

The long-term health of a longitudinal survey is dependent on retaining the ability to contact sample members at each wave. This requires an administrative system and a programme of operations to be put in place. Such operations might include regular mailings or e-mails, the collection and maintenance of contact details of close friends and relations of sample members, and linking sample details to administrative data sources that indicate address changes. Some of the challenges in tracking sample members and the outcomes of some commonly used techniques are discussed in Chapter 11.

Modes

There exists a wide range of methods of collecting survey data. Respondents may be interviewed either face to face or by telephone, or self-completion questionnaires may be administered. In each of these cases the instrument itself may or may not be computerised. In the face-to-face case, visual prompts such as show-cards may or may not be used. In the self-completion case, options exist for the method by which the questionnaire is delivered to, and collected from, respondents. All of these options have implications for costs, timetables, response rates and for data quality and survey error. The choice of data collection modes is therefore important. In the case of longitudinal surveys, this choice has some extra dimensions.

First, a greater variety of ways of combining modes is available. Sample members can be switched between modes from one wave to the next, in addition to any combination that may be done within a wave (for example, a self-completion component within a face-to-face interview). A particularly important consequence of this is that it may affect measures of change between waves if the modes in question have different measurement properties. This issue is the focus of Chapter 8.

Second, information can be collected at one wave that opens up extra mode possibilities at future waves. For example, respondents can be asked to provide e-mail addresses, allowing the possibility of an approach by e-mail to take part in a web survey at the next wave. However, these opportunities tend to lead to mixed-mode data collection (within a wave) as the proportion of sample members supplying a valid e-mail address or phone number, etc. rarely approaches 100 %.

1.6 QUALITY IN LONGITUDINAL SURVEYS

The ideas of survey quality (Biemer and Lyberg, 2003) provide a set of tools for maximising the usefulness of any survey. Within this, the concept of total survey error (Groves, 1989) can provide a useful framework for considering the statistical impacts of design and implementation considerations for longitudinal surveys. The total error in any survey-based estimate, $\hat{Y}$, of a population parameter, Y, is simply $\hat{Y} - Y$, the difference between the estimate and the true value of the parameter. Many factors can contribute to this error. The main components of error in any survey estimate are set out in Figure 1.1. We outline in turn key influences on each in the longitudinal survey context.

1.6.1 Coverage Error

Aside from undercoverage in the initial sampling frame, which is an issue common to all surveys, the main source of undercoverage on longitudinal surveys may be a failure to (adequately) include new population entrants in the sample. The extent to which this is a problem will depend on the definition of the study population as well as the survey sampling procedures.

Sources of Statistical Error in any Survey Estimate

Errors of Nonobservation	**Observational Errors**
Coverage	Measurement
Sampling	
Nonresponse	

Figure 1.1 Components of total survey error.

1.6.2 Sampling Error

The design issues that affect sampling error are not distinct for a longitudinal survey, other than the considerations regarding clustering and regarding differential selection probabilities, discussed above in Section 1.5.1. It should always be remembered, however, that the nature of the estimates for which sampling error is of interest is typically rather different from the nature of cross-sectional descriptive estimates. Knowledge of sampling error in one context cannot therefore be easily translated to the other context.

1.6.3 Nonresponse Error

In addition to the processes that can lead to nonresponse on any survey (Groves and Couper, 1998), there are some important influences on response propensity that are rather distinct in the longitudinal survey context.

Subsequent to the first wave, a major component of *noncontact* on longitudinal surveys is caused by geographical mobility of sample members. Between waves, some sample members will move home, change employment, change telephone number or change e-mail address. If the survey organisation is reliant upon any or all of these details to contact the sample member, then they will need to take extra measures to be able to make contact at the subsequent wave. Techniques for tracking sample members between waves are discussed in Chapter 11. For a face-to-face interview survey it is also necessary to have interviewers available to attempt contact with sample members in all the places to which they could move.

Participation in a longitudinal survey requires considerable commitment from sample members – not just a single interview, but several, over a period of time. In consequence, special incentives or motivation may be needed to compensate. Typically, sample members are offered a small payment for each interview, or some other form of gift, and particular effort is made to make the sample member feel an important, irreplaceable, component of the study and to persuade them of the value of the study. The use of incentives is discussed in Chapter 12.

A unique feature of longitudinal surveys relevant to *refusals* is that, after the first wave, sample members have already experienced the survey interview and therefore have a very good idea of exactly what it consists of, what kinds of questions will be asked, how difficult or sensitive they find it, and so on. This is very different from a typical survey situation, where a sample member will have only a rather vague and general impression of what they are being asked to do at the time when they are being asked to cooperate. Consequently, on a longitudinal survey it is very important to try to make the interview experience as pleasant as possible for the respondent. If a respondent finds the interview difficult, frustrating, embarassing, threatening, uninteresting or simply too long, they will be less likely to be willing to take part again at the next wave. Chapter 10 discusses sample attrition.

1.6.4 Measurement Error

Measurement error refers to the possibility that any individual survey observation might differ from the value that would be observed by a perfect measurement of the concept that the survey item was intended to measure. Such errors can arise for many reasons, associated with the response task, the respondent, the interviewer, the data collection mode and the interview setting (Biemer and Lyberg, 2003, chapter 4), but some features of measurement error are distinctive in the case of longitudinal surveys.

Of central interest to longitudinal surveys is the estimation of micro-level change over time. As measures of change draw upon data collected at different waves, any differences in measurement properties between waves or inconsistencies in the observed measures between waves can cause particular problems. Recall error in retrospective data can contribute to a phenomenon that is only observable in data from longitudinal surveys, known as *seam bias* or *seam effects* (Jäckle and Lynn, 2007; Jäckle, 2008). This refers to a common finding that the level of observed change between two consecutive time periods is much higher when the observations for each period come from two different interviews than when they come from the same interview. In other words, there appears to be an excess of transitions at the 'seam' between reference periods. This phenomenon can obviously introduce error to estimates of levels of change, but also to estimates of the correlates or causes of change, etc. Methods that reduce the extent of seam effects, and hence reduce error in estimates of change, are therefore desirable. Identification of such methods is, however, not entirely straightforward and requires understanding of the range of manifestations of recall error and their causes (Jäckle, 2008). *Dependent interviewing* encompasses a range of questioning techniques that take advantage of the bounding possibilities of longitudinal surveys described in Section 1.3.2 above and is discussed in Chapters 5 and 6.

When the study of change is based not on recall data but simply on comparisons of 'current status' indicators collected at each wave of a longitudinal survey, measurement error can still cause considerable problems. Even random misclassifications, for example, can bias estimates of change. Some analytical solutions to this problem are offered in Chapter 21.

Many aspects of survey methods can affect the measurement properties of a survey item, including whether the question and response options are presented visually or aurally, whether an interviewer is present, the setting of the interview, the context of the item, and so on. Any change between waves in survey methods therefore has potential to alter the measurement properties and hence to introduce error to estimates of change. Researchers should be particularly wary of design changes that have been shown to influence measurement for relevant topics or question types. Examples might include the effect of interviewer presence for items that have a social desirability connotation (Aquilino, 1994; Hochstim, 1967) or for which respondents may show a tendency to satisfice, the effect of aural vs. visual presentation for cognitively complex response construction tasks (Krosnick, 1999), and the effect of question context for attitude questions (Schuman and Presser, 1981; Tourangeau and Rasinski, 1988). Some of these measurement issues are explored in Chapter 8.

Even if survey methods are held completely constant between waves, there is still a risk that the measurement properties of an item for a given respondent can change between waves due to panel conditioning, as introduced in Section 1.4.2 above (and see Chapter 7). Methods should therefore be sought either to minimise the negative aspects of panel conditioning or to control for it in analysis. The latter approach is likely to require information from either a split panel or rotating panel design (see Chapter 2).

1.7 CONCLUSIONS

Longitudinal surveys are subject to all the methodological considerations that apply to any survey (Groves *et al.*, 2004) but they also have an added dimension that brings a wealth of extra opportunities but also some extra complexities and problems. This volume aims to provide an overview of those issues with the hope of promoting best practice in the design and administration of longitudinal surveys and encouraging clear thinking regarding the many issues that a longitudinal survey researcher must deal with. Close reading of many of the chapters will, however, reveal many areas in which the research evidence to inform good design decisions is rather thin. A secondary hope is therefore that this book will stimulate researchers to further investigate some of the many fascinating methodological issues surrounding longitudinal surveys.

REFERENCES

American Statistical Association (2000). *ICES: Proceedings of the Second International Conference on Establishment Surveys: June 17–21, 2000 Buffalo*, New York. Alexandria, VA: American Statistical Association.

Aquilino, W. S. (1994). Interview mode effects in surveys of drug and alcohol use: a field experiment. *Public Opinion Quarterly*, 58, 210–240.

Biemer, P. P. and Lyberg, L. E. (2003). *Introduction to Survey Quality*. New York: John Wiley & Sons, Inc.

Binder, D. (1998). Longitudinal Surveys: why are these surveys different from all other surveys? *Survey Methodology*, 24, 101–108.

Börsch-Supan, A. and Jürges, H. (2005). *The Survey of Health, Ageing and Retirement in Europe – Methodology*. Mannheim: MEA. Retrieved October 7, 2008, from http://share-methods.notlong.com

Cox, B. G., Binder, D. A., Chinnappa, B. N., Christianson, A., Colledge, M. J. and Kott, P. S. (1995). *Business Survey Methods*. New York: John Wiley & Sons, Inc.

De Graaf, P. M. and Kalmijn, M. (2006). Divorce motives in a period of rising divorce: evidence from a Dutch life-history survey. *Journal of Family Issues*, 27, 483–505.

Dex, S. and Joshi, H. (2005). *Children of the 21st Century: From Birth to Nine Months*. London: Policy Press.

Duncan, G., Hofferth, S. L. and Stafford, F. P. (2004). Evolution and change in family income, wealth, and health: The Panel Study of Income Dynamics, 1968–2000 and beyond. In J. House, *et al.* (Eds), *A Telescope On Society* (Chapter 6). Ann Arbor: Institute for Social Research, University of Michigan.

Ferri, E., Bynner, J. and Wadsworth, M. (2003). *Changing Britain, Changing Lives: Three Generations at the Turn of the Century*. London: Institute of Education.

Groves, R. M. (1989). *Survey Errors and Survey Costs*. New York: John Wiley & Sons, Inc.

Groves, R. M. and Couper, M. P. (1998). *Nonresponse in Household Interview Surveys*. New York: John Wiley and Sons, Inc.

Groves, R. M., Fowler, F. J., Couper, M. P., Lepkowski, J. M., Singer, E. and Tourangeau, R. (2004). *Survey Methodology*. New York: John Wiley & Sons, Inc.

Hochstim, J. R. (1967). A critical comparison of three strategies of collecting data from households. *Journal of the American Statistical Association*, 62, 976–989.

Jäckle, A. (2008). The causes of seam effects in panel surveys. *ISER Working Paper* 2008–14. Colchester: University of Essex. Retrieved October 13, 2008, from www.iser.essex.ac.uk/pubs/workpaps/pdf/2008–14.pdf

Jäckle, A. and Lynn, P. (2007). Dependent interviewing and seam effects in work history data, *Journal of Official Statistics*, 23, 529–551.

Jaspers, E., Lubbers, M. and De Graaf, N. D. (2008). Measuring once twice: an evaluation of recalling attitudes in survey research. *European Sociological Review*. Advance access published online 13 October 2008.

Jenkins, S. P. and Rigg, J. A. (2001). The dynamics of poverty in Britain. *DWP Research Report*, No. 157. London: Department for Work and Pensions.

Kalton, G. and Anderson, D. W. (1986). Sampling rare populations. *Journal of the Royal Statistical Society Series A*, 149, 65–82.

Kalton, G. and Citro, C. F. (1993). Panel surveys: adding the fourth dimension. *Survey Methodology* 19, 205–215.

Krosnick, J. A. (1999). Survey research. *Annual Review of Psychology*, 50, 537–567.

Roe, S. and Ashe, J. (2008). Young people and crime: findings from the 2006 Offending, Crime and Justice Survey. *Home Office Statistical Bulletin* 09/08. London: Home Office. Retrieved October 7, 2008, from http://www.homeoffice.gov.uk/rds/pdfs08/hosb0908.pdf

Rubin, D. and Baddeley, A. (1989). Telescoping is not time compression: a model of the dating of autobiographical events. *Memory and Cognition*, 17, 653–661.

Schuman, H. and Presser, S. (1981). *Questions and Answers in Attitude Surveys*. New York: Academic Press.

Smith, T. (1984). Recalling attitudes: an analysis of retrospective questions on the 1982 General Social Survey. *Public Opinion Quarterly*, 48, 639–649.

Tourangeau, R. and Rasinski, K. A. (1988). Cognitive processes underlying context effects in attitude measurement. *Psychological Bulletin*, 103, 299–314.

Tourangeau, R., Rips, R. and Rasinski, K. (2000). *The Psychology of Survey Response*. Cambridge: Cambridge University Press.

CHAPTER 2

Sample Design for Longitudinal Surveys

Paul Smith
Office for National Statistics, UK

Peter Lynn
University of Essex, UK

Dave Elliot
Office for National Statistics, UK

2.1 INTRODUCTION

In this chapter we review the considerations that influence sample design for longitudinal surveys, the range of sample design options available and research evidence regarding some of the key design issues. We attempt to provide practical advice on which techniques to consider for different types of longitudinal surveys, where they are most likely to be effective and what pitfalls to avoid.

2.2 TYPES OF LONGITUDINAL SAMPLE DESIGN

Surveys that collect data from the same units on multiple occasions vary greatly in terms of the nature of the information collected, the nature of the population being studied and the primary objectives (Duncan and Kalton, 1987; Binder, 1998). They also face a variety of practical constraints, from the level of financial resources available, to regulations on respondent burden. Consequently, there is also considerable variation in the choice of appropriate sample design. We identify five broad types of longitudinal survey design:

- *Fixed panel.* This involves attempting to collect survey data from the same units on multiple occasions. After the initial sample selection, no additions to the sample

Methodology of Longitudinal Surveys P. Lynn

are made. In principle, the only loss to the eligible sample is through 'deaths'. The 1970 British Cohort Study (BCS70) is an example of a fixed panel design (Plewis *et al.*, 2004).

- *Fixed panel plus 'births'*. This is like a fixed panel, except that regular samples of recent 'births' to the population are added. Typically, at each wave of data collection a sample of units 'born' since the previous wave is added. This may be preferred to a fixed panel if there are non-trivial numbers of births in the population during the life of a panel and there is a desire to represent the cross-sectional population at the time of each wave as well as the longitudinal population of wave 1 'survivors'. Most household panel surveys have this design, as a sample of 'births' into the eligible age range is added at each wave (Rose, 2000, chapter 1).
- *Repeated panel*. This design involves a series of panel surveys, which may or may not overlap in time. Typically, each panel is designed to represent an *equivalent* population, i.e. the same population definition applied at a different point in time. The England and Wales Youth Cohort Study (YCS) is an example of such a design: the panels consist of samples of the age 16–17 one-year age cohort selected in different years, each panel involving at least three waves over at least three years (Courtenay and Mekkelholt, 1987).
- *Rotating panel*. Predetermined proportions of sample units are replaced at each fieldwork occasion. Typically, each unit will remain in the sample for the same number of waves. A wide variety of rotation patterns have been used. As Kalton and Citro (1993) note, a rotating panel is in fact a special case of a repeated panel with overlap. It is special because the overlap pattern is fixed, and is typically balanced, and because each panel that is 'live' at one point in time is designed to represent the *same* population, allowing combination of the panels for cross-sectional estimation. Rotating panel designs are often used when the main objectives are cross-sectional estimates and short-term estimates of net and gross change. Labour Force Surveys have a rotating panel design in many countries (Steel, 1997).
- *Split panel*. This involves a combination of cross-sectional and panel samples at each fieldwork occasion. A common design described by Kish (1987, pp. 181–183) involves one panel sample from which data are collected on each occasion, plus a supplemental cross-sectional sample on each occasion. A series of cross-sectional surveys in which a proportion of sample elements are deliberately retained in the sample for consecutive surveys – referred to by Kalton and Citro (1993) as an *overlapping survey* – can also be thought of as a type of split panel. A variation is a series of repeated cross-sectional surveys but with a $P = 1$ stratum, so that in practice a subset of the sample units is included on every occasion. In this case the objective is a series of cross-sectional estimates, but longitudinal estimates are serendipitously possible for the $P = 1$ stratum. Many statutory business surveys have this characteristic.

This is of course only a broad typology and it does not fully describe the range of possible designs. For example, each panel in a repeated panel design may or may not include additional regular samples of births. In general, we would expect to use the fixed panel designs where we will select a single sample, and follow them to provide detailed information on changes in the characteristics. Repeated and rotating panels are useful for assessing changes in longitudinal estimates since they represent comparable populations (unlike a fixed panel), each treated with equivalent methodology (unlike a fixed panel plus births). In most respects, the various designs are asymptotically equivalent, the differences being ones of emphasis rather than fundamental ones.

One motivation for considering a panel design is to be able to construct measures derived from data collected at several time points, where these data could not reliably be collected retrospectively at one time point (e.g. totals or averages of quantities that may change frequently, such as income or physical activity, or indicators of whether certain low-salience events were ever experienced). But perhaps the prime motivation is to be able to measure gross change at the element level. This is useful for the study of transitions and stability in many fields. To study phenomena that change only slowly, or that may have long-term effects, long panels are needed. Where change is more rapid and only short- or medium-term effects are of interest, short panels or rotating panels may be preferred.

Often, cross-sectional estimates are required in addition to longitudinal ones. In principle, a fixed panel plus births can achieve this, but in many situations adding regular samples of births to a panel is a complex and expensive task. A more efficient way to achieve cross-sectional representativeness can be to select a fresh cross-sectional sample. A split panel design may be used in this case or, if the gross change of interest need be observed only over relatively short periods, a rotating panel design.

Repeated panels are often used where the population of interest is naturally thought of as a series of cohorts defined in time, each of which may be expected to have distinct experiences or where policy interventions are likely to apply at the cohort level. Examples might include cohorts defined by age, year of graduation, period of immigration, year of recruitment, etc.

2.3 FUNDAMENTAL ASPECTS OF SAMPLE DESIGN

2.3.1 Defining the Longitudinal Population

As in any survey it is necessary to define both the target population about which we will want to make inferences, and the frame population of units that are available to be sampled. Ideally these populations will coincide, but in many situations they are not completely congruent, and in these cases some appropriate weighting system may be necessary to adjust (approximately) the sample (from the frame population) to the target population.

However, population definition is complicated additionally in longitudinal surveys because the populations are dynamic. Some method of defining a fixed population on which to base a design is needed, and there are three broad approaches:

(1) *A static population based on the population at the time the first wave sample is selected.* There can be no 'births', but 'deaths' are allowed in this population as members emigrate, change their defining characteristics, or cease to exist. For practical purposes we can consider the population size to be fixed, but with units moving to 'dead', from which it may be possible or not to return. Therefore the accuracy of more interesting transitions will reduce with sample size, as sample units also move to 'death'. The big advantage of this approach is that the population information is fixed for all waves, which makes weighting and analysis of subsequent waves relatively straightforward.

A fixed panel design is the most natural design to study a population defined in this way. Such a population may become increasingly less relevant to policy decisions over time as it ages. For example, estimates from the early waves of the BCS70 in the 1970s – when the sample members were young children – are unlikely to be good predictors of equivalent parameters relating to young people in the 2000s. There is a danger that users may interpret results as applying to population definition (3) below, in respect of which estimates of change will obviously be biased if new units are more or less likely to change than established units, or if there is a relationship between the variable being measured and the probability of becoming a 'death'.

(2) *The population is defined as the intersection of the cross-sectional populations at each wave.* With this approach, both births and deaths over the life of the study are excluded from the study population. This creates the same problems as option (1) and has the added disadvantage that auxiliary population totals that are needed to calibrate sample estimates may be very difficult to construct. Also, the intersection, and hence the population definition, changes when a new wave is added.

A variant of this approach is to select the sample from the intersection, but then to provide some appropriate weighting to compensate for differences between the sample and the target population. This latter approach is used on the longitudinal dataset from the UK Labour Force Survey. Note that it only gives unbiased estimates for the target population if there is no relationship between the probability of being in the intersection and the survey variable(s).

(3) *The population is defined as the union of the cross-sectional populations at each wave.* This approach seems best suited to a genuinely dynamic population but obviously requires a correspondingly dynamic sample design to introduce births and so avoid bias. This approach is used on most household panel surveys, such as the British Household Panel Survey (Lynn, 2006) and the UK Household Longitudinal Study. There can be some difficulty in saying what a sample actually represents ('the population of all people in a household in the UK at any time during a five-year period', for example). Also, when a further wave is added, the union changes and may require revision of previous estimates.

In general therefore option (3) seems to be the design most likely to fulfil general requirements for longitudinal information, but has difficulties of both methodology and interpretation. In using option (1) or (2) in its place, there is a clear need to explain the limitations of what can be inferred to users of the information – noting however that there is a risk that such limitations may be ignored by users, particularly if no other information is available.

2.3.2 Target Variables

Optimal sample design cannot be determined in the abstract, but only relative to some estimate(s). Typically, sample designers identify a small set of key survey variables and attempt to optimise sample design for estimates of means, or possibly other functions, of those variables. Choice of strata, allocation to strata and the degree of sample clustering will be chosen to minimise the variance of those estimates. In the Office for National Statistics (ONS), the key estimates are typically means in the case of cross-sectional surveys, but for rotating panel designs there is a move towards using estimates of change in means. For a longitudinal survey, the key estimates will typically be rates of change, levels of change, durations, hazard rates or other dynamic parameters. Estimates of variance of these quantities are therefore needed (see Section 2.3.7) and knowledge of how the quantities are related to strata or primary sampling units (PSUs). Such information can be difficult to estimate as, unlike the case with cross-sectional surveys, there is often no similar recent survey to draw upon.

2.3.3 Sample Size

It is necessary to plan for the eventual sample size required for longitudinal analysis by taking account of any population births and deaths that are reflected in the sample but also specifically taking account of the expected attrition during the life of the survey. Such attrition may arise for several reasons. Sample members who respond at one wave of the survey may not be contacted in the next, either because they have changed their address and cannot be located (see Couper and Ofstedal, Chapter 11 of this volume) or because they are temporarily absent when an interview attempt is made; or they may refuse to take part in the survey (see Watson and Wooden, Chapter 10 of this volume). Some movers may of course be relocated at their new address, where this is known, but some loss of sample numbers can usually be expected from this cause. The treatment of movers in longitudinal surveys is considered further below.

If the proportions of the wave $i-1$ responding sample who are expected to respond at wave i can be estimated as p_i (with p_1 indicating the proportion of the initial sample responding at wave 1), then the initial sample size, n_0, needed to achieve (in expectation) a target number, n_k, after k interview waves can be simply calculated. With a fixed panel design, we have $n_0 = n_k \Big/ \prod_{i=1}^{k} p_i$. More generally, if response at each occasion is independent, then $n_k \sim B\left(n_{0,} \prod_{i=1}^{k} p_i\right)$, and using a Normal approximation for the confidence interval (under the reasonable assumption that n_0 is large) allows us to solve for the value of n_0 sufficient to be 95 % confident that n_k will be achieved,

$$n_0 = \frac{1}{\prod_{i=1}^{k} p_i}\left[n_k + 2\left(1 - \prod_{i=1}^{k} p_i\right) + 2\sqrt{\left(\prod_{i=1}^{k} p_i\right)^2 - (2+n_k)\prod_{i=1}^{k} p_i + (1+n_k)}\right].$$

In a rotating panel design with k waves and an initial sample size of n_0 in each panel, we have $n_0 = n_k \Big/ \sum_{j=1}^{k} \prod_{i=1}^{j} p_i$ (in expectation), where n_k is the target responding sample at each survey occasion, consisting of the sum of the respondents from each panel. Note that these all assume that once a unit is lost to the panel it is never recontacted. In a situation where units may be missed on one occasion but subsequently participate again, we would expect to achieve at least n_k interviews.

Experience of net attrition rates in both quarterly and annual panel surveys suggests that the greatest loss occurs between the first and second waves. For example, on the UK

Labour Force Survey, which uses a five-wave quarterly rotating address panel, $\mathbf{p}^T = \{0.728, 0.878, 0.963, 0.936, 0.956\}$ in Q1 2006.[1]

These approximations assume the same attrition across the whole population, but this is unlikely to be true in practice, so instead we would expect to calculate n_0 separately within groups of relative homogeneity (a longitudinal extension of the ideas presented in Pickery and Carton, 2008). In a sound sample based on good auxiliary information, this will mean that the most difficult cases will be heavily over-represented at the first wave, to allow for later attrition, which is likely to make the survey more expensive. Cutting back on this expense risks having estimates with high variances and biases at later waves.

2.3.4 Clustering

Decisions about the nature and extent of sample clustering should depend on slightly different criteria in the case of a longitudinal survey. Specifically, the relationship between clustering and data collection cost may change over the waves of a survey due to the mobility of sample units and sample attrition. The impact of clustering on the variance of estimates (design effects) may also change. To take this into account at the sample design stage requires knowledge of the number and timing of waves, and estimation of the changing effects of clustering over those waves.

For example, the General Household Survey in Great Britain has, since 2005, used a four-wave rotating panel design with annual interviews. Unit cost is reduced by clustering the initial sample within a sample of postal sectors. But unit cost increases at each subsequent wave as attrition and mobility in the sample reduces the number of interviews in each postal sector. One way to partly ameliorate this would be to retain the same postal sectors in the sample and perform the rotation of addresses within these sectors, but that could lead to biases in cross-sectional estimates over time. A careful compromise is therefore needed between selecting a sample size per sector that is large enough at wave 1 to still constrain unit costs at wave 4 and controlling the variance of estimates by limiting the wave 1 sample size per sector.

Identification of the optimal level of sample clustering is even more difficult with long-term panels, especially if the eventual life-span of the panel is unknown at the time of initial design. This was the case with the British Household Panel Survey, which was conceptualised as a survey with an indefinite number of annual waves but with initial funding only for five years. The decision to use a clustered design was based on crude estimates of the impact over the first few years of the survey. After many years of the survey, the unit cost of data collection is only marginally less than would be the case with a completely unclustered design, while design effects are still considerably greater than with an unclustered design. Therefore for a long-running longitudinal survey it may be better to have an unclustered sample, as the accuracy in later years may offset the additional cost in the earlier years.

2.3.5 Treatment of Movers

Movers create two types of problems in longitudinal surveys. First, there is a need to obtain new contact information and an associated increased risk of failure to contact the

[1] January–March 2006 figures for Great Britain: www.statistics.gov.uk/StatBase/Product.asp?vlnk=10675&Pos=&ColRank=1&Rank=192.

sample unit. Second, there are additional costs if face-to-face interviewing is used. Most social surveys that employ face-to-face interviewing use some form of cluster sampling or multistage sampling to control costs, so following a mover to a new address may incur considerable extra cost if the new address is not in one of the original sample areas. Also, there is a risk that no interviewer may be available to visit a mover if the move is only discovered during the field period, so procedures need to be put in place to track and record moves among the sample members. Problems are fewer on business surveys, as they tend to use remote methods of data collection, but there is still the challenge of identifying movers before the survey, because of lags in the administrative systems that are often used to create the frame.

However, omitting movers may create an obvious bias in some surveys. One way of bypassing this problem is to define sampling units that do not move. In the UK Labour Force Survey, the ultimate sampling unit is an address, so that movers out of a sample address during the study period cease to be part of the sample but are balanced by movers in. It is important in such a design to retain vacant addresses identified in any wave as part of the sample to be issued to interviewers in subsequent waves. Such a design provides a balanced sample for cross-sectional estimates and for estimates of net change, but not for estimates of gross change.

2.3.6 Stratification

Ideally, sampling strata should be defined that are relatively homogeneous in terms of a survey's target variables. In practice, of course, the ideal strata often cannot be determined in advance, cannot be identified in the sampling frame or are quite different for different target variables. In longitudinal surveys, the main targets for inference will often be measures of individual or gross change, both overall and for domains, for which the standard cross-sectional stratifiers may be of little use. Special studies may be required to identify suitable candidates, by analysing the relationships between survey-based estimates of change and potential stratification variables. Even obtaining survey-based estimates of change may be difficult if there are no previous longitudinal data to work with.

2.3.7 Variances and Design Effects

There is a substantial literature on estimating changes in totals and means from longitudinal and particularly rotating samples. Kish (1965) was one of the first to discuss the problem. His analysis applies in a static population when the finite population correction can be ignored. Holmes and Skinner (2000) also give useful general results for variances of poststratified and calibrated estimators of the change in means and totals for this case. In surveys with large sampling fractions the finite population correction is both important and open to a variety of different definitions – see Tam (1984), Laniel (1988), Nordberg (2000) and Berger (2004), where the analysis is extended to dynamic populations and cases where the finite population correction cannot be ignored.

Design effects for complex statistics such as these are often found to be less than those for simple linear statistics from a single sample (Kish and Frankel 1974) but Skinner and Vieira (2005) found unusually high design effects for regression coefficients based on a longitudinal sample, with the design effects growing with the number of occasions that were included in the analysis.

2.3.8 Selection Probabilities

Any longitudinal survey that includes a mechanism for sampling births must also allow the estimation of the selection probabilities of units that are added to the sample subsequent to the initial selection. This is not always simple. For example, some household panel surveys use an approach that adds to the sample the members of households into which members of initial sample households move. The selection probabilities of such people are a complex function of their household membership history over the life of the panel, but fortunately it is not required to know these probabilities. Instead, the 'weight share' method (Ernst, 1989; Lavallée 1995) can be used to produce unbiased estimates as long as people not in the initial households are not followed if they later leave the household of a person from the initial sample.

2.4 OTHER ASPECTS OF DESIGN AND IMPLEMENTATION

2.4.1 Choice of Rotation Period and Pattern

As with most aspects of sample design, the rotation pattern is a compromise between how frequently and how many times we are prepared to resurvey a sample unit and which characteristics we most want to analyse. Where changes over short periods are important, as often occurs in business surveys, we will typically use a pattern in which a sample unit is revisited on every survey occasion (for example monthly) for *T* occasions. Steel (1997, p. 20) characterised these designs as 'in-for-*T*', and contrasted them with designs where changes over longer periods than the frequency of the survey are also important, typically annual changes within a survey with a subannual period. These types of designs usually have a period in sample, a period out of sample and then a further period in sample, and are denoted *T-O-T*(*n*), with *n* giving the number of periods in sample – as in the US Current Population Survey (CPS), which has a 4-8-4 design (a sampled household is interviewed for four consecutive months, left out for eight months and then interviewed for four months again). This type of design ensures some annual overlap, and therefore takes advantage of sample autocorrelation in estimating annual flows (Figure 2.1).

The number of occasions on which a unit can be sampled is typically a balance between the length of time over which changes are to be estimated, the number of occasions on which it is reasonable to expect a sample unit to contribute to a survey, and operational challenges. In the case of business surveys, periods are often quite long to ensure some annual overlap. Many ONS monthly business surveys had an in-for-15 design in the mid-1990s to help support the management of response burden while allowing some annual overlap. However, respondents who are newly included in the sample require more effort to obtain a usable response (that is, the start-up costs are high relative to the ongoing costs), and replacing $\frac{1}{15}$ of the sample on each occasion proved relatively expensive, so in several surveys the rotation pattern has lengthened to 27 months, giving an in-for-27 design, although this has a much greater impact on individual respondents (Smith, Pont and Jones 2003). Statistically the gain is small, since the monthly overlap was already high, and has gone from 93 % to 96 %. There is much more impact on annual overlap, which has gone from 20 % to 56 %, but this is not the main

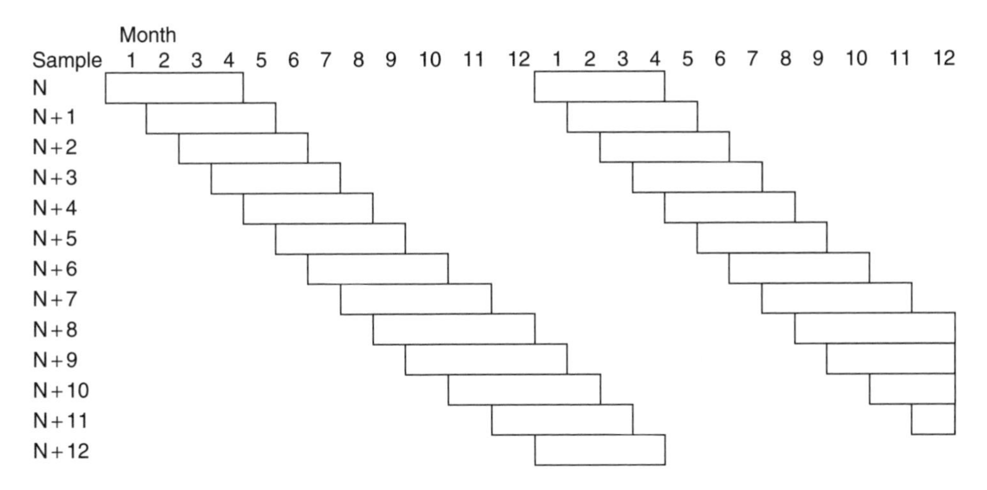

Figure 2.1 A 4-8-4 rotating panel design.

focus of design for short-period business surveys. The big change is in the number of new respondents each month, which is almost halved.

In contrast the costs in longitudinal surveys of households are often concentrated in the follow-up because of the need to follow people moving house (that is, start-up costs are small relative to ongoing costs), in which case the operational and burden constraints act together to keep the number of follow-ups small.

From this discussion it is clear that the objective of the survey has an impact on the choice of rotation period. The longer the period over which changes are to be analysed, the longer the period of surveys required. Over relatively long periods, however, the respondents are more likely to become jaded and drop out of the survey.

One strategy for reducing the impact of a large number of visits while maintaining relatively high sample overlaps for fixed periods is to have a discontinuous pattern, such as the 4-8-4 of the US CPS. This gives 75 % overlap between consecutive months and 50 % overlap between months a year apart, but low overlaps for other changes relative to an in-for-T design.

A stylised version of the decision tree for deciding on the rotation pattern for a longitudinal survey might be as in Figure 2.2.

2.4.2 Dealing with Births (and Deaths)

Most populations are not fixed, and are affected by births (immigration) and deaths (emigration). How to deal with these in a longitudinal study is important because of the effect on the validity and interpretation of the estimates. First let us take the easier challenge – how to deal with deaths. As long as these can be identified and distinguished from nonresponse, they are easily incorporated in analyses by using a code for dead units (and zeros for any numeric variables). If they are not identifiable, the challenge is greater. Imputation based on the valid responses will give biases in flows between states, underestimating flows to deaths and overestimating other flows. But from a design perspective, future deaths are already present in the sample, and so no special action is required.

Births, however, are another matter. With any nonstatic population new births will occur continuously, and if the sample is not updated it will quickly become deficient in

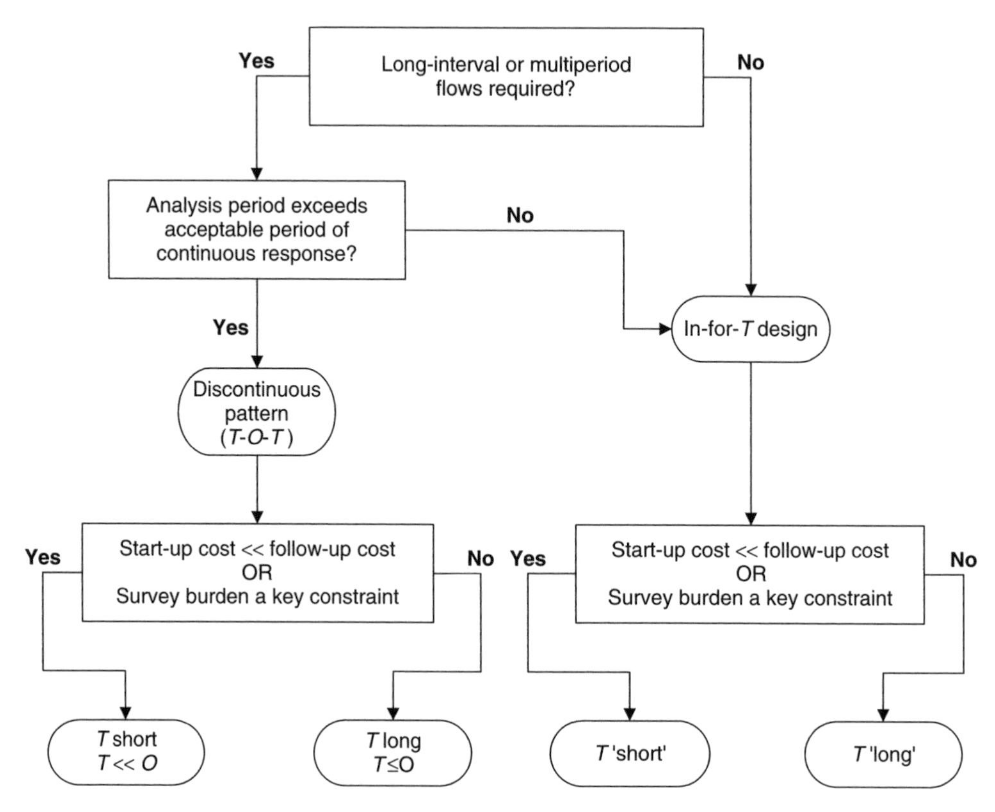

Figure 2.2 Stylised decision tree for selecting a rotation pattern.

this kind of observation. This means having a mechanism for \dentifying births to the population, and sampling from them. Appropriate weighting will be needed if the sampling fraction is different from the initial sample or differs from occasion to occasion if births are added to the sample periodically.

Where there is a good frame for the population that is kept up to date, as in many business surveys, the problem of keeping the sample up to date while retaining the longitudinal linkage is facilitated by using permanent random numbers (PRNs) as the basis of sampling (Ohlsson 1995; Smith, Pont and Jones 2003). Here, each unit on the register (including new births) is assigned a PRN when it is first added to the register. To sample, units in the population are ordered by their PRN and then a range of the available PRNs is used to define the sample; new births are included in the sample when their assigned random number falls in the same sample range. Varieties of PRN sampling can be used to manage rotating panel designs, as well as for strictly longitudinal surveys.

2.4.3 Sample Overlap

Where longitudinal information is produced as a byproduct rather than the main purpose of a panel or rotating panel design, there will be the need occasionally to redesign or reclassify the main survey. In this instance we have an interest in maximising the overlap between the two designs, and there are methods for doing this – see Kish and Scott (1971) and Hidiroglou *et al.* (1991, section 2.2.4).

2.4.4 Stability of Units and Hierarchies

There are two forms of instability over time that can affect sample design for longitudinal surveys. The first is inherent instability in the sample units. In some cases, the sample units (e.g. people) are well-defined and unlikely to change identification during the period of the survey. But in other cases (e.g. businesses), the units are not stable and can split or merge with other units. In these cases it is necessary to define clearly at the start what the approach will be – to follow up all parts or only certain parts, and how they will be treated in analysis.

The options of which parts of a sample to follow up are, from most restrictive (and cheapest) to least (and most expensive):

(1) only the entities initially sampled (e.g. households), so any change results in the unit dropping out of the survey. This is so restrictive that it would only really be useful in the study of a characteristic that is of the household, and not of the individuals within it, and we do not know of any examples of this kind of longitudinal study;
(2) all the components of the initial sample, wherever they end up, for example following all people in initially sampled households;
(3) all components of the initial sample, and any components with which they join, for example all households which contain *any* people from the initially sampled households;
(4) all components of the initial sample, and any components with which they have joined since the start of the study, for example all households that contain any people who have lived in a household with one of the initially sampled people. This last example is really a version of network (snowball) sampling (Berg 1988) and has the potential to lead to very large samples for later waves, and so is not normally adopted.

Options (2) and (3) are therefore the most practical. For a face-to-face interview survey there is little difference in travel costs between them, since the same households (or other survey units) must be visited in either case; the main difference is therefore whether a whole household is interviewed or only part of it. Whole households may be best where some context of the new household will be a useful explanatory variable in analysing the longitudinal outcomes, whereas for more personal data (for example, medical history) it may be sufficient to follow up the original sample people only.

Another possibility is that the units themselves may be stable but the classification of units may change over time. In other words, the characteristic that determines the classification is potentially transient. Examples might be persons with low income or schools with fewer than 100 pupils. This can be problematic if the intention is to restrict the sample to units in a particular category of the classification or to sample different categories at different rates. The first approach (restricting the sample) is only appropriate if the population of interest is genuinely a static one (e.g. people who had a low income on a particular date). The second approach (unequal sampling fractions) may improve the precision of (some) estimates initially but may reduce precision for equivalent estimates after several waves, depending on the speed at which changes in classification occur and the length of the panel. These factors can be difficult to predict in advance, making it difficult to determine the optimum allocation to strata (categories of the classification).

2.5 CONCLUSION

The range of issues for consideration when constructing a longitudinal sample design is large. Longitudinal surveys often need the complexity of cross-sectional designs with additional constraints and considerations for the longitudinal element. All these elements must be balanced to arrive at a final design achieving the best combination of properties within financial constraints. The major steps are to:

- define the population of units, including in time;
- define the units to be followed, and any rules to deal with changes to those units;
- ensure that the sample is representative (measurable) for the population (including any cross-sectional populations) required;
- make use of design procedures for the initial sample (stratification, clustering, etc. for *longitudinal* outcomes) to optimise the use of the available resources;
- consider the need for subsequent sampling techniques, such as to regularly add samples of 'births';
- implement the design.

Some aspects of sample design for longitudinal surveys are not yet well understood. An example is the effect of sample mobility over time on design effects and how this should be taken into account in the initial design. Other aspects – even if well-understood in principle – are difficult to assess empirically at the sample design stage due to uncertainty about the future of the survey. A common example is that funding is rarely assured for more than the first few waves of long-term longitudinal surveys at the time when the initial sample must be designed. Often, as few as one or two waves are assured. This makes it difficult to predict the eventual longevity of the panel and hence to make good assessments of the balance between costs and statistical considerations where these change over waves. For this reason, many longitudinal surveys have sample designs that may not, in the longer term, be the most efficient. There is plenty of scope for research and evaluation to improve the design of future longitudinal surveys.

REFERENCES

Berg, S. (1988). Snowball sampling. In S. Kotz and N. L. Johnson (Eds), *Encyclopedia of Statistical Sciences* (Vol. 8, pp. 528–532). New York: John Wiley & Sons, Inc.

Berger, Y. G. (2004). Variance estimation for measures of change in probability sampling. *Canadian Journal of Statistics*, 32, 451–467.

Binder, D. A. (1998). Longitudinal surveys: why are these surveys different from all other surveys? *Survey Methodology*, 24, 101–108.

Courtenay, G. and Mekkelholt, P. (1987). Youth Cohort Study handbook: the first ten years, *DfEE Research Series* No. 22. London: The Stationery Office.

Duncan, G. J. and Kalton, G. (1987). Issues of design and analysis of surveys across time. *International Statistical Review*, 55, 97–117.

Ernst, L. R. (1989). Weighting issues for longitudinal household and family estimates. In D. Kasprzyk, G. Duncan, G. Kalton and M. P. Singh (Eds), *Panel Surveys* (pp. 139–159). New York: John Wiley & Sons, Inc.

Hidiroglou, M. A., Choudhry, G. H. and Lavallée, P. (1991). A sampling and estimation methodology for sub-annual business surveys. *Survey Methodology*, 17, 195–210.

Holmes, D. J. and Skinner, C. J. (2000). Variance estimation for Labour Force Survey estimates of level and change, *Government Statistical Service Methodology Series* No. 21. London: The Stationery Office.

Kalton, G. and Citro, C. F. (1993). Panel surveys: adding the fourth dimension. *Survey Methodology*, 19, 205–215.

Kish, L. (1965). *Survey Sampling*. New York: John Wiley & Sons, Inc.

Kish, L. (1987). *Statistical Research Design*. New York: John Wiley & Sons, Inc.

Kish, L. and Frankel, M. R. (1974). Inference from complex samples (with discussion). *Journal of the Royal Statistical Society Series B*, 36, 1–37.

Kish, L. and Scott, A. (1971). Retaining units after changing strata and probabilities. *Journal of the American Statistical Association*, 66, 461–470.

Laniel, N. (1988). Variances for a rotating sample from a changing population. *Proceedings of the Business and Economic Statistics Section, American Statistical Association*, pp. 246–250.

Lavallée, P. (1995). Cross-sectional weighting of longitudinal surveys of individuals and households using the weight share method. *Survey Methodology*, 21, 25–32.

Lynn, P. (2006). *Quality Profile: British Household Panel Survey*, Version 2.0: Waves 1–13: 1991–2003. Colchester: University of Essex.

Nordberg, L. (2000). On variance estimation for measures of change when samples are coordinated by the use of permanent random numbers. *Journal of Official Statistics*, 16, 363–378.

Ohlsson, E. (1995). Co-ordination of samples using permanent random numbers. In B. Cox, D. Binder, B. Chinnappa, A. Christianson, M. Colledge and P. Kott (Eds), *Business Survey Methods* (pp. 153–169). New York: John Wiley & Sons, Inc.

Pickery, J. and Carton, A. (2008). Oversampling in relation to differential regional response rates. *Survey Research Methods*, 2, 83–92.

Plewis, I., Calderwood, L., Hawkes, D. and Nathan, G. (2004). *National Child Development Study and 1970 British Cohort Study Technical Report*, 1st edn. London: Institute of Education.

Rose, D. (2000). *Researching Social and Economic Change: The Uses of Household Panel Studies*. London: Routledge.

Skinner, C. and Vieira, M. (2005). Design effects in the analysis of longitudinal survey data. *S3RI Methodology Working Papers*, M05/13. Southampton, UK: Southampton Statistical Sciences Research Institute.

Smith, P., Pont, M. and Jones, T. (2003). Developments in business survey methodology in the Office for National Statistics, 1994–2000 (with discussion). *Journal of the Royal Statistical Society Series D*, 52, 257–295.

Steel, D. (1997). Producing monthly estimates of unemployment and employment according to the International Labour Office definition (with discussion). *Journal of the Royal Statistical Society Series A*, 160, 5–46.

Tam, S. M. (1984). On covariances from overlapping samples. *The American Statistician* 38, 288–289.

CHAPTER 3

Ethical Issues in Longitudinal Surveys

Carli Lessof

National Centre for Social Research, UK

3.1 INTRODUCTION

This chapter reviews some of the ethical issues that are raised by the design and implementation of longitudinal surveys. The objective is to summarise ways in which these differ from ethical considerations in cross-sectional surveys and to highlight aspects of current practice and thinking about how to deal with specific challenges. We begin with a brief review of the history of research ethics, to provide the relevant context, and then set out some of the issues that are particular to longitudinal surveys. Four key topics are discussed – informed consent (Section 3.3), free choice regarding participation (Section 3.4), avoidance of harm (Section 3.5) and participant confidentiality and data protection (Section 3.6). The final section discusses ethical overview and processes for ensuring adherence to ethical standards and then takes a forward look to likely emerging issues.

3.2 HISTORY OF RESEARCH ETHICS

Concerns about the ethics of medical research were raised – albeit unsuccessfully – in the 18th century when Jenner developed vaccination against smallpox. But a much more widespread concern developed in the aftermath of the Second World War and with the rapid growth of medical research in the 1950s, as a substantial number of experiments of very doubtful ethical standards were carried out without informed patient consent. Papworth (1962, 1967) wrote evocatively about these ethical failures. For the first time in the UK, medical ethics and patients' rights were put to the forefront of medical education and practice. The criticisms were not directed at surveys, but the concerns that were expressed led to the development of codes of practice and guidelines that must be met by medical and social scientists. These need to be adhered to by all investigators

Methodology of Longitudinal Surveys P. Lynn

and are key documents for Ethics Committees (or Institutional Review Boards) that are now the gatekeepers for much research.

There are many publications on professional ethics but in the survey context one with a particularly broad and lasting influence is the Code of Professional Ethics developed and published by the International Statistical Institute (1986). This is concerned with maintaining ethical collaborations and professional standards of objectivity. It sought to extend the scope of statistical enquiry for the benefit of the community, while guarding against predictable misinterpretations or misuse of various types of data. Recognising the intrusive potential of some studies, the Institute also emphasised the need for freely given informed consent whenever the active participation of human subjects was involved. In considering the need for confidentiality, dealing with information provided by proxies and in the secondary use of records, the Code provided standards that still apply. Since that time, the (UK) Social Research Association has produced a set of ethical guidelines that drew heavily on the ISI code and have since been updated in both 2001 and 2003 (Social Research Association, 2003). In addition, the British Government has produced its own code of practice for statistics and social research (Government Statistical Service, 1984), the Market Research Society (2005) has revised its code of conduct that was first produced in 1954 and, most recently, the Economic and Social Research Council (2005) has published a detailed framework of research ethics. The ESRC ethics framework provides a valuable contemporary summary:

(1) Research should be designed, reviewed and undertaken to ensure integrity and quality.
(2) Research staff and subjects must be informed fully about the purpose, methods and intended possible uses of the research, what their participation in the research entails and what risks, if any, are involved. Some variation is allowed in very specific and exceptional research contexts for which detailed guidance is provided in the guidelines.
(3) The confidentiality of information supplied by research subjects and the anonymity of respondents must be respected.
(4) Research participants must participate in a voluntary way, free from any coercion.
(5) Harm to research participants must be avoided.
(6) The independence of research must be clear, and any conflicts of interest or partiality must be explicit.

Practitioners designing longitudinal surveys face many of the same issues that present themselves in cross-sectional surveys. By definition, however, longitudinal surveys face an additional dimension, since they are centrally interested in measuring change or specific outcomes over time. The time dimension in longitudinal surveys means that study teams carry out certain activities that extend and sometimes deviate from standard practice for cross-sectional surveys. For example, longitudinal surveys have to concern themselves with panel maintenance (Couper and Ofstedal, chapter 11 of this volume) and attrition (Watson and Wooden, chapter 10 of this volume), and they increasingly use techniques such as dependent interviewing (Jäckle, chapter 6 of this volume), which involves referring to responses given at one wave at a future wave. The time dimension has to be explicitly factored into issues of consent, for example for data linkage. Even the concepts of avoiding harm or of informed consent (to participate generally and to agree to specific activities such as data linkage), which have been so central to the development of medical ethics, may need to be considered in a different light. Similarly, efforts to maintain participant confidentiality and to protect respondents' data may also be more challenging.

If it were possible, this chapter might set out some key guidelines for the ethical conduct of longitudinal social surveys that could be applied to any new survey about to begin. However, the complexity of many surveys prevents such a simplistic approach. Paradoxically, moreover, the fact that many longitudinal surveys continue over a long time period means that it is impossible to provide a definitive statement about what will and will not be ethical for the life of a particular survey. For many longitudinal surveys, it is also impossible for the investigators to know in advance exactly what data collection protocols will be used throughout the life of the survey. In medical research, it is acknowledged that the balance between the known benefits and risks of a study may change over time and may lead to a study being stopped. The examples are less stark in social science research that involves personal interviews, but various circumstances may still affect the ethical balance of a study or of an individual's participation in it:

- Over time the study investigators or funders can change and new substantive research issues and hypotheses may emerge. As a result, the nature or focus of a study can evolve. For example, the National Child Development Study (the British birth cohort study that began in 1958) started life as a study of mothers and infant health, then evolved to concern itself with the acquisition of basic skills and later extended the focus to the later life 'outcomes' associated with those skills.
- The relationship of respondents to the survey can change, for example when children in cohort surveys become adult, when adult participants age and experience cognitive decline, or when members of a household panel survey leave home, split from a partner or form a new partnership. Also, a participant's interest or confidence in the survey can grow or subside as personal experiences affect its salience.
- What is considered best practice in ethical research can itself change over the course of a long-term longitudinal survey. In recent years there has been a trend towards regulation, with increasing demands for formal procedures such as written consent from survey respondents. At the same time, there has been a growth in new research opportunities such as developments in technologies for genetic analysis and for linkages to administrative data. These have raised new ethical questions that could not have been foreseen at the outset of surveys that began two decades or more ago. Similarly, surveys starting now will no doubt encounter major contextual changes that will affect ethical considerations but cannot be predicted at the outset.

To give an example, we can imagine that the Millennium Cohort Study (MCS) – whose sample members were born in 2000–2001 – will continue to follow the children into their 60s, as the National Survey for Health and Development, which started to collect data in 1946, is doing currently. A simple thought experiment of this kind shows how challenging it would be to predict all the ethical considerations that will be relevant in the 2060s. The consequence of this is that surveys must be subject to regular review to ensure that appropriate ethical controls are in place, and must be given some protection from the 'changing winds' of contemporary preferences.

The remainder of this chapter sets out some of the ethical issues that are faced specifically by longitudinal surveys. We review four of the key ethical principles identified by the ESRC – informed consent, voluntary participation, avoidance of harm and confidentiality (points 2, 4, 5 and 3 in the list above) – and provide illustrations based on particular aspects of the longitudinal survey research process. The first and last issues raised by the ESRC (integrity and quality; resolution of conflicts of interest) are considered only briefly and more generally in the conclusions.

3.3 INFORMED CONSENT

> Research staff and subjects must be informed fully about the purpose, methods and intended possible uses of the research, what their participation in the research entails and what risks, if any, are involved. (ESRC, 2005, p. 23)

A fundamental requirement of ethical research is to obtain the informed consent of the research participants. There have been occasions when consent was not gained, for example in the notorious case of the Tuskegee Syphilis Study, which began in the 1930s and studied the natural history of syphilis, and which breached every notion of what is ethical in clinical research (Katz *et al.*, 2006). There are a few legitimate exceptions where informed consent is not needed and these are discussed further below. In the great majority of longitudinal studies, however, informed consent must be acquired, and at several points in time.

3.3.1 Initial Consent

First, a survey requires consent for initial participation. In some respects this is the same as seeking consent for a cross-sectional survey. However, the questions raised are whether, when and how eligible individuals should be told that they have been selected to be a member of a longitudinal survey and how much they should be told about what this is likely to involve. The scope of a study is not always known at the outset of the survey. For example, in some instances, what turned out to be the first wave of a longitudinal survey was only planned as a cross-sectional survey and the idea of returning to the sample came later. In other cases only a general intention to carry out further waves exists at the time of the first wave, without any specific plans or funding being in place. Most would argue that the longitudinal nature of the survey should be revealed to sample members at the first wave where possible. Where this is done, it is important to stress the voluntary nature of participation in future waves.

That said, in surveys where the extent to which a sample is representative of the population is crucial, researchers can fear 'putting off' respondents who need only make a commitment to the first stage of a survey for their input to be valuable, and they may prefer to downplay the longitudinal nature of the study at that stage. The researchers hope, of course, that commitment will grow once the participant has become involved with the survey. In other surveys, the extent to which the sample is representative of the total population is less important. Here, the key issue is to recruit participants willing to remain in the study indefinitely (Coleman *et al.*, 1999) and this may mean that stating clearly the long-term goals of the project, in a way that is compatible with the most open dialogue with respondents, is crucial to ensure maximum commitment at the outset.

3.3.2 Continuing Consent

Having gained initial agreement, further issues arise about informed consent for future waves or stages of data collection. Data protection legislation has meant that it has become increasingly common practice in cross-sectional surveys to ask for permission to revisit the household in the future (and in some instances for personal information to be given to another research organisation should a funder wish to transfer fieldwork to another contractor). But once individuals have been recruited to a longitudinal survey, practitioners are understandably reluctant to seek positive consent for a future visit. Instead, they may include a final questionnaire item expressing the intention to revisit the respondent in one or more years in the future, and asking for contact details of a friend or relation who would be likely to know the sample member's whereabouts should they move. This provides the respondent with a clear opportunity to spontaneously report that they do not want to take part again, but without explicitly inviting the respondent to decline further contact. Such an approach leaves the decision about participation for most sample members to be made at the time of the next interview. The rationale is that even if someone gave a full commitment at the end of the current interview, this could only be for permission to be reapproached in the future. And an individual who feels they have 'had enough' at the end of an interview in one year may feel quite differently before the next stage of fieldwork arrives. In summary, explicitly seeking consent to participate twice – at the end of one wave and the start of the next – may be unnecessary and unhelpful.

Once respondents have agreed to take part in the initial data collection, they are more likely to agree to further interventions and requests, in part because they trust the investigators and the survey. The investigators' interests may seem to be aligned with those of participants, but this may not always be the case, or this may change over time. Continuing consent should not therefore be taken for granted. It should be acknowledged that however committed a study member may be, longitudinal surveys with a number of data collection activities or waves inevitably involve the respondent in a series of separate decisions rather than a continuous obligation. Consent to participate can only, therefore, be seen as valid for a particular wave of activity. Each time a sample member is asked to take part in a new wave of data collection their agreement to take part is again required. In some surveys, such as the Millennium Cohort Study, the Multi-Centre Research Ethics Committee (MREC) asks that each sample member should sign a written document agreeing to participate. Written agreements of this kind may become a norm in future social surveys but it goes beyond the current standard in large-scale surveys in most countries. Indeed many would argue that it is unnecessary and disregards the relationship of trust that is usually established in the early stages of the survey.

3.3.3 Consent to Trace Respondents

A key objective of all longitudinal surveys is to maintain response over waves. A vital part of achieving this involves tracing any respondents who move address (Couper and Ofstedal, chapter 11 of this volume). There is some debate about whether sample members should be asked to consent to their (new) address being sought by various means if they move. Few surveys currently ask for such consent, arguing that activities of this kind use the data provided by respondents in ways that respondents would accept. Furthermore, seeking someone's address cannot directly harm that person as they are given an opportunity to consider whether they want to continue to cooperate with the survey once they have been found.

With any sample of named persons it is common practice to make discreet enquiries with near neighbours without releasing information about the study or respondent, and also to search phone books or other public sources. In longitudinal surveys, pre-emptive efforts are made to encourage respondents to tell the study team of any planned move or to inform them of any completed move. Where these measures fail, a diverse range of techniques may be used to trace people who change address. Increasingly, longitudinal surveys collect and use telephone numbers (both landline and mobile) and e-mail addresses that may remain constant even when a household moves. A very effective method is to ask each respondent at every wave to provide or update a stable contact address through which they could be approached in the event of a future move. The benefit of this kind of approach is that respondents have implicitly consented for this tracing activity to take place. Administrative sources may also provide information about new addresses. For example, in the UK it is possible for surveys, via the Office for National Statistics, to use information from the National Health Service Central Register. Via the Health Authority and general practitioner, letters can be sent to study members, asking them to renew their contact with the study organisers. While such efforts to trace individuals are both ethical and necessary – both for the study and to ensure that the participant is given proper opportunities to continue – there remains the risk of invading what has been the intended privacy of the respondent. For this reason, the possibility of asking for consent to trace when the survey is first introduced to the respondent should be considered.

3.3.4 Consent for Unanticipated Activities or Analyses

A longitudinal survey will often have additional elements – involving either extra data collection or extra uses of the data collected – that were not identified at the time when participant consent was obtained. Although some argue that all elements of the survey should be revealed to sample members at the outset and that it is unethical to reveal further elements at a later date, this is often unrealistic for longitudinal social surveys as to anticipate opportunities brought about questions and funding may be uncertain and it may not be possible to anticipate opportunities brought about, for example, by technological advances. The longer the time frame over which a longitudinal survey operates, the greater the likelihood that additional unanticipated elements will be proposed at some point. Luckily, the very nature of longitudinal surveys provides regular opportunities to return to participants to seek additional consents if necessary. This is certainly an advantage over cross-sectional surveys, for which it is also possible that unanticipated later activities could be proposed – for example, linking microdata from an administrative system that was not available at the time of the survey.

3.3.5 Implications for Consent of Changing Circumstances of Sample Members

In some instances, the issue of informed consent can be complicated by changes in the circumstances of sample units subsequent to granting initial consent.

For example, in birth cohort surveys, the sample units are typically newborn or very young children. A parent is asked to provide the data and therefore gives initial consent and answers questions about themselves and their child. As the child grows old enough to exercise some preferences, the child assents to physical or cognitive measurements but is not considered able to give informed consent. When the child becomes older, decision making shifts from the parent to the child and the child is asked to respond on his or her own behalf. This transition can be a difficult one and is exacerbated by the fact that it takes place at a time when the young person is establishing a separate identity. It is not entirely clear to what extent these children are able to disregard the commitment offered by their parents and refuse to take part. Nor is it clear whether the parents understood at the outset that their role would diminish. Additionally, the situation can be complicated by the possibility that only one of the two parents initially gave consent (a situation for which clear guidelines need to be established in advance).

There is an equivalent problem at the other end of the age spectrum where individuals who have participated in a survey may lose the ability to give informed consent as they become physically, cognitively or mentally impaired. At this stage, it is important to find the correct balance between representing these individuals in the survey – by reducing the burden of the questionnaire or by involving proxy informants – and accepting their loss to the panel with the consequent increase in nonresponse bias. The English Longitudinal Study of Ageing (ELSA) acknowledges this issue by inviting respondents to nominate a relative, friend or carer to act as an informant on their behalf in the future. This person may either provide a proxy interview for a living ELSA respondent who cannot take part themselves, or may act as an informant for an End of Life Interview after their death. An ethical issue raised here is whether the request should go further and, rather than ask for a preferred nominated proxy, ask people explicitly whether they consent to having a proxy interview or End of Life Interview conducted on their behalf.

Another particular case is provided by longitudinal surveys where multiple members of a household become sample members and are asked to take part in data collection. Household panel surveys represent one important example, where typically all members of initially-sampled households become sample members. In other surveys, such as ELSA, the focus of the study is on partnerships, with both members of each partnership considered as sample members. In either case, all sample members are followed at subsequent waves, even if a split occurs. This makes it important that each individual sample member should give informed consent and that they should understand that they personally are part of the study sample, not just by virtue of being part of their current household or partnership. Splits between related sample members, as illustrated here, also introduce ethical concerns regarding the use of previously collected data. This is discussed further in Section 3.6.1.

3.3.6 Consent for Linkage to Administrative Data

Increasingly, the investigators of longitudinal surveys want to link micro-level administrative data to the survey data (Calderwood and Lessof, chapter 4 of this volume). This too requires the informed consent of respondents. The administrative data need not be limited to data current at the time of the survey interview. It can include historical information stretching back many years. Alternatively, the administrative data can continue to be collected and linked indefinitely, even after an individual leaves the survey or after the survey stops collecting primary data. In either case, the exact nature of the proposed data linkage must be explained clearly to sample members so that the researchers can be clear that they have received consent for the proposed linkage. It may be necessary to allow for some sample members to consent to part but not all of what is proposed (for example, to linkage of present and past data but not to further linkage in the future).

Throughout the history of using administrative data to supplement survey research, the role of consent has changed. For example, the influential study of doctors' smoking habits by Doll and Hill (2004) did not seek informed consent to link death certificate data to survey data. At the time there were no ethical committees that scrutinised research designs. Now, there is general agreement that consent to link administrative data to survey data is needed except in the most exceptional circumstances. Researchers must explain to respondents what the administrative dataset is, the information that will be used and the purpose to which the data will be put. Further assurances about confidentiality and the protection of the data are normally added.

Though linking survey and administrative data has become much more widespread in recent years, this has raised a number of practical questions about consent, some of which are specific to longitudinal surveys. For example, how long can consent of this kind last? Can it be collected once, or should there be a cool-off period, for example in the form of a check question at the next wave (as is normally the case with ELSA respondents)? Or, more stringently, should consent be collected on every occasion that the respondent is interviewed, as occurs with consent for linkage to the Hospital Episodes database for MCS? Or perhaps it should be collected once with respect to each planned repeat of the linkage? A final question – whether consent for linkage is automatically nullified if the participant leaves the study, and whether the answer depends on the nature of their exit – is discussed further below.

3.3.7 Using Administrative Data without Full Consent

Some longitudinal surveys draw the survey sample from an administrative data source. This means that administrative data are immediately available for all sample members, before the researchers have had an opportunity to ask for consent. It may also mean that the data source can be used to track eligible individuals over time, regardless of their participation and consent. This raises questions about the ethics of using the data in these ways. For example, in the evaluation of the (UK) New Deal for Lone Parents, approximately 70 000 individuals were identified from benefit records and their outcomes (in terms of exits from benefit) were observed (Lessof *et al.,* 2003). Explicit consent was only collected from those who were interviewed face to face at the first wave of the evaluation survey, so that their survey data could be linked to the administrative records that had been identified. This method is very effective in counteracting some of the problems thrown up by nonresponse and attrition, particularly in disadvantaged populations such as this one. It does, however, require 'silent' observation of a large group using administrative data only. But if permission to do this kind of activity is denied, some useful data collection will be impossible, while some survey estimates will be severely biased. In considering the ethical aspects (as in clinical studies) there is therefore one important question. This is whether an individual's health, interests or confidentiality could be affected negatively. The view reached by the researchers involved in the New Deal survey was that there was no negative impact. This conclusion may be easier to reach when no survey data have been collected from respondents, as in the case of longitudinal studies that only link together multiple administrative data sources, such as the Office for National Statistics (ONS) Longitudinal Study and the Work and Pensions Longitudinal Study.

3.3.8 Can Fully Informed Consent be Realised?

The previous sections of this chapter illustrate that there are many aspects of informed consent that may need to be addressed by a longitudinal survey. But informed consent can be a theoretical ideal that proves illusive. When defining informed consent, a Canada Council Consultative Group on Ethics (1977) stated that:

> No research involving human subjects should be undertaken without their freely-given, informed consent, if possible in writing.... The information given should be complete and presented in a way which takes into consideration the level of (their) comprehension. An exact description should be provided of all aspects of the research project... Subjects should always be apprised of any considerations which might lead them to refuse to participate.... Those participating in a research project should never, either before or after the experiment, have any reason for saying that they did not fully understand what was involved.

In discussing these guidelines Jowell (1986) pointed out that, clear and comprehensive though they may be, their fulfilment is not always a simple matter. As an example, he suggests that both consent and coercion can be informed, uninformed or disinformed – that 'informed coercion' is the condition under which many censuses are conducted and that 'uninformed coercion' is the condition under which many observation studies are conducted (since the subjects are unaware of their participation). Furthermore, when the aim is to investigate antisocial or unlawful practices by those in positions of influence, informed consent could not be obtained. It could be argued that asking for it might change the behaviour that the study was designed to investigate, which could in some sense render the study unethical. Important though it is, it cannot therefore be argued that informed consent is the unquestionable cornerstone of all research practice.

In considering informed consent, we need to understand what study participants actually believe with regard to the use of their data, and what they believe they do and do not need to know to make informed decisions about participation. The complexity of many – perhaps most – longitudinal surveys make it impossible to ensure that every sample member has been fully informed of, and has understood, every aspect of the research project. Rather, respondents should be informed of the key aspects about which they can reasonably expect to be informed and should be able to request further information if desired (and should receive full responses to those requests). Above all it is important for respondents to know that safeguards are provided. In practice, some respondents may take more assurance from the fact that there has been an ethical review of the study than from detailed information which they may feel is too detailed or even overwhelming. That overview is often provided in the UK by research ethics committees, which involve lay people from a range of backgrounds working alongside independent academics and others with an interest in ethical issues. In other countries, institutional review boards or other bodies perform a similar function.

3.4 FREE CHOICE REGARDING PARTICIPATION

> Research participants must participate in a voluntary way, free from any coercion. (ESRC, 2005, p. 25)

Another important principle in carrying out ethical social research, distinct from informed consent but related to it, is ensuring that potential study participants have a genuine choice about whether or not to take part. An individual should not be coerced into participating in any study, and interviewers need to be encouraging and persuasive to maintain response while avoiding being forceful. But free choice relates not only to the option to refuse but also to the option to participate. It implies that the research team must make reasonable efforts to ensure that all sample members have the opportunity to (continue to) participate in a survey.

There are various ways in which cooperation can be encouraged, for example by providing information leaflets or newsletters that demonstrate the value of the survey or by keeping in touch between waves to show that the respondent is valued by the study team. More specifically – though the British Birth Cohort Studies and some other longitudinal surveys reject this approach – it is now common for incentives to be given to respondents 'as a token of appreciation'. UK examples of longitudinal surveys providing such incentives include the Families and Children Study (FACS), the British Household Panel Survey (BHPS) and ELSA. The incentives used are quite modest payments in the region of £10 per wave. Informal testing with respondents during pilot studies has shown that such payments are rarely perceived as coercive, but the ethical issues associated with any payments still need to be considered. In some studies such as ELSA and the Health and Retirement Study (HRS), which is carried out in the USA, the incentive is prepaid by cheque and is therefore unconditional on participation. The HRS also makes small payments as 'finders fees' to family members or neighbours who provide follow-up information, and it has occasionally increased incentives to study members who initially refuse to participate. ELSA has similarly tried using increased incentives for cases that, for a variety of reasons, have been returned by interviewers as unsuccessful. These have been issued for a second attempt to a different interviewer, while at the same time a letter is sent to the household offering a higher incentive. Study teams need to make sensitive choices about when it is appropriate to ask an interviewer to reapproach a reluctant respondent.

Longitudinal surveys are distinctive in that respondents who continue to take part wave after wave show a very personal commitment and may be hesitant to refuse even if they feel that they are being asked for too great a commitment. As a result, it beholds the investigators to consider respondents' well-being when making decisions about the content and frequency of future waves of data collection. In the BHPS and FACS the interview has been relatively short and self-contained, but even so the respondents give a lot of their time. In other cases a survey may continue over many years, with relatively long and demanding interviews. Some involve several members of the household, as in the MCS, and include physical or cognitive measurements (for example the British Birth Cohort Studies and ELSA). It is important that investigators plan opportunities to listen to the views of respondents through feedback from pilot studies and through other means such as specially arranged meetings of panel members. If investigators lack a good understanding of what is and is not acceptable, there is a risk that many sample members will stop cooperating with the study. This could threaten the validity and value of the study.

Continuing participation by a large proportion of the sample members in a longitudinal survey is of course at least as important as response at any given wave. Once the first wave is complete, it is important to approach all participants again, unless they have explicitly asked to withdraw or have refused to allow any future approach. Clearly, a balance must be struck between inviting individuals who have opted-out because of a set of temporary circumstances and, on the other hand, persisting with an invitation that is unwelcome. Studies employ various approaches to manage this issue. The FACS, which involves an annual face-to-face interview, gives respondents a 'study holiday' before returning to check whether they might be persuaded to rejoin at a future wave.

A balance needs to be reached between not coercing respondents and failing to provide equality of access and information. Either could be considered unethical. Survey research practice shows that there are alternative routes to maintaining a participant's involvement in a study, for example by allowing them to provide data by another mode or to carry out a shorter interview. This may be appropriate for sample members who are reluctant or sick. The use of proxy informants may also be considered, in which case the most ethical approach may be to seek the respondent's permission in advance to allow a nominated informant to carry out a proxy interview.

Allowing free choice implies that respondents must be genuinely able to withdraw from a survey. From time to time, a respondent may ask to do this. Though this happens only occasionally, and for a variety of reasons, it is a circumstance that is challenging for many investigators as longitudinal research projects involve a composite of data collection and analytical activities, and it is not immediately obvious what the implications of such a withdrawal should be, and what set of decisions is both practical and ethical.

It is clear that an individual who has asked to withdraw from a study should not be invited to take part in any primary data collection activity in the future, and systems are needed to put this into effect reliably. However, though few studies would deliberately risk irritating a study member in this way, it is less obvious whether that person would choose to receive correspondence from the study team, for example telling them about the latest findings.

It also seems likely that someone who asks to withdraw from a study would not want to be included in the next stage of any ongoing administrative data collection exercise. One example of this is the proposed biennial downloads of National Insurance Contributions and benefits, to which roughly three-quarters of ELSA respondents agree. However, this assumption may be wrong; it might be the case that they would not object to future downloads of administrative data as long as it did not involve them personally in any action.

The appropriate response is even less clear with another administrative data linkage. This is the agreement to be flagged on the National Health Service Central Register, which, at some time in the future, will result in information from the person's death certificate being passed to the study team so that their age and cause of death can be identified. At the point when an individual withdraws from a study, it is again unclear whether the 'one-off' consent that they gave for this flag, perhaps many years earlier, is necessarily void.

As well as these concerns about collecting data, the withdrawal of an individual from a study raises difficult questions about what analysis can be done in the future and how their past data should be treated. It may be that any data already collected about an individual, but not yet made available for analysis, should not be released. It is also possible, though some would say undesirable, that analysts who request survey data for the very first time should be offered new versions of all past datasets with that case removed. Some would argue, further, that any analysts holding past datasets should be notified of any withdrawals and that the relevant data records should be deleted or the whole dataset should be recalled. If this approach is to be implemented effectively then very strict controls are needed when new data are released so that updated versions can be provided and monitored. Surveys that never release general datasets – and those that reissue the complete dataset rather than issuing datasets incrementally – are in a much better position to manage this issue than others.

The contrary position is that analysts who already hold specific datasets should be allowed to complete their current work and even carry out any new analyses with that case under the confidentiality restrictions that they have already agreed. Furthermore, that past datasets should be kept intact so that analyses are consistent. The resistance to deleting data from past datasets is both practical and principled. Deleting past data might mean that the analyst would need to rerun their analyses; it would mean that published findings could not be perfectly replicated; it would make one analysis project slightly incompatible with another and it would have a marginal effect on any weighting variable that had been constructed. In principle, it can be argued that the respondent gave their data at a specific time, it was carefully anonymised so that it contained no personal details about them and on that basis it was analysed in good faith. With that in mind, some analysts would argue that a withdrawal from a study represents a clean break for future data collection, but not for the analysis of records of past activities.

In practice, it is highly unlikely that a particular respondent understands the choices that are available to them or the principal investigator, or the implications these choices will have. It is conceivable that a study member who is withdrawing, however disgruntled, will feel more positive about a study that has disappointed them if they are able to express these preferences. They may, in fact, be happy to know that the data they have contributed in the past continue to help analysts. On the other hand, if they genuinely want to withdraw without reservation, then it is important that they understand what can and will be done, and if necessary what cannot.

This issue has several implications. Researchers may hesitate to approach sample members who have asked to withdraw, but there is certainly a lot to learn from people who no longer want to be part of the study. Furthermore, to act ethically, longitudinal studies should ask individuals who wish to withdraw what their wishes are in relation to these issues and to clarify what is and is not possible. All longitudinal surveys may need to be more explicit in the future (as some already are) about the actions they will take in this situation.

3.5 AVOIDING HARM

> Harm to research participants must be avoided. (ESRC, 2005, p. 25)

All medical and social researchers have an ethical obligation to avoid causing harm. Social surveys rarely risk harming participants, but there are four possible ways in which this might happen:

(1) In surveys that involve anthropometric measurement – whether by an interviewer or a nurse – there may be instances when direct harm could result. For example, taking a blood sample could cause bruising, a measurement of walking speed could lead to a fall or an assessment of grip strength could cause pain to a respondent with arthritis. Though problems of this kind are rare, measurements like these are only

carried out where safety precautions are in place, genuine benefit is anticipated and protocols are closely followed.

(2) Even in a standard face-to-face interview where there is minimal likelihood of physical harm, some questions may cause distress or embarrassment. Asking about depression, life satisfaction or incontinence are examples, and even apparently mundane questions can occasionally upset respondents with particular sensitivities. One of the objectives of pilot studies is to identify problem questions, which can then be removed, asked more sensitively or included in a self-completion instrument to provide additional privacy.

(3) There may also be a risk of 'social harm', for example exposing to neighbours the fact that an individual is involved in a study of welfare recipients, or to carers that their clients are answering questions designed to identify abuse of the vulnerable old. Risks of this kind are managed in various ways, for example by training interviewers not to release information to nonparticipants and by framing the survey to take account of possible sensitivities. In certain circumstances, this may partially conceal the genuine purpose of the study and so reduce the possibility of fully informed consent.

(4) Finally, there may be indirect effects that could be construed as harmful to respondents, for example if the findings of a study inform a national change in pension policy that has a negative effect on an individual. Few would argue that consequences of this kind would make a project unethical, as the harm is not caused by the survey itself, but by the policy decision, and is general, not personal (in principle the same effect would have occurred even if a particular respondent had not been a participant).

The nature of potential harm in social surveys is generally much less dramatic than in medical research and these risks rarely prevent either cross-sectional or longitudinal surveys from progressing. For longitudinal surveys, however, there is an additional consideration relating to the possible effects of either intentional or accidental feedback to respondents.

There is an ethical obligation to feed back results to avoid harming the respondent,[1] for example if blood samples are taken and analysis in a laboratory reveals an abnormality such as leukaemia. Less extreme findings of anaemia or high blood levels of cholesterol are more common, and feedback of this kind can have an effect on a longitudinal survey if respondents change their behaviour during the course of a study. It can only be left to investigators to consider whether individual changes of this kind can affect the reliability of data provided in a particular study, but there is general agreement that the ethical need to provide the information in order to avoid harm usually over-rides concerns about conditioning respondents' behaviour and therefore potentially affecting survey results.

[1] There are some exceptions. For example, providing feedback from genetic analysis could inappropriately influence a respondent's assessment of health risks or of heritable disease. Part of the consent for this kind of sample collection is therefore to ensure that respondents understand they will not receive any results. A study team may also resist feeding back measures that are not considered reliable at the individual level – for example, where assessments of cognitive function in a research setting might mistakenly be assumed to be of diagnostic value.

The concern, whenever feedback is given, is that care must be taken to ensure that details provided should not be misconceived or influence a participant to take actions that are inappropriate. Consistent with this, interviewers are trained to refer respondents to other agencies rather than to provide specific feedback, for example if financial questions reveal that a respondent is not claiming a benefit to which they might be entitled. In such circumstances, investigators are encouraged to give helpline numbers as appropriate.

There is also a risk that some participants may be influenced by feedback that is not intended. For example:

- Interviews can throw a spotlight, whether healthy or unhealthy, on a participant's life. In longitudinal surveys some harm may come from being made more aware of one's own decline – for example, in walking speed or in cognitive function – though this effect can be minimised by the customary precaution that the assessments are carried out in private. It is also possible that feeding back study findings can encourage comparisons with others, which may show the respondent's life in an unfavourable light. However, it must be accepted that influences of this kind arise in the course of everyday life and can never be totally avoided.
- In very rare instances, individual data from a survey can have a more powerful influence on a respondent's life. A child, for example, who was involved in a survey but subsequently separated from her birth parents can seek information about her childhood from the survey. The ethical position is that all participants have a right to the information held about them.

The best means of avoiding harm in any social survey is to maintain the best possible relationship between respondents and the study. This involves communicating effectively and providing clear explanations of the study's objectives and findings so that the participant has a positive response to being a participant:

- Providing information and updates about the progress of the survey is ethically desirable in longitudinal surveys (though, as stated above, it may occasionally encourage comparisons that could be unhelpful) and achieveable. Fortuitously, such feedback will also tend to encourage participants to keep in contact and so will minimise attrition (Watson and Wooden, chapter 10 of this volume).
- The reverse of this positive approach is reflected in the harm that can arise when study managers fail to respond appropriately to specific enquiries or fail to feed back to respondents promptly. Participants may feel that the commitment and time that they give to a longitudinal survey should command respect, so bureaucratic errors or a failure to respond might be seen as a serious betrayal.

On the other hand, participants in longitudinal qualitative research studies can at times have a very positive feeling that they themselves are the beneficiaries. Investigators sometimes comment about participants who perceive themselves to be members of a group that is special and irreplaceable and, for them, the experience is meaningful. There is a reciprocal relationship between respondent and interviewer that grows over time, so there are benefits in sending the same interviewer to a household whenever possible. Interviewers in quantitative surveys give the same feedback, though there is little research that systematically explores the effects of this. That said, respondents will sometimes drop out if faced with an interviewer they do not like, so it is possible that respondents should, from time to time, be given the choice to have an alternative interviewer. In practice, this is often done for quantitative surveys through the practice of 'reissuing' unsuccessful cases to different interviewers.

3.6 PARTICIPANT CONFIDENTIALITY AND DATA PROTECTION

> The confidentiality of information supplied by research subjects and the anonymity of respondents must be respected. (ESRC, p. 25)

The final principle underlying any ethical study that we consider here is that of participant confidentiality and the protection of a respondent's data. There are two main areas where this needs to be considered in relation to longitudinal research. The first is within the interview itself and relates to confidentiality issues that arise as a result of dependent interviewing. The other is the treatment of research data – an issue that also preoccupies cross-sectional research studies.

3.6.1 Dependent Interviewing

In recent years, longitudinal surveys have incorporated a technique called dependent interviewing (Jäckle, chapter 6 of this volume).

Dependent interviewing increases the complexity of any questionnaire and increases the resources needed to prepare each wave of a longitudinal survey. However, it is generally perceived positively by respondents and reduces the burden placed on them. Though it does no harm, the technique can occasionally cause frustration or a loss of confidence if a previous response has been miscoded or misrecorded. Issues of this kind are, however, rare. Arguably, reactive dependent interviewing is more likely to cause offence since it might be construed as checking up on or contradicting a respondent. On the whole, this does not appear to create problems in practice. But there are two more significant issues that are introduced by dependent interviewing that need to be addressed when considering the ethical balance of a study.

The first issue that requires care is to avoid revealing confidential data inappropriately. For example, it seems reasonable that we should only feed back information to the individual who originally gave the information in order to avoid providing other household members with information that was given in confidence. This rule is not strictly adhered to if the interviewer is trying to ascertain something as straightforward as who lives in the household and is eligible for interview. But it is particularly important if, for example, a proxy informant completes all or part of an interview on behalf of someone who has previously responded to the survey in person. In that event it would not be appropriate to repeat back to the proxy informant any information that the sample member had reported in private in the past. Similarly, in surveys such as ELSA, where concurrent interviewing is allowed, it would not be appropriate to feed information given at one wave back to a respondent in the presence of a new partner. More generally, it can be argued that it is safest never to feed back information in the presence of any other person who was not present at the time the information was given.

Another scenario that may present difficulties is when the interviewer changes. It can be argued that information given in confidence to one interviewer should not be reported back to the respondent by another. However, few surveys avoid this as it is generally felt that respondents are giving their information to the study rather than to a specific interviewer.

There are therefore arguments against feeding back some answers, particularly the most sensitive information and particularly when other people may be present. A possible solution is to avoid complex situations such as concurrent interviewing, at least where the household composition has changed. Another solution is to provide respondents with a consent question that would 'switch off' dependent interviewing if the respondent was not comfortable with past information being introduced into the current interview. However this would affect the consistency and quality of data collected. Furthermore, feedback from piloting and in-depth discussions suggests that respondents expect us to remember what they have told us in the past and appreciate the fact that we acknowledge and update this information. That said, in some instances this is less clear cut, for example where the initial interview was carried out for one study and the follow-up interview (which may not have been anticipated at that time) was carried out for another. This was the situation when ELSA drew its sample from respondents to the Health Survey for England. In theoretical terms at least, this situation needs to be explicitly considered and defended.

3.6.2 The Treatment of Research Data

Researchers involved in collecting primary data have an obligation to safeguard the information provided by study participants and to ensure that their identities remain confidential. This obligation is equally great whether the study is cross-sectional or longitudinal. Arguably, however, longitudinal surveys may present higher disclosure risks.

The main criterion used for assessing the potential disclosure risk is that an analyst who has no knowledge about who has taken part in a survey should not be able to identify an individual (Office for National Statistics, 2004, pp. 7–8). In a longitudinal survey, observations of change across several waves might be used to generate unique combinations that would provide clues to an individual's identity. For example, the movement of an individual from one area of the country to another, perhaps in combination with a transition from one occupational classification to another, may create virtually unique scenarios that could lead to identification. Tight controls on disclosure risk are therefore needed for longitudinal surveys.

Assessments of disclosure risk usually set aside concerns about disclosure that might take place when a rogue analyst knows that an individual has taken part in a study or knows something about an individual that is not in the public domain. But the examples below show that ignoring this type of abuse may not be appropriate for longitudinal surveys (and perhaps some cross-sectional surveys as well):

- A number of longitudinal surveys have recruited their samples using selection criteria that are unusually transparent to the outside world. For example, three of the four British birth cohort studies selected all births in a given week of a given year, in 1946, 1958 and 1970. Similarly, the Avon Longitudinal Study of Parents and Children (ALSPAC) sampled a slightly broader birth cohort, but within a tight geographical location. This means that if one knew that an individual fitted these criteria it would be possible to search for them.

- Even without these tell-tale criteria, participation in a longitudinal survey, almost by definition, means that there will be repeated contacts between the study team and the study members that may increase the likelihood that attention will be drawn to an individual's participation. For example, ALSPAC invites children to visit clinics where measurements are taken, and the density of study children in local schools means that discussion at the school-gate about participation is almost inevitable (and indeed is welcomed). In these and other similar cases, an outside observer might relatively easily identify a study member if very strict controls were not imposed. As a result, ALSPAC have traditionally had stringent data control arrangements.
- Many longitudinal surveys have multiple sample members within a household (for example, MCS, ELSA and BHPS). In some cases, these individuals may even give some of their responses in each other's presence. An individual with the right set of skills may be able to find their own record on a dataset, and by association is quite likely to find that of their partner, child or other household member. It is unlikely that someone would wish to do this and have the wherewithal to do so, but it is not inconceivable, particularly when we remember that households split and that some longitudinal surveys (such as BHPS and ELSA) will continue to follow several parties.

To some extent, the complexity involved in carrying out longitudinal analysis may provide a natural barrier to casual or accidental abuse. Nevertheless, the arguments above suggest that our approach to the protection of confidentiality and the avoidance of disclosure could, perhaps, be based on stricter criteria. This is something that is implicitly or explicitly acknowledged by many longitudinal survey teams. The various levels of controls exercised by some of the main longitudinal surveys are illustrated in the following examples:

(1) Signed agreement from the study investigators is needed before data from surveys such as NCDS and BCS70 can be released from the social science data archive by the (UK) Economic and Social Data Service (ESDS).
(2) The ESDS has recently introduced 'special licenses' to provide additional controls for more sensitive datasets (see www.esds.ac.uk/orderingdata/speciallicence.asp).
(3) Some longitudinal surveys only allow analysis to take place in a secure data enclave, or by using special tools that facilitate secure remote analysis. This is true of the ONS Longitudinal Study and the Scottish Longitudinal Study, which do not allow any data to be released externally.
(4) ALSPAC and NSHD provide approved researchers with special datasets for each research project, each of which is kept separate and must be destroyed when the project is completed.
(5) Other surveys use a combination of approaches depending on the sensitivity of the data being used. For example, ELSA provides access to core data through ESDS, but imposes stringent controls on analysis, which involves geographical identifiers (producing special reduced datasets for that purpose), and insists that researchers work in a recognised safe setting when using administrative or highly sensitive data.

It is also important that longitudinal surveys maintain effective in-house security measures. These tend not to be the main focus of attention in discussions of disclosure risk but since some part of the study team is likely to hold personal details the potential for disclosure is actually greatest in this hub of activity. The ALSPAC study appears to provide an excellent example of carefully prescribed arrangements.

3.7 INDEPENDENT ETHICAL OVERVIEW AND PARTICIPANT INVOLVEMENT

After World War II, public concern was aroused in the UK when a number of clearly unethical medical research projects came to light. This led to the establishment and continuing development of new codes of practice and to the establishment of Ethics Committees, concerned initially with clinical research but subsequently with cross-sectional and longitudinal surveys of individuals that include any medical component, including simple anthropometric measures such as height or weight. All surveys share some similar ethical issues, but the time scale of longitudinal surveys adds to these, especially if there are changes in the nature of a study or when, with time, children become adults or adults age both physically and mentally. What is considered to be best ethical practice in research can also change as the years pass.

In this chapter we have considered four fundamental principles that underlie ethical research practice (informed consent, voluntary participation, avoidance of harm and confidentiality) and seen how they should be – and are – applied in longitudinal surveys. In addition, the ESRC identify two general principles that must be adhered to: that the 'research should be designed, reviewed and undertaken to ensure integrity and quality', and that 'the independence of research must be clear, and any conflicts of interest or partiality must be explicit'.

On the whole, Ethics Committees have demanded formal arrangements to ensure that these principles are applied. Scientific teams scrutinise content and methods and oversight committees ensure that there are appropriate governance controls in place, but the informal dimensions of ethical research also need to be given consideration. To take one small example, in securing informed consent, Ethics Committees place emphasis on the content of the advance letter and (increasingly) on the content of any study information leaflet or consent form. In practice, many respondents pay little attention to the content of these documents, and the act of giving consent requires trust, a concept that may sometimes be neglected or dismissed. In addition to the formal documentation that is designed to elicit informed consent, activities such as the training and support given to interviewers may be as or more important in providing assurance and key information to the respondent. Figure 3.1 presents a conceptual model of both formal and informal survey procedures for handling the ethical issues of informed consent, free choice and harm.

Of course, study participants also rely on the survey team to ask only what is reasonable and to make good use of their contribution. As a result, investigators have a particular responsibility to identify and respond to any concerns that a participant might have, and to consider issues that a vulnerable participant may not identify for himself or herself. Also, investigators have an obligation to consider the more detailed ethical dilemmas that their own surveys are likely to raise, which may not be noticed by an Ethics Committee viewing the survey from a distance. They should also be encouraged to consult with independent experts and representatives of their participant group, as ALSPAC has done so effectively.

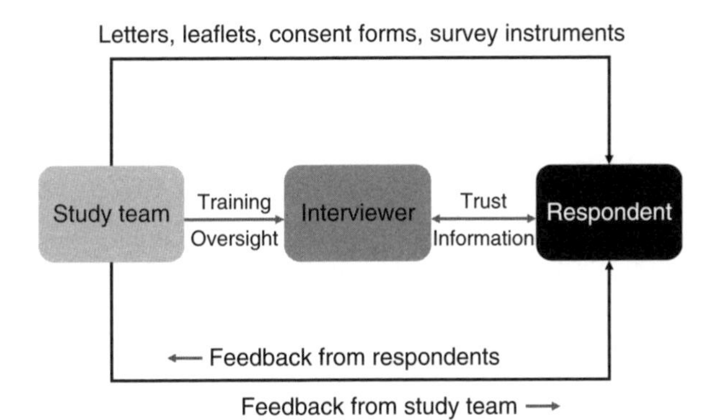

Figure 3.1 Formal and informal dimensions to ethical issues (trust).

Important though all of these ethical considerations are, it has been suggested that the actual conduct of social scientists must depend largely on the scientist's own set of values rather than the imposition of formal or legal controls (Douglas, 1979; Beauchamp *et al.*, 1982). In practice, the ethical attitude of the investigator remains as vital as any formal or legal requirements.

Attitudes change, and the meticulous standards we now require would have seemed excessive a generation ago. In applying the ethical values we now adopt, there is nevertheless a need to avoid unnecessary bureaucratic requirements that have nothing to do with ethical values. The question that needs to be asked, both by investigators and by ethics committees, is: can the use of these data in the way that is proposed harm the interests or the confidentiality of those who gave permission for their use? If the answer is no, the conscientious investigator – and the members of ethics committees – should conclude that the survey is ethical.

ACKNOWLEDGEMENTS

I am grateful to Borbala Nagy for research assistance, to Maurice Lessof for perceptive comments on the draft text, and to the editor, whose suggestions greatly improved this chapter.

REFERENCES

Beauchamp, T. L., Faden, R. R., Wallace, R. J. and Walters, L. (1982). *Ethical Issues in Social Science Research*. Baltimore: Johns Hopkins University Press.

Canada Council Consultative Group on Ethics (1977). *Ethics: Report of a Consultative Group on Ethics*. Ottawa: The Canada Council.

Coleman, M. P. *et al.* (1999). *Cancer Survival Trends in England and Wales 1971–1995: Deprivation and NHS Region*, Studies in Medical and Population Subjects No. 61. London: The Stationery Office.

Doll, R. and Hill, A. B. (2004). The mortality of doctors in relation to their smoking habits: a preliminary report. *British Medical Journal*, 328, 1507.

Douglas, J. D. (1979). Living morality versus bureaucratic fiat. In C. B. Klockars and F. W. O'Connor, (Eds), *Deviance and Decency* (pp. 13–33). London: Sage.

Economic and Social Research Council (2005). *Research Ethics Framework (REF)*. Swindon: ESRC. Retrieved September 20, 2008, from www.esrc.ac.uk/ESRCInfoCentre/Images/ESRC_Re_Ethics_Frame_tcm611291.pdf

Government Statistical Service (1984). *Code of Practice on the Handling of Data Obtained from Statistical Enquiries*. London: HMSO.

International Statistical Institute (1986). Declaration on professional ethics. *International Statistical Review*, 54, 227–242.

Jowell, R. (1986). The codification of statistical ethics. *Journal of Official Statistics*, 2, 217–253.

Katz, R. V., Russell, S. L., Kegeles, S. S. and Kressin, N. R. (2006). The Tuskegee Legacy Project: Willingness of minorities to participate in biomedical research. *Journal of Health Care for the Poor and Underserved*, 17, 698–715.

Lessof, C., Miller, M., Phillips, M., Pickering, K., Purdon, S. and Hales, J. (2003). *New Deal for Lone Parents Evaluation: Findings from the Quantitative Survey*, National Centre for Social Research DWP Report WAE147. London: Department for Work and Pensions.

Market Research Society (2005). *Code of Conduct*. Retrieved September 22, 2008, from www.mrs.org.uk/standards/codeconduct.htm

Office for National Statistics (2004). *National Statistics Code of Practice: Protocol on Data Access and Confidentiality*. London: ONS. Retrieved September 20, 2008, from www.statistics.gov.uk/about_ns/cop/downloads/prot_data_access_confidentiality.pdf

Papworth, M. H. (1962). Human guinea-pigs: a warning. *Twentieth Century Magazine*, Autumn.

Papworth, M. H. (1967). *Human Guinea Pigs*. London: Routledge & Kegan Paul.

Social Research Association (2003). *Ethical Guidelines*. London: SRA. Retrieved September 20, 2008, from www.the-sra.org.uk/documents/pdfs/ethics03.pdf

CHAPTER 4

Enhancing Longitudinal Surveys by Linking to Administrative Data

Lisa Calderwood
Centre for Longitudinal Studies, Institute of Education, UK

Carli Lessof
National Centre for Social Research, UK

4.1 INTRODUCTION

Linkage of survey data to administrative data has become increasingly common in the UK and elsewhere. This growth has been encouraged by the expansion of what is available and technically possible. At the same time, there has been a rise in concern about linkages and an increase in the ethical and regulatory constraints that surround the process.

The primary motivation in linking to administrative records has been to enhance the survey data in order to provide greater opportunities for research. Administrative data can be used to supplement and/or validate survey data. However, in order to fully exploit such opportunities there are substantive issues about how survey and administrative data should be used together. There are also practical issues that need to be considered; these encompass technical issues regarding linkage methodology as well as ethical and legal issues about consents, legislation and disclosure risk. Administrative data can also be integrated into the design of the survey, as a sampling frame or to adjust for nonresponse bias.

Many of the potential uses of administrative data and the issues associated with their use apply to any type of survey. However, there are also particular considerations that are unique to longitudinal surveys or different in the case of longitudinal surveys.

This chapter begins by discussing the value of administrative data as a research resource (Section 4.2) and summarising record linkage methodology (Section 4.3).

The main focus is on discussing the different motivations for linking survey data to administrative data and how these differ for longitudinal surveys compared with

Methodology of Longitudinal Surveys P. Lynn

cross-sectional surveys (Section 4.4). Ethical and legal issues involved in linkage are also discussed (Section 4.5).

The final section (Section 4.6) summarises and concludes.

4.2 ADMINISTRATIVE DATA AS A RESEARCH RESOURCE

Administrative data are routinely collected by organisations, institutions, companies and other agencies so that the organisation can carry out, monitor, archive or evaluate the function or service it provides. Unlike survey data, they are not primarily generated as a research resource.

Administrative data have potential as both a cross-sectional and longitudinal research resource, either alone or in combination with data from other sources such as surveys and censuses (e.g. Nathan, 2001). Administrative data may be either longitudinal or cross-sectional in nature. Many administrative datasets collect and store information by spell, e.g. period of welfare benefit receipt or time spent in hospital. Such datasets are inherently longitudinal. Successive spells for a given individual may be linked with each other so that change can be observed over time. It is also possible to create a longitudinal data resource by linking a cross-sectional survey to administrative data – either to cross-sectional administrative data collected at a different time point than the survey data or to longitudinal administrative data. Longitudinal data resources can also be created by linking different administrative datasets to each other or linking census data to longitudinal administrative data.

Linkage of administrative records to survey data or linkage between different administrative datasets may take place at either individual level or aggregate level. Linkage at individual level involves matching an administrative data record for a particular person to interview data collected from that person in a survey or linking together different administrative records for the same person. Aggregate-level information from administrative records can be linked to individual-level survey data. Most commonly this involves matching ecological data about the area in which survey respondents live, such as the unemployment rate in the local area or deprivation index of the local area. One example of this is a project that linked survey data from the Department for Work and Pensions' longitudinal Families and Children Study (FACS) with local area information (McKay, 2003). Administrative data may also be used as a sampling frame for social surveys and, in the case of longitudinal surveys, to subsequently keep track of respondents for future follow-up.

Using administrative data for research has many advantages in comparison with survey data. Many administrative data resources are censuses of the population of interest in that they, in theory, have 100 % coverage. Typically surveys are samples of populations. As a result administrative data sources often contain very large sample sizes that would be too costly to achieve in a survey. These large sample sizes also facilitate detailed analysis of population subgroups and small areas that would be costly to sample efficiently through a survey. Administrative data may be more accurate than survey data as certain types of measurement error associated with surveys, such as recall error and misreporting, may be overcome. In particular, administrative datasets may contain historical records over a period of many years about which it would be difficult or impossible to collect accurate retrospective information using traditional survey methods. It is also possible to extract administrative data relatively quickly, easily and cheaply in comparison to the time, effort and cost involved in carrying out a survey. A benefit of using administrative data is the avoidance of intrusion in individuals' lives – study members are not involved in any additional effort and so burden is reduced. Also, administrative data typically have the benefit of minimal attrition over time as administrative systems track and trace people more successfully than survey organisations.

Administrative data are generated by many organisations but it is data held by government departments and other public sector organisations that have perhaps the greatest potential for social and economic research. Most government departments keep records of the services they deliver and the processes they register. As a result, they hold information about varied and fundamental aspects of people's lives. This includes key demographic information such as birth and death, data about experience of illness and receipt of health services, participation in and outcomes from the education system, as well as employment spells, income earned, benefits received, tax paid and National Insurance contributions made.

Administrative data themselves are increasingly being used as a research resource in order to provide evidence for civil servants and public bodies. The design and development of administrative databases is increasingly influenced by their potential value as an aid to research in their own right or in combination with linked survey data. For example, the recent creation of the National Pupil Database in England resulted from a decision to centralise records in order to allow the Department for Education and Skills (now the Department for Children, Schools and Families) to monitor the performance of individual schools and Local Education Authorities. This database has since been linked to a variety of surveys and used extensively as a research resource for longitudinal analysis in its own right. Similarly, the UK Department for Work and Pensions (DWP) has undertaken a programme of work to link the records used to administer each of a number of separate benefits (welfare payments). This has enabled cross-domain analysis – both current and longitudinal (Department for Work and Pensions, 2008). The DWP have also used administrative data on benefits longitudinally in combination with small-area data to look at benefit trajectories for Income Support and Job Seekers Allowance claimants in particular areas (McLennan, 2004). Administrative data linkage is also a strategy that the DWP has considered for longitudinal investigation of the link between child poverty and child outcomes (Plewis *et al.*, 2001).

Given their nature and content, it is unsurprising that administrative data held by government departments and public sector organisations are among those most often linked to social surveys in the UK. However, administrative data held by private companies may also be exploited in this way. Some social surveys have made linkages to employer-held information about earnings and occupational pensions. Market research surveys already link to commercial data, for example about spending patterns and debt, and there is great potential for using such data to address social research questions such as the relationship between consumption, diet and health, and between consumption and poverty.

Historically, the use of administrative data for research in the UK has been limited. One notable exception is the Office for National Statistics (ONS) Census Longitudinal Study (LS), which was begun in 1974 and covers about 1 % of the population (500 000 individuals) in England and Wales. It involves linking individual records from different decennial censuses for all individuals born on particular birth dates. These longitudinal census records are then linked to ONS vital event administrative data about mortality, births, deaths and cancer registration. The LS is an important large longitudinal research resource. An equivalent study in Scotland – the Scottish Longitudinal Study (SLS) – has recently been established (Hattersley and Boyle, 2007). This study is based on a 5 % sample (20 birth dates) of Scottish census data and will cover about 274 000 people. As well as linking to vital event data, the SLS will also link to Hospital Episode Statistics (HES) administrative data on hospital admissions.

In recent years, there has been greater recognition in the UK of the potential of administrative data as a research resource. In 2003, the ONS published a discussion paper about proposals for an Integrated Population Statistics System that would bring together individual-level data from census, surveys and administrative records to create a research resource (ONS, 2003). A report to the Research Resources Board (RRB) of the Economic and Social Research Council (ESRC) in 2004 outlined the value of linked administrative data for longitudinal research (Smith *et al.*, 2004). In 2006, the ESRC published a report of the potential for administrative data to be used as a research resource, which included a detailed review of the availability of administrative data in five different substantive areas: education, labour market, health, business and demography (Jones and Elias, 2006). In 2007, the ONS set up a centre for combining data sources within their Methodology Directorate (Cruddas, 2007).

In other countries, there is a much longer history of using administrative data as a research resource. Some European countries like Denmark and Sweden have an integrated population statistics system and in other countries like USA and Canada administrative data have been integrated into survey design much more commonly than in the UK.

The exploitation of administrative data for research purposes has been greatly facilitated by computerisation of administrative records and by technological advances. It is now easier, cheaper and quicker to generate new research resources from existing sources than it is to create new data sources.

4.3 RECORD LINKAGE METHODOLOGY

This section outlines the development of record linkage methodology, methods of automated record linkage and practical considerations in relation to linkage. These issues are discussed more extensively elsewhere (Jaro, 1995; Gill, 2001; Blakely and Salmond, 2002; Herzog *et al.*, 2007). We consider here methods for linking two or more different records that are believed to relate to the same person. Usually the records come from separate sources and so are in separate data files, though record linkage may also be attempted within data files to identify duplicate records. Record linkage methodology involves comparing individual identifiers that appear on each record in order to determine which pairs of records relate to the same person.

The aim of record linkage is to maximise the number of pairs of records that are correctly linked together, i.e. to maximise the number of linked records that have been correctly linked (true positives). However, it is possible for pairs of records relating to different people to be incorrectly linked (false positives). Similarly, true matches may not be linked (false negatives). Minimising false positive links will tend to lead to an increase

Table 4.1 Possible outcomes of record linkage.

Outcome of linkage	Records relate to same person	Records relate to different persons
Linked	True positive	False positive
Unlinked	False negative	True negative

in false negative nonlinks and vice versa. A common compromise involves aiming to minimise the sum of both of these. Table 4.1 summarises the possible outcomes of record linkage for each pair of records compared.

There are two types of individual-level record linkage: exact (or deterministic) matching and probabilistic matching. A third type of linkage often referred to in the literature is statistical data linkage – also referred to as data fusion (Baker *et al.*, 1989; Rässler, 2002). This is different in that the aim is not to bring together records for the same person. Rather it is to bring together records for people who are similar in terms of the characteristics recorded on the different datasets. This type of matching is used in propensity score matching and evaluation methodology, particularly in order to generate real or virtual refreshment samples. Evaluation methodology is discussed further in the next section.

In deterministic matching, pairs of records are only accepted as a link if there is exact agreement on all matching variables. This type of linkage requires a variable to be present on each file that is universally available, fixed, easily recorded, uniquely identifying and readily verifiable (Gill, 2001). In general, this means that an ID number is required, as other types of identifying characteristics such as name, sex and date of birth do not fulfil these criteria. In the UK, the most common examples of these are National Insurance Number (NINO) and National Health Service Number. In some other countries, individuals have a unique ID number that is used across administrative databases.

If such a variable is not available, an alternative matching method must be used. In practice data files will contain error, often due to clerical error when variables are transcribed or input. The existence of error means that, even if an ID number is available, deterministic matching may lead to false positives and false negatives. Also, the ID number may be missing for some cases. Probabilistic matching allows matches even where there is some disagreement between matching variables. Probabilistic record linkage uses a set of matching variables and allocates a weight to each. These weights are used to calculate a total weight for each comparison pair depending on which of the variables they match on – often referred to as the 'agreement weight'. A large, positive total weight indicates that all or most of the matching variables agree and a large negative weight indicates that all or most do not agree. Upper and lower threshold values for this total weight are established depending on the trade-off between false positives and false negatives. Record pairs with a total weight above the upper threshold are categorised as links and pairs with a total weight below the lower threshold are categorised as nonlinks. Pairs with a weight between the threshold values are compared manually to ascertain whether or not they should be linked.

The weight for each match variable is determined by the probability that the variable will agree by chance (u probability) and the probability that the variable will agree if the comparison pair is a true match (m probability). The m probability is one minus the error rate in the field. The more reliable a field is, the higher the m probability.

The *u* probability can be estimated either from the files to be matched or from external data sources at population level and depends on the distribution of the values of the variable in the file or population. It is generally different for different variables and can be different for different values of the same variable. For example, sex can only take two values – male and female – so the probability that sex will agree purely by chance is very high – there is a 1 in 2 chance (assuming the distribution of males and females in the file mirrors the general population). Month of birth has a 1 in 12 chance of matching purely by chance so a greater weight should be attached to matches on month of birth than sex. There are also variables, like surname, for which the *u* probability will depend on the value of the variable. A match on a very common surname will carry less weight than a match on a less common surname.

In large files, blocking is used to limit the number of comparisons that need to be made. Blocking involves choosing one or more of the match variables and only making comparisons between pairs that match exactly on this variable. Before any linkage can be attempted there is often substantial work required to prepare the files to ensure that matching variables are in the same format in each file and that each file has the same structure. This is discussed in detail by Gill (2001).

Hierarchical matching combines elements of deterministic and probabilistic matching. It involves specifying multiple matching variables in a priority order. Each record is matched on as many of the variables (starting with the top priority variable) for which an exact match can be found. Consequently, there will be several 'match types' – records matched only on the first two variables, records matched on the first three variables, and so on. Match types involving fewer variables have a higher probability of agreement purely by chance and this probability can be calculated as described above.

A variant is to use multiple nonhierarchical match types. For example, some records may match on variables A, B and C, some on A, B and D and some on A, C and D. A benefit of this approach is that it can lead to a greater proportion of records that match on at least one type (e.g. Gomatam *et al.*, 2002; Jenkins *et al.*, 2008). With both hierarchical and nonhierarchical matching, information about which cases match on which match types can help to evaluate the linkage. Those pairs of records that agree on multiple match types may be less likely to be false positive matches than pairs of records that match on fewer or only one match type.

Record linkage methodology, as outlined above, was developed in the 1950s and 1960s (Fellegi, 1997). The advent of computers enabled automated methods for record linkage to be developed and applied to large data files. Automated methods for record linkage using weights and probabilities were first developed by Howard Newcombe and colleagues at Statistics Canada. Their methods were further developed and formalised into a statistical model by Felligi and Sunter (1969), who also presented an optimal procedure for deciding on the threshold weights depending on the acceptable probabilities of false positives and false negatives. The 1980s saw the development of general purpose matching and linking software – particularly by Statistics Canada and the US Census Bureau – and refinements and development of record linkage methodology by various authors (Jabine and Scheuren, 1986; Jaro, 1995; Larsen and Rubin, 2001; Winkler, 2001; Blakely and Salmond, 2002).

In the UK, an early example of research that linked survey data and administrative records was provided by Doll and Hill (1954) in their longitudinal survey of doctors' smoking habits and their eventual cause of death. A postal survey of doctors' smoking habits was supplemented in later years by obtaining death certificates. The study established, beyond doubt, that smoking was a major cause of death from diseases such as cancer of the lung and coronary thrombosis.

Another early example of record linkage – but on a much larger scale involving automated matching – was the Oxford Record Linkage Study (ORLS). This began in 1962 and involved linking together medical records from different sources for 350 000 people living in the Oxford area. This study was extended to all of the Oxford Health Region in 1985 with a population of 2.5 million and has been used in many substantive applications, including looking at associations between diseases (Goldacre *et al.*, 2000).

4.4 LINKING SURVEY DATA WITH ADMINISTRATIVE DATA AT INDIVIDUAL LEVEL

There are many different motivations for linking survey data with administrative data.

4.4.1 Sampling, Sample Maintenance and Sample Evaluation

Administrative datasets can be used as sampling frames for surveys. The administrative dataset can then be used to provide evidence on nonresponse bias, to develop weighting strategies for dealing with nonresponse and to provide contact information for sample members.

In the context of a longitudinal survey, it is particularly important to understand potential bias that can occur as a result of attrition between waves or sweeps of data collection. One example of the use of administrative records to understand nonresponse bias to a longitudinal survey is the MCS in the UK. The sample for this study was drawn from Child Benefit records held by the DWP. As information was available from the administrative records for all sampled families, it was possible to compare the characteristics of respondents with nonrespondents in order to develop nonresponse weights (Plewis, 2004).

Integrated survey designs of this kind have been common practice in the US for many years, especially those focused on welfare recipients that are often sampled from social security records (Cohen, 2004). There are many examples of longitudinal studies in the USA that are sampled through administrative records. In the Medical Expenditure Panel Survey, administrative records are used to calculate nonresponse weights (Cohen, 2004). The New Beneficiary Survey uses information from mortality records to inform analyses of the correlates of attrition (Antonovics *et al.*, 2000). The Medicare Current Beneficiary Survey (MCBS) uses administrative data as a sampling frame to provide substantive data items on healthcare, to provide targeted refreshment samples to correct for initial undercoverage of certain population groups and to adjust for losses due to attrition. The Survey of Income and Program Participation (SIPP) was designed from the beginning to incorporate administrative data linkages and development work for this integrated survey design started in 1973, with the actual survey starting in 1983 (Herriot *et al.*, 1988).

Administrative data can also be used to help with sample maintenance, as they may contain home addresses of respondents and other contact information. Obtaining such information can help greatly with tracing respondents and thereby minimising panel attrition due to mobility. In the MCS, address updates are provided before each sweep by the DWP. Other administrative datasets that have been used to provide addresses for respondents in longitudinal surveys in the UK are the National Health Service Central Register, which includes every person registered with a General Practitioner in England and Wales, and the Driver and Vehicle Licensing Authority records, which include the home address of everyone licensed to drive in the UK and the registered address and owner of every licensed motor vehicle in the UK.

Administrative datasets can also be used to help evaluate the representativeness of survey samples, both cross-sectionally and longitudinally. Even in the absence of linkage, this can be done by comparing the responding sample with the complete population represented in the administrative data. However, such comparisons are limited to comparable variables available in both the survey and administrative data and may be limited to comparison of marginal distributions for which administrative-based statistics are published. Comparisons based on linked data are generally far more powerful, as multivariate microdata methods can be used and the responding sample can be compared with the selected sample and/or the total population. Such comparisons can shed light on the nature of any undercoverage and/or loss from attrition.

4.4.2 Evaluation Methodology

Another particular use of administrative data is in evaluation methodology. Administrative data have been used extensively in the UK and the USA in the evaluation of active labour market polices, which are designed to get benefit recipients into work. UK examples include the New Deals for Young People, Disabled People and Lone Parents (Lessof *et al.*, 2003; Phillips *et al.*, 2003), and US examples include the Project Network for disabled people (Rupp *et al.*, 1999). Longitudinal administrative data have been used alone to evaluate such programmes by looking at data from different cohorts – before and after policy implementation (Knight and Lissenburgh, 2004) – but more commonly they are used in combination with survey data.

Administrative data on benefit claimants are often used to select samples of participants or potential participants in a government programme, who are then invited to take part in a longitudinal survey that aims to interview individuals before, during and after their participation in the programme. However, it is likely that there will be attrition in the longitudinal survey, i.e. not all of the individuals interviewed before the programme will be interviewed after the programme. If this attrition is nonrandom, this may lead to incorrect conclusions about the outcome of the policy. In particular, differential attrition rates by the outcome in question – in this case entry to employment – will lead to biased estimates of the effectiveness of the policy. For example, if respondents who drop out of the survey are more likely than the sample as a whole to become employed as a result of the programme, the survey will under-estimate the effectiveness of the programme. However, if the sample is selected from benefit records, it is possible to observe the outcome for all sampled individuals, regardless of whether or not they continue to take part in the survey.

A further issue in relation to evaluation methodology is estimating the counterfactual. To evaluate the effect of a programme it is necessary to estimate what would have happened to the participants if they had not taken part in the programme, e.g. what proportion would have become employed anyway? This is usually done using comparison or control groups, sampled from administrative records, who do not take part in the programme. However, it is rarely possible to randomly assign persons to participate or not participate in the programme (but see Purdon and Skinner, 2003) and it is likely that those who do participate are different from those who do not. If these differences are related to the likelihood of gaining employment, then even sophisticated control group methods can produce biased estimates of the effect of the programme. For this reason, some evaluations have used methods such as propensity score matching to identify comparison or control groups from administrative data records (Bryson *et al.*, 2002; Lessof *et al.*, 2003; Phillips *et al.*, 2003; Dorsett, 2004; Dolton *et al.*, 2006). This is a form of statistical matching as discussed above.

4.4.3 Supplementing and Validating Survey Data

Administrative data can be used to supplement survey data by providing additional substantive data items. These may be items that could not be collected by normal survey means or they may replace the need to ask questions directly of respondents. In the latter case, collecting information from administrative records potentially frees up interview time for other questions or reduces the number of questions that need to be asked and hence the burden on the respondent. The former case may arise because respondents are not able to provide the desired information, because there is likely to be recall bias in respondents' reports or because the recall task is too burdensome to be feasible (Dex, 1995). Data on pupil attainment levels on Standard Assessment Tests (SATs) can be obtained from the National Pupil Database, data on hospital admissions from Hospital Episode Statistics or data on benefit receipt from DWP records. In examples like these, it may be more reliable to get this information from administrative data because respondents may not know or be able to report accurately SAT scores, exact dates and reasons for hospital admissions and exact dates and amounts of benefit receipts. Survey responses to questions of this kind may be more incomplete or inaccurate than administrative records.

For these reasons, collecting data from administrative records may be a useful and effective way to supplement survey data and provide opportunities for research that may not have otherwise been possible. For example, in the USA proposals are under consideration to link the Current Population Survey (CPS) to state-level administrative data about participation in the food stamp programme in order to explore the dynamics of programme participation (Wittenburg and Alderson, 2004).

In addition, linking to administrative data may reduce respondent burden, either by reducing the number of questions that need to be asked or by allowing some complex, detailed and uninteresting questions to be replaced by questions that respondents find more interesting or salient. An example of this in a longitudinal survey is the Canadian Survey of Labour and Income Dynamics (SLID). SLID aimed to obtain estimates of income from tax data rather than asking respondents directly. Respondents were offered a shorter interview if they consented to data linkage. Overall, Michaud *et al.* (1995) concluded that this method has reduced respondent burden and improved data quality – in particular, reduced the spikiness of income data – although they had some concerns that some 'underground economy' income items that appeared in the survey data did not appear in the administrative data. Analysts should be aware that mixed-method approaches such as this could lead to very different error structures in the data from each source (administrative data and survey data).

One problem with supplementing survey data with administrative data is that the definitions and concepts used in the administrative data are designed for administrative purposes rather than research purposes and may not correspond well with the scientific concepts that the survey is designed to measure. In a longitudinal survey, stability of measurement over time is extremely important in order to be able to estimate change. Relying on administrative data sources is particularly risky for longitudinal surveys as the definitions used in the administrative data may change over time.

However, the major problem with supplementing survey data with administrative data is that it is not usually possible to obtain the administrative data for all survey respondents. This will result in bias in analyses using the administrative data if respondents for whom the data are not available differ from respondents for whom they are available. There are three main reasons why it is not usually possible to obtain administrative data for all survey respondents. The first is related to the consent to linkage; the second is related to the success of the linkage; and the third is related to completeness of the administrative data. These are discussed in turn below.

In the UK, written consent is usually required in order for the information held about a person on the administrative record to be released to a third party. Surveys typically ask respondents to sign a data release form during the interview. Consent rates for the use of administrative data are usually less than 100 % of survey respondents. Nonconsenters may therefore introduce a potential source of nonresponse bias if there are differences in the characteristics of each group. Evidence from the UK MCS in relation to birth registration and hospital maternity data shows that although consent rates were high (92 %), some groups were less likely to consent than others. In particular, consent was lower among mothers living in Northern Ireland and mothers from ethnic minority groups. Education, age and lone parent status were also related to consent (Tate *et al.*, 2006). Two UK studies both reported consent rates of 78 % for linkage to DWP benefit and tax credit data, but both found evidence of consent bias. The first, based on a national general population sample, found that respondents who were aged 40–49, single householders, had shorter previous interviews or had problems with that interview were less likely to consent (Jenkins *et al.*, 2006). The second, based on a sample of persons in England aged 50 or over, found that respondents aged 75 or over and those in the lowest income quintile were less likely to consent (Marmot *et al.*, 2003; Lessof *et al.*, 2004). Both studies also requested consent to link to other data sources and found that the nature of consent bias was different for the different consents, though there were correlations between them. Jenkins *et al.* asked for consent to link to employer records while Lessof *et al.* asked for consent to link to the National Health Service Central Register, to hospital episode data and to employer pension data. A review of seven UK epidemiological surveys that took place between 1996 and 2002 with over 25 000 respondents also found evidence of consent bias (Dunn *et al.*, 2004) and there is evidence from the USA of consent bias in relation to record linkage (Woolf *et al.*, 2000).

Longitudinal studies have an advantage over cross-sectional studies in that they are able to return to their respondents at multiple time points in order to try to boost consent rates, with the aim of reducing consent bias. Example of studies in which this has been done include the Health and Retirement Study (HRS) in the USA (Olson, 1999) and the English Longitudinal Study of Ageing (ELSA) (Lessof *et al.*, 2004) and the third sweep of the MCS in the UK.

However, respondents who give consent may not be successfully matched to the administrative data, and again it is possible that there will be systematic biases in the characteristics of nonmatched cases (for example, if matches cannot be found for women who have reverted to their maiden names following a divorce). The process of linking survey data with administrative records usually involves a form of probabilistic matching using identifying variables that are available in both the survey and the administrative datasets. Jenkins *et al.* (2008), reporting a methodological project based on a subsample of British Household Panel Survey respondents, found that there were some differences between cases who were matched to DWP data and those who were not matched. Men and single householders were less likely to be matched and those with lower education and lower income were more likely to be matched. However, as the authors discuss, this differential matching rate may reflect different probabilities of benefit and tax credit receipt. Respondents who had not received benefits or tax credits would not appear on the DWP records, so it was not expected that there would be a match in the administrative data for all respondents. In the MCS, match rates were very high – 99 % for birth registration data and 84 % for Hospital Episode Statistics (HES) data. There was evidence of differences between those who were matched and those who were not matched to hospital records and the factors associated with successful matching were different in different countries. In England, matching was less likely if the mother was not born in the UK, was a lone parent, was Black Caribbean or Black African or if the child was not the first born. In Wales, matching was less likely if a language other than English was spoken at home and the child was not the first born. In Scotland and Northern Ireland, there were no significant differences between matched and nonmatched cases (Hockley *et al.*, 2008).

In the USA, the longitudinal Health and Retirement Survey (HRS) has been linked to administrative data on earnings histories, social security benefit histories and supplemental security income payment histories. Due to a combination of refusal to linkage and unsuccessful matching, HRS has linked data for about 75 % of respondents. Olson (1999) finds evidence that the characteristics of respondents with linked data are different from those without.

The success of matching to administrative data will depend in part on the quality of the match variables. Longitudinal surveys have advantages over cross-sectional surveys as they should be able to provide match variables that are more complete (because there are multiple opportunities to collect the data) and potentially more unique (as there are typically more potential match variables, including longitudinal ones).

Commonly used match variables are demographic information (such as sex and date of birth) and name and address. In longitudinal surveys these variables are likely to be more accurate and more complete than in a cross-sectional survey as they can be verified on more than one occasion. In addition, some of these variables – such as surname and postcode – can change over time. Longitudinal surveys are able to collect updates to these variables at each wave. This may be important if repeated linkages or historic linkages are to be attempted. This is often necessary as repeated information over time is needed for longitudinal analyses. Data linkage may be carried out at multiple time points that can be before, during and after the survey has taken place.

For nonmatched cases, longitudinal surveys are able to go back to respondents specifically in order to try to improve the accuracy of the match variables. An example of this is the SIPP in which social security numbers (SSNs) are collected from respondents to facilitate linkage. If it is discovered that there is error in the SSN, respondents are asked for this number again at a later sweep (Herriot *et al.*, 1988).

The final reason why administrative data may not be available for all survey respondents is that there may be missing data or undercoverage in the administrative records. For example, in the UK, databases such as the National Pupil Database and Hospital Episode Statistics only cover state-provided services and do not include the minority of people who receive private education or private health care. Hence, surveys relying on obtaining estimates of pupil attainment or hospital care from administrative records must either accept that this information will not be available for this nonrandom group of respondents, and hence there will be bias in the survey estimates, or attempt to collect this information in another way. Most commonly this is done by collecting data directly from this subsample of respondents, which is only a partial solution to the problem of bias as it means that there is nonrandom variability in the source and reliability of the data.

However, the problem of nonresponse bias affects all estimates based on survey data. All surveys suffer from initial unit nonresponse and particular data items suffer from item nonresponse. Longitudinal surveys also suffer from attrition over time. As a result, the problems of potential bias in linked administrative data may be small compared with overall nonresponse bias, especially on longitudinal surveys. Moreover, much information is typically available from the survey data about respondents for whom administrative data cannot be obtained and this information may be used to correct for potential linkage bias through, for example, weighting. Again, this is especially true for longitudinal surveys, which may have very extensive information about the characteristics of respondents from multiple prior interviews. Moreover, because survey questions are usually a scarce and expensive resource (particularly in longitudinal surveys), collecting data through linkage may still be an attractive and cost-effective prospect despite this problem of potential bias. In addition, if data are collected from administrative records rather than from respondents the resulting decrease in respondent burden may lead to an improvement in response rates and a reduction in sample loss through attrition, which could reduce overall nonresponse bias.

A potential statistical problem with using administrative data as well as survey data is that the different design properties of each data source may mean that the linked data have a particularly complex design structure, which can require non-standard statistical manipulations to analyse appropriately (Chesher and Nesheim, 2006).

Another problem with using administrative data instead of survey data is that the substantive variables available from administrative data sources may contain measurement error. If there is measurement error in the administrative record, then it may not be advisable to rely on record linkage to provide supplementary data items. Perhaps respondents are more reliable than records after all?

Unfortunately neither respondent reports nor administrative records are error-free. For this reason, administrative data are often used not to supplement or substitute for information collected directly from respondents but as a way of validating survey data. Particular substantive variables are collected both by putting questions to respondents and from administrative data, and the results are compared. If there is agreement between the administrative record and the survey record, this may give greater confidence in the estimate. If there is not agreement between the administrative record and the survey record, then further investigation is needed in order to improve the accuracy of the data. In longitudinal studies, it may be possible to cross-reference information from more than one sweep of survey data and from more than one sweep of administrative data. It is also possible in the context of a longitudinal survey to ask particular questions to respondents again, particularly if there is disagreement between the administrative and survey records, and to attempt to improve the survey questions.

In the UK, the DWP has used administrative data to evaluate the reliability of benefit claims as reported in the longitudinal Labour Force Survey (LFS). They estimated that there was under-reporting of benefit claims in the LFS (Brook, 2004). In the MCS, data on birth weight obtained from birth registration data were compared with mothers' own reports. This analysis revealed a high level of consistency between the different sources. The birth weight reported in the survey was within 100 g of the registration weight in 92 % of cases (Dezateux *et al.*, 2005) but certain characteristics of the respondents were related to agreement between the survey and administrative data. In particular, mothers from non-White ethnic groups, long-term unemployed mothers and those living in disadvantaged or ethnic wards had lower levels of agreement (Tate *et al.*, 2005). Also, a mother's report of mode of delivery was compared with administrative records of delivery from Hospital Episode Statistics. It was found that overall agreement levels were very high – 94 % using a six-group classification of delivery mode and 98 % using a three-group classification – but that disagreement between survey data and administrative records was more common for women whose babies were not first born and in women who were not born in the UK (Quigley *et al.*, 2007).

There are also examples from the USA of using administrative data to validate survey data. The Medical Expenditure Panel Survey (MEPS) uses administrative data from medical providers both to evaluate the accuracy of survey data and to adjust or recalibrate survey estimates of medical expenditure (Cohen, 2004). The SIPP compared survey-reported levels of benefit receipt with estimates from social security records. They found that about three-quarters of respondents reported an amount that was within 10 % of the administrative record value but that some groups of respondents – including more recent benefit recipients – were more likely than other groups to report amounts that were consistent with the administrative record (Olson, 2002).

Administrative data may also be used to impute information that is missing in the survey data, either directly from the administrative record (if a variable that is missing in the survey is available in the administrative record for the respondent) or by using the administrative data as auxiliary variable(s) in statistical imputation, such as conditional hot decking.

In addition, comparing administrative data to survey data may be useful for evaluating how administrative variables have been used in analysis based solely on administrative data. For example, the National Pupil Database has been extensively used as a research resource in its own right and the Free School Meal (FSM) status of pupils is commonly used as a proxy for socio-economic status. By linking the administrative data to survey data from the Avon Longitudinal Survey of Parents and Children (ALSPAC) containing more detailed information on socio-economic status, it is possible to evaluate the extent to which FSM status proxies socio-economic status (Hobbs and Vignoles, 2007).

4.5 ETHICAL AND LEGAL ISSUES

There are ethical and legal considerations to be resolved in linking longitudinal surveys to administrative data.

4.5.1 Ethical Issues

Most linkages require the written consent of respondents for the administrative data to be released. This usually involves asking respondents to sign a consent form. It is important from an ethical perspective to ensure that this consent is fully informed (Lessof, Chapter 3 of this volume). This means that it is essential to explain to respondents what the administrative dataset is, the nature of the information that will be obtained and the reasons for requiring this linkage. In longitudinal surveys a relationship of trust is built up between respondents and the survey (mediated through the interviewer) and this may mean that respondents in longitudinal studies are more likely to be compliant than in cross-sectional surveys, so arguably longitudinal surveys have a greater responsibility to take extra measures to ensure that consent is fully informed.

An example of this is the approach taken by the English Longitudinal Study of Ageing. At the second wave of this study, respondents were reminded of the permissions they had given at the first wave and the implications of these permissions, and were given an opportunity to withdraw consent. Following the same approach, it could be argued that when respondents withdraw from a longitudinal survey they should be reminded of the permissions they have previously given, especially if this allows the survey to collect information about them into the future, in order to give them the opportunity to withdraw these permissions as well.

In addition, as longitudinal surveys often wish to carry out repeated linkages over time, there are ethical issues about the longevity of the consent that is obtained. In particular, the issue is whether the survey should attempt to collect consent for unlimited future linkage or whether the consent should be time-bounded. The former has advantages in terms of securing as much data as possible but it may not be considered ethical to ask respondents to make such an unlimited future commitment.

4.5.2 Legal Issues

It is, of course, imperative that any data linkage is lawful. However, the legal framework around data linkage is not well-defined in the UK. There have been attempts to clarify this in relation to data sharing for official statistics (ONS, 2004, 2005). The main legal considerations are the Data Protection Act (DPA) 1998 and the Freedom of Information (FOI) Act, which both contain regulations about the release of personal data but from opposite perspectives; the DPA provides safeguards against the release of these data whereas the FOI Act specifies when such data are able to be released. There is also a UK common-law right to confidentiality and, from the EU Data Protection Directive, a requirement for an individual to actively signify consent to the release of their data.

In addition, before seeking to gain consent from respondents, researchers must first agree with the holders of the administrative data and their legal advisors on a form of words to be used for the consent form. For surveys in which consent is being given by parents for the release of data about their children, it is also necessary to ensure that consent is collected from a parent who is legally able to do so, for example, consent from a step-parent or even a nonmarried natural father may not be legally valid.

4.5.3 Disclosure Control

The provision of additional administrative data could also increase the risk of disclosure, especially in longitudinal surveys where the cumulative nature of data about individuals means that disclosure risk is already higher than in a cross-sectional survey. This is one reason why administrative data are so safely guarded. In some cases, administrative data may contain information that respondents do not know themselves and may not wish or need to know. Appropriate access arrangements for linked administrative data must be made. It may be that only limited kinds of linked administrative data should be publicly available and that access to certain data should be restricted, for example through the use of special downloads or safe settings/data enclaves. The practicalities of setting up and supporting a safe setting or data enclave poses organisational challenges in relation to establishing systems and rules to monitor and evaluate outputs and providing the staffing resources to make this possible (Ritchie, 2005).

4.6 CONCLUSION

Linking longitudinal survey data to administrative data can enhance longitudinal research studies in many ways. By validating and supplementing survey data, administrative information can improve data quality and provide additional research opportunities for analysts. The ability to use administrative records (particularly if used as sampling frames) to track respondents and to understand and adjust for attrition are important benefits that can improve the overall quality of longitudinal data.

The procedures involved in linkage do, however, present challenges. These concern the methods to be used in the matching process, the minimisation of potential bias and measurement error and the simultaneous need to manage ethical issues and disclosure risk.

Finally, there is very little evidence about respondents' views of administrative data linkage and how this might affect their willingness to cooperate with a longitudinal survey. Perhaps asking permission to link to administrative databases that are perceived as 'important' or 'official' may enhance the status of the survey in the eyes of respondents and hence increase their loyalty or commitment to the survey. However, the reverse may also be true; data linkages may have a negative effect on their perception of the survey and their future cooperation, especially if they feel that collecting data from these administrative sources is intrusive or inappropriate. This would be a useful avenue for further research.

REFERENCES

Antonovics, K., Haveman, R., Holden, K. and Wolfe, B. (2000). Attrition in the New Beneficiary Survey and Follow-up, and its correlates. *Social Security Bulletin*, 61, 40–51.

Baker, K., Harris, P. and O'Brien, J. (1989). Data fusion: an appraisal and experimental evaluation. *Journal of the Market Research Society*, 31, 153–212.

Blakely, T. and Salmond, C. (2002). Probabalistic record linkage and a method to calculate the positive predictive value. *International Journal of Epidemiology*, 31, 1246–1252.

Brook, K. (2004, September 27). Linkage of labour market survey data with administrative records. Talk presented at *ESRC Research Methods Programme Seminar: Linking Survey Responses and Administrative Records*, London. Retrieved January 2, 2008, from www.ccsr.ac.uk/methods/events/linkage/

Bryson, A., Dorsett, R. and Purdon, S. (2002). The use of propensity score matching in the evaluation of active labour market policies. *Department of Work and Pensions Working Paper*, No. 4. London: Department of Work and Pensions.

Chesher, A. and Nesheim, L. (2006). Review of the literature on the statistical properties of linked datasets. *Department of Trade and Industry Occasional Paper*, No. 3. Retrieved January 2, 2008, from http://www.berr.gov.uk/files/file24832.pdf

Cohen, S. B. (2004). Integrated survey designs: a framework for nonresponse bias reduction through the linkage of surveys, administrative and secondary data. *Agency for Healthcare Research and Quality Working Paper*, No. 04001, http://www.ahrq.gov

Cruddas, M. (2007, 11 May). Combining data: developing a centre in Methodology Directorate to meet the challenges. Paper presented to the *12th Meeting of the UK National Statistics Methodology Advisory Committee*, London.

Department of Work and Pensions (2008). *Work and Pensions Longitudinal Study – Background Information*. Retrieved January 2, 2008, from www.dwp.gov.uk/asd/longitudinal_study/ic_longitudinal_study.asp

Dex, S. (1995). The reliability of recall data: a literature review, *Bulletin de Méthodologie Sociologique*, 49, 58–89.

Dezateux, C., Foster, L., Tate, R., Walton, S., Samad, L., Bedford, H., *et al.* (2005). Children's health. In S. Dex and H. Joshi (Eds), *Children of the 21st Century: From Birth to Nine Months* (pp. 133–158). Bristol: The Policy Press.

Doll, R. and Hill, A. B. (1954). The mortality of doctors in relation to their smoking habits. *British Medical Journal*, 4877, 1451–1455.

Dolton, P., Smith, J. and Azevedo, J. P. (2006). The economic evaluation of the new deal for lone parents. Paper presented at the *2006 Latin American Meeting of the Econometric Society (LAMES)*, Mexico City. Retrieved January 2, 2008, from www.webmeets.com/files/papers/LACEA-LAMES/2006/111/paper_lames.pdf

Dorsett, R. (2004). Using matched substitutes to adjust for nonignorable nonresponse: an empirical investigation using labour market data. *Policy Studies Centre Research Discussion Paper*, No. 16. Retrieved January 2, 2008, from www.psi.org.uk/docs/rdp/rdp16-refreshment-samples-matching-and-attrition-bias.pdf

Dunn, K. M., Jordan, K., Lacey, R. J., Shapley, M. and Junks, C. (2004). Patterns of consent in epidemiological research: evidence from over 25,000 responders. *American Journal of Epidemiology*, 159, 844–854.

Fellegi, I. (1997). Record linkage and public policy – a dynamic evolution. In *Record Linkage Techniques*. Washington, DC: Federal Committee on Statistical Methodology, Office of Management and Budget.

Fellegi, I. P. and Sunter, A. B. (1969). A theory for record linkage. *Journal of the American Statistical Association*, 64, 1183–1210.

Gill, L. (2001). Methods for automatic record matching and linkage and their use in national statistics. *National Statistics Methodology Series*, No. 25. London: Her Majesty's Stationery Office. Retrieved January 2, 2008, from www.statistics.gov.uk/downloads/theme_other/GSSMethodology_No_25_v2.pdf

Goldacre, M., Kurina, L., Yeates, D., Seagroatt, V. and Gill, L. (2000). Use of large medicinal databases to study associations between diseases. *Quarterly Journal of Medicine*, 93, 669–675.

Gomatam, S, Carter, R., Ariet, M. and Mitchell, G. (2002). An empirical comparison of record linkage procedures. *Statistics in Medicine*, 21, 1485–1496.

Hattersley, L. and Boyle, P. (2007). The Scottish Longitudinal Study: an introduction. *LSCS Working Paper 1.0*. St Andrews: Longitudinal Studies Centre Scotland.

Herriot, R., Bowie, C., Kasprzyk, D. and Haber, S. (1988). Enhanced demographic-economic data sets. *Survey of Income and Program Participation Working Paper*, No. 82, US Department of Commerce Bureau of the Census.

Herzog, T. N., Scheuren, F. J. and Winkler, W. E. (2007). *Data Quality and Record Linkage Techniques*. New York: Springer-Verlag.

Hobbs, G. and Vignoles, A. (2007). Is free school meal status a valid proxy for socio-economic status (in schools research)? *Centre for the Economics of Education Discussion Paper*, No. 84. London: London School of Economics and Political Science.

Hockley, C., Quigley, M. A., Hughes, G., Calderwood, L., Joshi, H. and Davidson, L. (2008). Linking Millennium Cohort data to birth registration and hospital episode records. *Paediatrics and Perinatal Epidemiology*, 22, 99–109.

Jabine, T. B. and Scheuren, F. J. (1986). Record linkage for statistical purposes: methodological issues. *Journal of Official Statistics*, 2, 255–277.

Jaro, M. A. (1995). Probabalistic linkage of large public health data files. *Statistics in Medicine*, 14, 491–498.

Jenkins, S. P., Cappellari, L., Lynn, P., Jäckle, A. and Sala, E. (2006). Patterns of consent: evidence from a general household survey. *Journal of the Royal Statistical Society*, 169, 701–722.

Jenkins, S. P., Lynn, P., Jäckle, A. and Sala, E. (2008). Feasibility of linking household survey and administrative record data: new evidence for Britain. *International Journal of Social Research Methodology*, 11, 29–43.

Jones, P. and Elias, P. (2006). *Administrative Data as Research Resources: a Selected Audit, Draft Version 2.0*. Warwick: Warwick Institute for Employment Research and Economic and Social Research Council.

Knight, G. and Lissenburgh, S. (2004). Evaluation of lone parent work focused interviews: final findings from administrative data analysis. *DWP Research Report*. London: Department of Work and Pensions.

Larsen, M. D. and Rubin, D. B. (2001). Iterative automated record linkage using mixture models. *Journal of the American Statistical Association*, 96, 32–41.

Lessof, C., Banks, J., Taylor, R., Cox, K. and Philo, D. (2004, September 27). Linking survey and administrative data in the English Longitudinal Study of Ageing. Talk presented at *ESRC Research Methods Programme Seminar: Linking Survey Responses and Administrative Records*, London. Retrieved January 2, 2008, from www.ccsr.ac.uk/methods/events/linkage/

Lessof, C., Millar, M., Phillips, M., Pickering, K., Purdon, S. and Hales, J. (2003). *New Deal for Lone Parents Evaluation: Findings from the Quantitative Survey*, WAE 147. Sheffield: Department of Work and Pensions. Retrieved January 2, 2008, from www.dwp.gov.uk/jad/2003/wae147rep.pdf

Marmot, M., Banks, J., Blundell, R., Lessof, C. and Nazroo, J. (2003). *Health, Wealth and Lifestyles of the Older Population in England: the 2002 English Longitudinal Study of Ageing*. London: Institute for Fiscal Studies.

McKay, S. (2003). *Local Area Characteristics and Individual Behaviour: FACS Data Linking Project*. London: Department of Work and Pensions on behalf of the Controller of Her Majesty's Stationary Office.

McLennan, D. (2004, September 27). Claiming matters: changing patterns of benefit receipt across Wales, 1995 to 2000. Talk presented at *ESRC Research Methods Programme Seminar: Linking Survey Responses and Administrative Records*, London. Retrieved January 2, 2008, from www.ccsr.ac.uk/methods/events/linkage/

Michaud, S., Dolson, D. and Renaud, M. (1995). Combining administrative and survey data to reduce respondent burden in longitudinal surveys. *Survey of Labour and Income Dynamics Research Paper*, No. 95-19. Ottawa: Statistics Canada.

Nathan, G. (2001). Models for combining longitudinal data from administrative sources and panel surveys, *Proceedings of the 53rd Session of the International Statistical Institute: Invited Papers*, pp. 350–355.

Olson, J. A. (1999). Linkages with data from Social Security administrative records in the Health and Retirement Study. *Social Security Bulletin*, 62, 73–84.

Olson, J. A. (2002). Social security benefit reporting in the Survey of Income and Program Participation and in social security administrative records. *ORES Working Paper*, No. 96.

London: Social Security Administration, Office of Policy, Office of Research, Evaluation and Statistics, http://www.ssa.gov/

ONS (2003). *Proposals for an Integrated Population Statistics System*. London: Office for National Statistics. Retrieved January 2, 2008, from www.statistics.gov.uk/downloads/theme_population/ipss.pdf

ONS (2004). *National Statistics Code of Practice: Protocol on Data Matching*. London: Office for National Statistics. Retrieved January 2, 2008, fromwww.statistics.gov.uk/about/national_statistics/cop/downloads/NSCoPDatamatching.pdf

ONS (2005). *Data Sharing for Statistical Purposes. A Practitioner's Guide to the Legal Framework*. London: Office for National Statistics. Retrieved January 2, 2008, from www.statistics.gov.uk/downloads/theme_other/NSDataSharing.pdf

Phillips, M., Pickering, K., Lessof, C., Purdon, S. and Hales, J. (2003). *Evaluation of the New Deal for Lone Parents: Technical Report for the Quantitative Survey*, WAE 146. Sheffield, Department of Work and Pensions. Retrieved January 2, 2008, from www.dwp.gov.uk/jad/2003/wae146rep.pdf

Plewis, I. (2004). *Millennium Cohort Study First Survey: Technical Report on Sampling*. London: Centre for Longitudinal Studies.

Plewis, I., Smith, G., Wright, G. and Cullis, A. (2001). Linking child poverty and child outcomes: exploring the data and research strategies. *Department of Work and Pensions Research Working Paper*, No. 1. London: Department of Work and Pensions.

Purdon, S. and Skinner, C. (2003). Survey design and estimation issues in the evaluation of labour market programmes in Great Britain. In *Bulletin of the International Statistical Institute 54th Session: Proceedings* (CD-ROM). Wiesbaden: NOK-ISI.

Quigley, M. A., Hockley, C. and Davidson, L. L. (2007). Agreement between hospital records and maternal recall of mode of delivery: evidence from 12 391 deliveries in the UK Millennium Cohort Study. *BJOG: An International Journal of Obstetrics and Gynaecology*, 114, 195–200.

Rässler, S. (2002). *Statistical Matching*. New York: Springer-Verlag.

Ritchie, F. J. (2005). Statistical disclosure detection and control in a research environment. Unpublished manuscript. Retrieved January 2, 2008, from doku.iab.de/fdz/events/2007/Ritchie.pdf

Rupp, K., Driessen, D., Kornfeld, R. and Wood, M. (1999). The development of the Project NetWork administrative records database for policy evaluation. *Social Security Bulletin*, 62, 30–42.

Smith, G., Noble, M., Anttila, C., Gill, L., Zaidi, A., Wright, G., *et al.* (2004). *The Value of Linked Administrative Records for Longitudinal Analysis*. Report to the ESRC Research Resources Board.

Tate, A. R., Calderwood, L., Dezateux, C. and Joshi, H. (2006). Mother's consent to linkage of survey data with her child's birth registration records in a multi-ethnic national cohort study. *International Journal of Epidemiology*, 35, 294–298.

Tate, A. R., Dezateux, C., Cole, T. J. and Davidson, L. (2005). Factors affecting a mother's recall of her baby's birth weight. *International Journal of Epidemiology*, 34, 688–695.

Winkler, W. E. (2001). Record linkage software and methods for merging administrative lists. *Bureau of the Census Statistical Research Division: Statistical Research Report Series*. Washington, DC: US Census Bureau.

Wittenburg, D. and Alderson, D. (2004). Linking the current population survey to state food stamp program administrative data: phase II report, data development initiatives for research on food assistance and nutrition programs – final report. United States Department of Agriculture, http://www.ers.usda.gov/ The reference number on the report is E-FAN-04-005-1

Woolf, S. H., Rothemich, S. F., Johnson, R. E. and Marsland, D. W. (2000). Selection bias from requiring patients to give consent to examine data for health services research. *Archives of Family Medicine*, 9, 1111–1118.

CHAPTER 5

Tackling Seam Bias Through Questionnaire Design

Jeffrey Moore, Nancy Bates, Joanne Pascale and Aniekan Okon
US Census Bureau

5.1 INTRODUCTION

This chapter examines the impact of dependent interviewing procedures on 'seam bias', a phenomenon peculiar to longitudinal panel surveys. Seam bias refers to the tendency for estimates of change measured across the 'seam' between two successive survey administrations to far exceed change estimates measured within a single interview – often by a factor of 10 or more. The presence of seam bias almost always signals measurement error. Much research over the past two decades has documented the existence of seam bias in longitudinal surveys, and has also shed light on its essential nature – too little observed change within the reference period of a single interview, and too much at the seam. Attempts to control seam bias have met with some success, but have been limited primarily to employment-related characteristics.

The US Census Bureau implemented new procedures in the 2004 panel of the Survey of Income and Program Participation (SIPP) in an attempt to significantly reduce seam bias for a wide variety of characteristics. The primary tool for accomplishing this was a more extensive and more focused use of dependent interviewing (DI) procedures, wherein 'substantive answers from previous interviews are fed forward and used to tailor the wording and routing of questions' in the next interview (Jäckle, Chapter 6 of this volume). This chapter describes those procedures, and examines their impact on estimates of month-to-month change across the initial waves of the new panel for reports of participation in government transfer programmes, school enrolment, employment, earnings and health insurance coverage, through a comparison with similar estimates derived from the most recent prior SIPP panel, the 2001 panel. We find evidence of significant improvement with the new procedures – estimates of month-to-month change from the initial waves of the 2004 panel are in general much

Methodology of Longitudinal Surveys P. Lynn

less afflicted with seam bias than their 2001 counterparts. Even with the improvement, however, much seam bias still remains.

The remainder of this chapter is organized as follows: Section 5.2 briefly describes the seam bias phenomenon, and summarizes work that has attempted to understand and ameliorate it. Section 5.3 provides a brief background on SIPP, and describes and contrasts its old and new DI procedures. Section 5.4 presents the primary research results, which consist of comparisons of 2004 SIPP panel seam bias results, across a variety of characteristics, with results for the same characteristics derived from the old questionnaire used in the 2001 SIPP panel. Section 5.5 offers our conclusions, including implications of the current findings for future research.

5.2 PREVIOUS RESEARCH ON SEAM BIAS

Seam bias began to draw the attention of survey methodologists in the early 1980s. Czajka (1983, p. 93), for example, describing data from a survey that was the precursor to the US Census Bureau's SIPP, notes 'a pronounced tendency for reported program turnover to occur between waves more often than within waves'; Moore and Kasprzyk (1984) document the effect quantitatively. Soon the phenomenon was identified in the SIPP itself (Burkhead and Coder, 1985; Coder *et al.*, 1987), and in other ongoing longitudinal survey programmes such as the Panel Study of Income Dynamics (Hill, 1987), and the US Census Bureau's quasi-longitudinal labour force survey, the Current Population Survey (CPS) (Cantor and Levin, 1991; Polivka and Rothgeb, 1993). In its subsequent panels, SIPP has continued to provide much evidence of seam bias (Weidman, 1986; Martini, 1989; Young, 1989 – see Jabine *et al.* (1990); Kalton and Miller, 1991; Ryscavage, 1993; Hill, 1994 and Kalton (1998) for summaries of SIPP seam bias research), so much so that Weinberg (2002) lists it as a key unresolved research issue for the survey. Michaud and colleagues have produced numerous papers documenting seam bias and attempts to ameliorate it in Statistics Canada's longitudinal surveys (e.g. Murray *et al.*, 1991; Grondin and Michaud, 1994; Dibbs *et al.*, 1995; Hale and Michaud, 1995; Michaud *et al.*, 1995; Brown *et al.*, 1998; Cotton and Giles, 1998); and in recent years researchers on the other side of the Atlantic have demonstrated that European longitudinal surveys are by no means immune (Holmberg, 2004; Hoogendoorn, 2004; Jäckle and Lynn, 2007). LeMaître (1992), in an excellent general review, summarizes the first decade of seam bias research in terms that still seem apt: 'seam effects would appear to be a general problem with current longitudinal surveys, regardless of differences in design [p. 5]'. Marquis and Moore (1990) confirm that seam bias severely compromises the statistical utility of estimates of change.

Since the very beginning, researchers have considered it almost axiomatic that the amount of change measured between interview waves is overstated. Collins (1975), for example, speculates that between two-thirds and three-quarters of the observed change in various employment statistics (as measured in a monthly labour force survey) was spurious; Polivka and Rothgeb (1993) estimate a similar level of bias. Michaud *et al.* (1995, p. 13) describe apparent change in income across successive survey waves as 'grossly inflated'; similarly, Lynn and Sala (2006, p. 507) label the amount of change they observe from one survey wave to the next in various employment characteristics as 'implausibly high'; see also Cantor and Levin (1991), Hill (1994), Hoogendoorn (2004) and Stanley and Safer (1997).

Other researchers have focused on the other side of the equation – the understatement of change within an interview wave, sometimes called 'constant wave responding' (Martini, 1989; Young, 1989; Rips *et al.*, 2003). Moore and Marquis (1989), using record check methods, confirm that both factors – too little change within the reference period of a single interview, and too much at the seam – operate in concert to produce the seam effect. Kalton and Miller (1991) offer supporting evidence for that assessment, as does LeMaître (1992). Rips, Conrad and Fricker (2003) tie these phenomena to a combination of memory decay over time and strategies that respondents invoke to simplify a difficult reporting task. In support of these positions they cite evidence of increasing seam bias with an increase in the interval between the interview date and the to-be-recalled change (see, for example, Kalton and Miller, 1991), and with increasing task difficulty in general (e.g. Lynn and Sala, 2006).

Along with a better appreciation of the pervasiveness of seam bias, and a better understanding of its underlying nature, came increased calls for possible remedies, among which DI procedures were often mentioned (e.g. Kalton and Miller, 1991; Corti and Campanelli, 1992). Excellent summaries of the pros and cons of DI can be found in Holmberg (2004), Murray *et al.* (1991), Mathiowetz and McGonagle (2000) and Jäckle (Chapter 6, this volume). For those concerned about seam bias, however, and the more general problem of accurate measurement of transitions, the need to control spurious change made DI very attractive. This has been especially true with regard to the measurement of employment-related phenomena. After tests of DI in the CPS showed great promise (e.g. Cantor and Levin, 1991), DI was introduced permanently into CPS procedures in the early 1990s, and has greatly reduced the overestimate of between-interview change in various labour force characteristics (Polivka and Rothgeb, 1993). Hill (1994), in a comparison of successive SIPP panels, one of which did not use DI for employment-related questions, the other of which did, reports similar results. Use of DI in Statistics Canada's Labour Market Activity Survey, and later its Survey of Labour and Income Dynamics, has virtually eliminated seam bias for employment characteristics, according to Brown *et al.* (1998), Cotton and Giles (1998) and LeMaître (1992). More recently, in Great Britain, Lynn and colleagues have experimented with different forms of DI for labour force and other types of questions; they find somewhat inconsistent effects in different circumstances for different forms of DI, but in all cases find the level of spurious change to be consistently highest under conditions of nondependent interviewing (Lynn and Sala, 2006; Lynn *et al.*, 2006; Jäckle and Lynn, 2007).

5.3 SIPP AND ITS DEPENDENT INTERVIEWING PROCEDURES

SIPP is a nationally representative, interviewer-administered, longitudinal survey conducted by the US Census Bureau. It provides data on income, wealth and poverty in the United States, the dynamics of programme participation and the effects of government programmes. Each SIPP panel consists of multiple waves (or rounds) of interviewing, with waves administered three times a year at four-month intervals. The SIPP sample is split into four equivalent subsamples, called 'rotation groups'; each rotation group's interview schedule is staggered by one month, in order to maintain a constant workload for field staff. All SIPP interviews are conducted with a computer-assisted questionnaire; the first interview is administered in person but subsequent interviews are generally conducted via telephone. The SIPP core instrument, which contains the survey content that is repeated in every survey wave, is detailed, long and complex, collecting information about household structure, labour force participation, income sources and amounts, educational attainment, school enrolment and health insurance over the prior four-month period. A typical SIPP interview takes about 30 minutes per interviewed adult. See the US Census Bureau (2001) for a more complete description of the SIPP programme.

5.3.1 SIPP's Pre-2004 Use of DI

Throughout its 20-year history prior to the 2004 panel, SIPP made much use of DI in its 'control card' questions about the household roster and the demographic characteristics of household members, but little in the main body of the questionnaire. In the survey's early panels this was in part a function of its paper-and-pencil interview mode, which is much less conducive to a smooth and accurate administration of dependent questions than is computer-assisted interviewing (CAI) (Corti and Campanelli, 1992; Brown *et al.*, 1998). However, even after the introduction of CAI in the 1996 SIPP panel, neither that panel nor those that followed made much more use of such questions than did their predecessors. In the 2001 SIPP panel, for example (the most recent SIPP panel before the 2004 redesign), some key subject matter areas, such as health insurance coverage, did not use any dependent procedures; each wave of the survey asked about health insurance without any reference to past reports.

Other areas of the 2001 questionnaire employed dependent-like procedures that offered respondents general reminders of their prior reports, but then fell back on completely nondependent wording for the actual question regarding the current wave. For example: 'Last time I recorded that you received Foster Child Care payments. Did you receive any Foster Child Care payments at any time between [MONTH 1] 1st and today?' Extending Jäckle's (Chapter 6, this volume) terminology, we might label this the 'remind, ignore' or 'remind only' approach. This form of DI offers one clear advantage over fully dependent questioning ('remind, continue' or 'remind, confirm/still' in Jäckle's terminology): it is simple to implement, because it does not require any restructuring of the initial questionnaire beyond the simple addition of the 'Last time I recorded...' introduction. Major drawbacks of the form, however, are that it only weakly anchors the respondent's current report to the known past, does little to invite consideration of whether that past state has continued or changed and – not unlike a nondependent question – leaves the respondent focused primarily on the immediate reporting period.

5.3.2 Development of New DI Procedures

In the mid-1990s, concerns about increasing nonresponse and attrition led the US Census Bureau to launch a research and development programme to redesign the SIPP questionnaire for the 2004 SIPP panel. The main focus of this effort was 'interview process' improvements that would yield a less burdensome interview. Data quality improvements were also targeted, however, including a reduction in seam bias, which was found not to have changed for the better with the introduction of CAI procedures in the 1996 panel (Moore *et al.*, 2004). Thus, new procedures were designed to reduce seam bias, primarily through an increased emphasis on the use of DI, as follows:

(1) With the advent of computer-assisted interviewing in 1996, SIPP expanded its traditional, strict four-preceding-calendar-months reference period to also include the current month, up to the date of the interview. This change was motivated more by aesthetic, rather than substantive, considerations – it permitted simpler question wording ('Since [MONTH 1] 1st...' rather than 'At any time between [MONTH 1] 1st and the end of [MONTH 4]...') and made for a more natural response process, since it allowed respondents to report on very recent events. The 'month 5' data were largely ignored, however. No attempt was made to exploit the fact that when an interview month event is reported, a basic fact about the *next* wave's four-month reference period is already known, because the interview month of one interview wave is the first month of the next wave's reference period. That situation changed in the 2004 panel questionnaire; interview month information from one survey wave is now used in the next wave to decide whether to ask a dependent question and, if so, the specific form of that question.

(2) We framed the new questionnaire's dependent questions in truly dependent language, explicitly linking the current wave report to what is known from the last interview, and focusing the cognitive task on whether or not the prior circumstances did or did not continue on into the current wave. The concentration on whether something *continued* from one interview's reference period to the next actually led us to be more restrictive about when to use dependent procedures. In SIPP's 2001 questionnaire, an event that occurred in any month of the previous interview's reference period was sufficient to trigger the 'Last time I recorded...' question introduction in the next wave – even if the event happened only early in the previous interview's reference period and was no longer appropriate to the notion of 'continuing'.[1] The new instrument, in contrast, only considers the previous interview's months 4 ('last month') and 5 (the interview month) in determining whether to ask a dependent question. Events that happened only before those months trigger a nondependent question in the subsequent interview wave, with no mention at all of pre-month-4 events or characteristics.

More specifically, we instituted the following new procedures, with some slight variations, throughout the 2004 SIPP questionnaire:

- An event reported in the interview month of the prior wave (i.e. the first month of the current wave's reference period) triggers an initial confirmatory question in the next interview, e.g: 'Last time I recorded that you received Food Stamps in April. Is that correct?' A 'yes' confirms the person's status for the current reference period, and a later question fills in the details about the remaining months of the reference period. If the respondent does not confirm the prior wave

[1] The 2001 instrument's dependent questions about prior-wave jobs and businesses were an exception to this rule, and in fact closely mirrored the procedures implemented in the 2004 questionnaire for other subject matter areas.

report, then the questionnaire asks about the remainder of the current reference period, e.g.: 'Did you receive any Food Stamps since May 1st?'

- A different strategy is used for events of interest reported in 'month 4' of the prior wave (the last month of that wave's reference period), but *not* in the prior wave's interview month. In almost all cases the interview month report covers only a portion – and often a very small portion – of that month, so a 'no' report actually could mean 'not yet'. Thus, where the 'month 4' report is a 'yes' and 'month 5' is a 'no' ('not yet'), the next wave's interview recalls the 'month 4' circumstances and asks whether they continued into the current wave: 'Last time I recorded that you received Food Stamps in March. Did you continue to receive Food Stamps after April 1st?' The response establishes the person's status for the current reference period; a 'yes' triggers later questions about each individual month.
- If an event or circumstance was not reported in the prior wave, or was only reported in a month other than month 4 or month 5, then the respondent is asked a nondependent question about the current wave.

(3) New DI techniques are also used in the 2004 questionnaire as a follow-up procedure, to reduce nonresponse to income amount questions. Questions about income amounts now begin as nondependent questions, exactly as before, but switch to a dependent format in the event of an initial nonresponse.[2] This 'reactive' form of DI (see Lynn *et al.*, 2006) is in place for all income amount questions in the 2004 questionnaire beginning in wave 2; no such procedures had been employed in any previous SIPP panel. Initial evidence suggests that these procedures have been quite successful at reducing item nonresponse (Moore, 2006a, 2006b). We also thought, however, that we might see some impact of these new procedures on income amount transitions, although perhaps to only a limited extent due to their more limited use. The results of this investigation, too, are summarized briefly in Section 5.4.

5.3.3 Testing and Refining the New Procedures

The project to develop the new questionnaire included a series of three field experiments to evaluate and refine the revised procedures. Doyle, Martin and Moore (2000) describe the design of the field experiments; Moore *et al.* (2004) also cover field test design issues, and provide information concerning the full array of changes implemented in the SIPP questionnaire. The results of these experiments were sufficiently positive (see Moore and Griffiths, 2003) that the new DI procedures for seam bias reduction and for the reduction of income amount nonresponse were implemented in the redesigned instrument used in the 2004 SIPP panel.

[2] Other reactive procedures were also implemented in the 2004 panel to identify possible keying errors in earnings amounts; even though they seem unlikely to have had any noticeable impact on amount transitions, we note this fact in the interest of completeness. Amount reports that deviated (up or down) from the previous wave's report by more than some threshold percentage were flagged for the interviewer's immediate but 'off-stage' attention, to be corrected if necessary or to be accepted as initially entered. Our pretest experience leads us to suspect that these flags were elicited only rarely, but there are no hard data to test this assertion.

5.4 SEAM BIAS COMPARISON – SIPP 2001 AND SIPP 2004

This section examines the impact of the new DI procedures on seam bias for programme participation and other 'spell'-type characteristics. We also look beyond the immediate goal of the DI nonresponse follow-up procedures and examine their impact on transitions in income amounts. We note that our analysis does not cover all characteristics included in SIPP; those selected were chosen with an eye toward breadth and importance, but primarily because they could be analysed reasonably easily with preliminary, internal data files. Our use of preliminary files means that the results presented here may differ from those obtained from future analyses using final, edited data. We use the best evaluation method available to us – a comparison of the 2004 seam bias results with those of the immediately preceding 2001 panel – recognizing that drawing conclusions from a 'natural experiment', as opposed to a designed one, requires additional strong assumptions (e.g. that sample design and field staff differences and the mere passage of time can be ignored). While we acknowledge these limitations, we have no reason to believe that they actually influence our findings in important ways, or affect overall conclusions.

5.4.1 Seam Bias Analysis for Programme Participation and Other 'Spell' Characteristics

Our analysis uses data from the first four interview waves of the 2001 and 2004 SIPP panels. Each panel started its wave 1 interviewing in February of the panel year, so corresponding waves of the two panels cover the same calendar months, three years apart. For each panel we carried out three separate seam bias analyses, one for each successive pair of waves (waves 1–2, 2–3 and 3–4), in effect treating each pair of waves as if it provided an independent set of eight months' worth of data, with one seam in the middle. We chose this approach for its simplicity, for the ease it offered with regard to linking sample cases across waves and also to avoid unnecessary loss of otherwise useful cases absent from only one or two of the four waves (e.g. attritors and in-movers). Each analysis excludes cases for which an interview was obtained in only one of the two waves, and within each characteristic we further exclude cases for which data are missing for either of the months at the seam.

Table 5.1 summarizes the results of these analyses, showing, by panel, the simple average of the three separate estimates for each statistic. The simple average accords equal weight to the three estimates, ignoring the fact that the number of cases from which they are derived generally declines slightly across the three tests, due primarily to attrition. We opted for this approach to avoid giving extra weight to the wave 1–2 pair, which differs from the others in that it includes the only completely nondependent interview (wave 1) in the entire panel, and to avoid giving extra weight to some periods of the calendar year at the expense of others just because of SIPP's arbitrary interview schedule. We doubt that the decision to treat the three wave-pairs equally has any important impact on our results or conclusions, primarily because the differences among the three for any of the characteristics examined are quite minimal.

Table 5.1 is in two parts. Part 1 summarizes the results for characteristics that were captured with very different procedures in the two panels – i.e. where the 2004 questionnaire used the new DI procedures. Part 1A presents the results for 'need-based' public-assistance-type programmes, and Part 1B for non-need-based income sources

Table 5.1 Seam bias analysis for various characteristics, 2001 and 2004 SIPP panels.

Characteristic	Panel	Analysis N for analysed interviews (total observed changes)			Average % of all changes that were at the seam	Average month-to-month change rates (%)			Change rate ratio: seam/off-seam	Average 'directional' change rates at the seam (% change in 2004 compared to 2001)	
		W1–W2	W2–W3	W3–W4		All	Off-seam	Seam		% 'Yes' that changed to 'No'	% 'No' that changed to 'Yes'
Part 1: Characteristics with new dependent interviewing procedures in the 2004 panel											
A. Need-based programmes											
'Public' health insurance coverage[a]	2001	45 560 (2208)	41 572 (1831)	40 799 (1627)	81.9 %[e]	0.6	0.13	3.6	27.7	27.1	1.8
	2004	67 211 (2008)	62 227 (1625)	60 473 (1493)	57.6 %[e]	0.4	0.21	1.6	7.6	5.0 (– 82 %)	1.2 (– 33 %)
Receipt of Federal SSI (Supplemental Security Income)[b]	2001	1596 (334)	1563 (352)	1587 (352)	84.6 %[f]	3.1	0.55	18.5	33.6	13.9	88.6
	2004	1826 (500)	1917 (618)	1929 (687)	77.6 %[f]	4.5	1.2	24.7	20.6	19.4 (+ 40 %)	86.1 (– 3 %)
Receipt of Veterans' Compensation/ Pensions[b]	2001	641 (118)	619 (103)	621 (94)	82.2 %[e]	2.4	0.50	13.7	27.4	10.4	82.4
	2004	1096 (118)	1041 (110)	1006 (96)	47.5 %[e]	1.5	0.90	4.9	5.4	3.8 (– 63 %)	52.1 (– 37 %)
Receipt of AFDC/ TANF[b,c]	2001	575 (345)	534 (326)	479 (267)	55.5 %[e]	8.4	4.4	32.6	7.4	29.0	52.8
	2004	835 (499)	786 (461)	734 (416)	36.3 %[e]	8.3	6.2	21.2	3.4	18.1 (– 38 %)	38.1 (– 28 %)

Receipt of WIC (Women, Infants, Children) benefits[b]	2001	1172 (546)	1130 (436)	1147 (443)	52.7 %[e]	5.9	3.3	21.7	6.6	17.0	68.3
	2004	1713 (672)	1703 (648)	1667 (635)	34.6 %[e]	5.5	4.2	13.3	3.2	9.0 (– 47 %)	49.0 (– 28 %)
Receipt of Food Stamps[b]	2001	2104 (976)	2033 (884)	2008 (823)	52.2 %[e]	6.2	3.5	22.7	6.5	18.1	62.7
	2004	3844 (1512)	3837 (1435)	3802 (1377)	34.4 %[e]	5.4	4.1	13.0	3.2	9.2 (– 49 %)	47.9 (– 24 %)
B. Other (non-need-based) characteristics											
School enrolment	2001	50 948 (6188)	47 458 (9629)	46 920 (5086)	28.9 %[e]	2.1	1.7	4.3	2.5	23.3	2.1
	2004	73 919 (8082)	67 729 (13 549)	65 758 (6189)	17.7 %[e]	1.9	1.8	2.7	1.5	8.7 (– 63 %)	2.1 (0 %)
Private health insurance coverage	2001	51 378 (5607)	47 856 (4971)	47 064 (4616)	73.4 %[e]	1.5	0.46	7.6	16.5	5.4	14.1
	2004	78 162 (6483)	73 217 (6627)	71 128 (5991)	62.8 %[e]	1.2	0.53	5.4	10.2	3.2 (– 41 %)	11.3 (– 20 %)
Receipt of Social Security[b]	2001	9929 (780)	9569 (753)	9442 (655)	79.4 %[e]	1.1	0.26	6.0	23.1	3.8	92.8
	2004	14 999 (1321)	14 610 (1378)	14 403 (1173)	62.2 %[e]	1.3	0.55	5.5	10.0	3.0 (– 21 %)	87.3 (– 6 %)
Receipt of Workers' Compensation[b]	2001	259 (222)	194 (151)	178 (128)	49.3 %[f]	11.2	6.7	38.4	5.7	37.3	41.8
	2004	352 (242)	323 (220)	326 (238)	36.1 %[f]	10.0	7.4	25.3	3.4	23.8 (– 36 %)	30.6 (– 27 %)
Receipt of child support payments[b]	2001	1460 (860)	1470 (817)	1479 (750)	41.6 %[c]	7.8	5.4	22.9	4.2	17.2	63.1
	2004	2476 (1403)	2408 (1298)	2289 (1177)	31.0 %[e]	7.7	6.4	15.5	2.4	10.6 (– 38 %)	48.6 (– 23 %)

Table 5.1 (Continued)

Characteristic	Panel	Analysis N for analysed interviews (total observed changes)			Average % of all changes that were at the seam	Average month-to-month change rates (%)			Change rate ratio: seam/off-seam	Average 'directional' change rates at the seam (% change in 2004 compared to 2001)	
		W1–W2	W2–W3	W3–W4		All	Off-seam	Seam		% 'Yes' that changed to 'No'	% 'No' that changed to 'Yes'
Receipt of alimony[b]	2001	172 (63)	158 (57)	157 (53)	56.3 %[f]	5.1	2.6	19.9	7.7	16.6	60.6
	2004	215 (84)	220 (93)	207 (85)	41.1 %[f]	5.8	4.0	16.8	4.2	14.7 (– 11 %)	33.8 (– 44 %)
Receipt of private pensions[b]	2001	3146 (612)	2894 (489)	2877 (480)	87.6 %[e]	2.5	0.37	15.5	41.9	11.9	87.4
	2004	4681 (417)	4570 (305)	4523 (287)	58.5 %[e]	1.1	0.44	4.7	10.7	3.5 (– 71 %)	71.2 (– 19 %)
Receipt of Federal Civil Service pensions[b]	2001	478 (104)	397 (65)	375 (54)	94.8 %[e]	2.5	0.15	16.6	110.7	14.0	97.2
	2004	761 (50)	745 (35)	728 (55)	70.2 %[e]	0.9	0.31	4.4	14.2	3.3 (– 76 %)	96.7 (– 1 %)
Receipt of 'wealth' income (annuities; estates/trusts)[b]	2001	271 (241)	199 (181)	185 (158)	67.7 %[e]	12.6	4.8	59.8	12.5	60.3	51.9
	2004	456 (320)	378 (159)	363 (165)	31.1 %[e]	7.5	6.1	15.9	2.6	13.6 (– 77 %)	28.8 (– 45 %)

Part 2: Characteristics with the same dependent interviewing procedures in both panels											
Medicare	2001	51 950 (706)	48 508 (429)	47 809 (404)	67.9 %[g]	0.15	0.05	0.71	14.2	0	0.86
	2004	75 713 (868)	70 152 (483)	68 067 (445)	69.5 %[g]	0.12	0.04	0.59	14.8	0.02	0.71
Employment at the same job[d]	2001	40 684 (15 410)	29 533 (11 811)	36 562 (13 140)	36.2 %[h]	5.4	4.0	13.7	3.4	8.8	33.6
	2004	52 799 (17 510)	50 862 (22 052)	48 873 (20 879)	37.7 %[h]	5.7	4.1	15.3	3.7	9.4	44.8

Notes: The data source is preliminary, internal, unedited ('TransCASES') questionnaire data files, unweighted. Table entries represent the simple mean of three estimates derived from individual analyses of each of three pairs of SIPP waves – waves 1–2, 2–3 and 3–4. Each individual analysis included only 'adults' (people age 15+) for whom a completed interview was obtained in both waves, and excluded cases with missing data in either seam month.

[a] Analysis restricted to people aged 20+.

[b] Analysis restricted to people who reported at least one month of receipt of this income type in either of the two successive survey waves.

[c] AFDC = Aid to Families with Dependent Children; TANF = Temporary Aid to Needy Families (which replaced AFDC as a result of the 'welfare reform' legislation of the mid-1990s).

[d] The unit of analysis is the job, rather than the person; the analysis includes all jobs held in at least one month in either of the two successive survey waves. Note that an error in the 2001 wave 3 rotation 1 instrument resulted in the inability to link wave 3 jobs with wave 2 jobs. The error only affected rotation group 1 – the instrument operated correctly for rotation groups 2, 3 and 4. Therefore, the data from rotation 1 are excluded from the 2001 W2–W3 analysis.

[e] According to a simple t-test of the difference between two proportions, the 2001–2004 difference was statistically significant in each of the three individual analyses.

[f] According to a simple t-test of the difference between two proportions, the 2001–2004 difference was significant in two of the three individual analyses, and in the same direction (but nonsignificant) in the third.

[g] All three individual 2001–2004 differences were nonsignificant.

[h] All three individual 2001–2004 differences were statistically significant, but the sign of the difference was inconsistent across the three comparisons.

and characteristics. Part 2 presents results for two characteristics whose measurement procedures did not differ across the two SIPP panels. In the case of Medicare (the US government health insurance programme for the elderly), once a person is eligible and enrolls he/she is covered for life. Thus in both 2001 and 2004 there were no DI procedures – a 'yes' response in one wave was simply carried over automatically to all subsequent waves without asking, and a 'no' simply caused the nondependent question to be re-asked in the next wave. For jobs, on the other hand, both panels used the same fully dependent, 'remind, confirm/still' DI procedures.

The summary statistics (column headings) in Table 5.1 are defined as follows:

- *Analysis* N *for analysed interviews*. This shows, for each characteristic and each wave-pair, the total number of cases included in the analysis and the total number of month-to-month transitions observed in the two waves. The analysis sample is limited to those who were interviewed in both waves of the pair, and for most (but not all[3]) characteristics it is also limited to those who provided at least one 'yes' value for any month in either wave.
- *Average % of all changes that were at the seam*. This column shows the number of month-to-month changes – in either direction – observed at the seam as a percentage of all observed changes (averaged across the three separate analyses). In the absence of seam bias, we would expect about one-seventh (14 %) of all changes to be seam changes, since the seam is one of seven month-pairs in a two-wave analysis. We also summarize here the results of significance tests comparing the 2001 and 2004 estimates in each of the three analyses.
- *Average month-to-month change rates*. These columns show the likelihood of observing a change from one month to the next, in either direction, for all month-pairs combined and separately for 'off-seam' and seam month-pairs. We have no absolute standard for assessing the quality of these rates, which are a function of both the particular 'volatility' of the characteristic in question and how the analysis universe is defined.
- *Change rate ratio: seam/off-seam*. This column divides the seam change rate by the off-seam rate to produce an estimate of how much the seam change rate is inflated relative to the off-seam change rate. (This statistic, unlike the others in Table 5.1, is not an average of the three separate analyses, but is calculated directly from the average change rates shown in the table.) Here we can apply an objective quality standard: in the absence of any seam bias the likelihood of a change across the seam would be about the same as for any other pair of months, and thus the change rate ratio would be close to 1.0.
- *Average 'directional' change rates at the seam*. This pair of columns displays the observed percentage of 'yes' (on a programme, enrolled in school, covered by health insurance, etc.) cases in the last month of one wave's reference period that changed to a 'no' (off, not enrolled/covered, etc.) in the first month of the next wave's reference period and, similarly, the observed percentage of 'no' cases that changed to 'yes'.

[3] The definition of the analysis sample for each characteristic simply follows SIPP's conventions for how the questionnaire data are stored (as a 'person' record, a 'job' record, a 'program' record, etc.). The only practical impact of this decision is whether cases with all 'no' values across all eight months are included or excluded.

These rates, too, are affected by differences in the volatility of each characteristic; the no-to-yes rate, in addition, is also highly sensitive to the definition of the analysis universe. To permit a comparison between the two SIPP panels in the likelihood of observing 'yes-to-no' and 'no-to-yes' changes at the seam, we also show, in the 2004 row, the change in the statistic from 2001 to 2004 as a percentage of the 2001 estimate.

Despite the large amount of information in Table 5.1, and the wide variations among the different characteristics in the levels of the estimates presented, we find the essential features of the results to be remarkably consistent. We see those essential features as follows:

(1) *Seam bias has declined substantially in the 2004 SIPP panel.* The nondependent or dependent-like ('remind, ignore') 2001 procedures were significantly less effective at controlling seam bias than are the fully dependent ('remind, continue' or 'remind, confirm/still') procedures introduced in 2004. This is readily apparent in the proportion of all month-to-month changes observed at the seam: across all 15 characteristics subject to the new DI procedures the 2004 panel estimate is lower than the 2001 estimate. As noted earlier, the 15 pairs of estimates in Part 1 represent 45 separate comparisons, of which 42 showed a statistically significant difference (see Table 5.1 footnotes) according to a simple *t*-test of the difference between two proportions; 36 of the 42 significant differences were significant at the $P < 0.001$ level or beyond. The same result can be seen in the 'Change rate ratio' column, where the ratio of the seam change rate to the off-seam change rate is always closer to 1.0 in 2004 than it was in 2001. In many cases the 2004 estimate is less than half its 2001 counterpart.

(2) *The decline is attributable to the new DI procedures.* As clearly as seam bias declined in 2004 where SIPP implemented new DI procedures, it did not decline where the interview procedures were the same in both panels, as shown in Part 2. For both Medicare coverage and employment at a particular job, the use of very similar interviewing procedures yields very similar seam bias results. This finding offers strong support for the notion that the differences shown in Part 1 are due to the new DI procedures, and not to different samples, different interviewing staffs, the different times that the measurements were collected or other artefacts.

(3) *DI shows positive seam bias effects across a wide range of characteristics.* Seam bias afflicts the measurement of characteristics associated with rich and poor alike, and the improved DI procedures introduced in 2004 were similarly unrestricted in their impact. We divided Part 1 of Table 5.1 into two categories of characteristics – 'need-based' programmes to assist the low income population, and other income sources and characteristics that apply to the general population (or that in some cases are skewed toward the wealthy) – primarily to make it easy to see that there was really no need to do so. The impact of DI appears to have been largely the same in both categories.

(4) *The positive effects of DI are a result of reduced change at the seam and increased change off the seam.* As noted, seam bias has been shown to be the net effect of too many changes observed at the seam and too few changes observed elsewhere (Moore and Marquis, 1989). The new DI procedures directly countered those tendencies. As shown in the 'Average month-to-month change rates' columns, for every characteristic the off-seam change rate is higher in the 2004 panel than in the earlier panel, and for 14 of the 15 characteristics the rate of change at the seam in

2004 is lower than in 2001. DI reduced spurious change reports at the seam, and reduced spurious *non*change reports across the months within a wave.

(5) *SIPP's new DI procedures acted primarily to reduce spurious 'yes-to-no' change at the seam.* As shown in the right-most columns of Table 5.1, across all but two of the characteristics subject to the new DI procedures in 2004 the decline in 'yes-to-no' changes at the seam is greater, in percentage terms, than the decline in 'no-to-yes' changes at the seam. This pattern makes sense because the dependent procedures employed in 2004 are 'asymmetrical' (Murray *et al.*, 1991) – they only apply to those who are in a 'yes' status (enrolled, covered, participating, etc.) at the end of the prior interview.

Given that focus, why did the new DI procedures have any impact at all on no-to-yes transitions? That they did is obvious: with the sole exception of school enrolment, no-to-yes change at the seam was consistently higher in 2001, before the new DI procedures were implemented. We suspect that the key is the careful targeting of DI in the new SIPP panel, as contrasted with the indiscriminate approach that it replaced. The new DI procedures were only triggered by spells known to be in progress at the end of the previous wave's reference period. In contrast to the old DI format, no mention was made of spells that were known to have ended before the end of the reference period. We suspect that the old format's irrelevant reminders to respondents about already-ended spells may have masked the fact that the respondent had already, in the prior wave, reported the spell's termination, thus subtly encouraging him or her to mis-recall a new spell as a continuation of an old one, resulting in a false no-to-yes change report at the seam. (LeMaître (1992) notes this flaw in the previous SIPP design, and its possible negative consequences.)

(6) *Despite the improvements due to DI, much seam bias still remains.* Improvement in seam bias in the SIPP 2004 panel is unmistakable; that bias is far from having been eradicated is equally unmistakable. With the single exception of school enrolment, every characteristic – notwithstanding its improvement relative to 2001 – still displays, in 2004, an overabundance of changes at the seam. For example: in the proportion of all changes that are seam changes, the best outcomes (again, excluding school enrolment) are still above 30 %, which is more than twice as high as would be expected if there were no seam bias. Similarly, in the 'change rate ratios', the best performing characteristics show a rate of change at the seam that is more than twice the rate observed between months within a single interview wave, and in most cases the improvement still leaves at least 3–4 times more seam changes than there should be.

Although our purpose here is to examine general trends, rather than the results for particular characteristics, we want to focus briefly on the school enrolment results, which stand out from the others in the much lesser extent to which they are afflicted with seam bias. Even 'pre-improvement', in 2001, the seam bias estimates for school enrollment are lower than for any other characteristic *after* the addition of improved DI procedures in 2004. And the 'postimprovement' results in 2004 arguably contradict the notion that 'much' seam bias remains in the DI-improved estimates. We suspect that the unique profile for school enrolment is due to its familiar seasonal patterns, which makes reporting in terms of calendar months a relatively easy task compared to other characteristics. (Interestingly, Moore and Kasprzyk (1984), in their very early seam bias investigation, report high levels of bias for every type of characteristic examined, save one – receipt of educational benefits.) This suggests that, given reasonable cues to begin with, respondents can report transitions with reasonable accuracy. And with the addition of other useful cues – specifically, carefully designed DI procedures – it is possible for respondents to produce reports of transitions that are largely devoid of error.

5.4.2 Seam Bias Evaluation for Income Amount Transitions

Beginning in the second interview wave, the redesigned 2004 panel SIPP questionnaire introduced dependent questions into its procedures for capturing income amounts. As noted, these new DI techniques are used in a 'reactive' manner (Lynn *et al.*, 2006), as a follow-up to an initial nonresponse to an income amount question. Although the main purpose of this change was to reduce item nonresponse (Moore, 2006a, 2006b), here we examine its impact on income amount transitions, focusing specifically on monthly earnings from a job. We use an arbitrary definition of an amount 'change' – namely, a difference in earnings between two adjacent months of plus or minus 5 %.[4] We again analyse each pair of waves separately, and we restrict the analysis sample to those interviewed in each wave of the pair who held the same job in each wave. Occasional job changes within a wave, and occasional nonresponse to the amount question in one or the other of adjacent months, result in some month-to-month fluctuation in the number of cases available for analysis.

The earnings change analysis is complicated by other questionnaire differences, specifically the new procedures in the 2004 panel, which encourage respondents to select the 'most convenient' method for reporting their earnings – as monthly totals (the traditional standard), as submonthly gross pay amounts, as an annual salary or by reporting hourly pay rates and hours worked. In the 2001 panel, most workers – about 85 % – reported their earnings using monthly amounts; in the new panel the proportions have flipped almost exactly, with about 85 % reporting their earnings using something other than monthly amounts. Both panels transform nonmonthly data into gross monthly earnings. Unfortunately, however, the transformation algorithms used in the 2001 and 2004 panels are not consistent, resulting in important differences in how monthly amounts are created from nonmonthly reports. Because we sought to assess the impact of DI on respondents' *reports* of change, apart from the impact of processing decisions on change patterns, we decided to exclude from the analysis those who reported in a nonmonthly fashion and to limit our analysis to those who reported monthly amounts, either by reporting monthly totals directly or by reporting all of their individual, submonthly payment amounts, which could then be summed to produce monthly totals.[5]

Figures 5.1a, 5.1b and 5.1c summarize the results of our analysis of seam bias for earnings in waves 1–2, 2–3 and 3–4, respectively. They offer stark visual evidence that DI significantly reduced the change 'spike' across the interview seam. Figure 5.1a, for example, shows that almost 70 % of earnings reportedly changed (according to our ± 5 % definition) across the wave 1–2 seam in 2001, compared to about half that

[4] Kalton and Miller (1991), using data from an early SIPP panel, apply this same definition to examine seam bias in Social Security payment amounts. They find a large seam effect consisting of almost nonexistent month-to-month change within a single wave, contrasted with change at the seam about two-thirds of the time.

[5] Concern about the possible impact of these analysis sample restrictions led us to repeat the investigation for three other income sources – Food Stamps, Social Security and child support – which did not require any such restrictions, since the survey procedures in both panels only permitted monthly-type reporting. These analyses produced very similar results to those reported here for job earnings (data not shown).

rate in 2004, when DI was available as a nonresponse follow-up procedure. Note also that when DI was not available in either panel, in the three month-pairs within wave 1 the line graphs are virtually identical, with a constant change rate of about 20 %. After the seam spike in 2001, the picture within wave 2 returns to a pattern almost identical to that of wave 1, in contrast to the postseam change rates in 2004, which are only about half of what they were before the wave 1–2 seam. Figures 5.1b and 5.1c present strikingly similar results. In all three analyses the percentage of cases falling outside the threshold at the seam is significantly lower in the 2004 panel than it was in 2001, in some cases reduced by over half,[6] and the change rate for off-seam month-pairs (with the exception of those in wave 1) is also consistently lower in the 2004 panel compared to 2001.

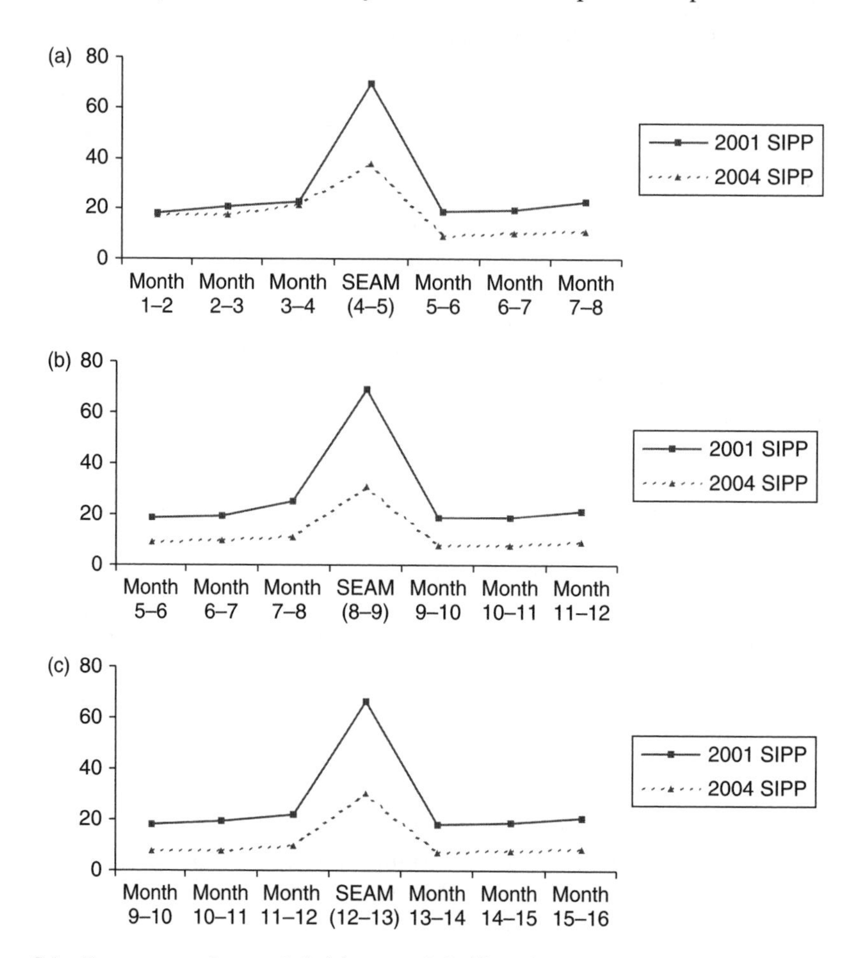

Figure 5.1 Percentage of wave 1–2 (a), wave 2–3 (b) and wave 3–4 (c) cases with a month-to-month 'change' ($\pm 5\,\%$) in job earnings. Average number of cases: (a) 18 000 (2001) and 7200 (2004); (b) 12 200 (2001) and 6300 (2004); (c) 15 800 (2001) and 5600 (2004).

[6] Waves 1–2: 69.5 % vs. 37.8 %, $t = 47.2$, $P < 0.001$; Waves 2–3: 69.1 % vs. 31.0 %, $t = 52.2$, $P < 0.001$; Waves 3–4: 66.4 % vs. 30.1 %, $t = 50.6$, $P < 0.001$.

Because nonresponse rates for earnings amount questions are not generally excessive (Moore *et al.*, 2000), we expected that use of the DI procedures would be a fairly rare event. Thus, the apparent magnitude of their impact on change at the seam is surprisingly large. In fact, use of the DI procedures was quite common – they were invoked over half the time in waves 2 and 3 when asking about job earnings, a rate that far exceeds typical rates of nonresponse to earnings amount items. Other evidence also suggests that the DI questions were not used strictly as a nonresponse follow-up tool. Several 2004 panel interview observation reports by Census Bureau staff note interviewers' tendency to use the DI follow-ups as a means to 'peek ahead' at the answers reported in the last interview, and to help respondents to answer the amount questions without even giving them a chance to report on their own. Thus, the greater than anticipated seam effect reduction may have been the result of interviewers' tendency to transform the intended reactive DI follow-ups into proactive-style questions.

The consistently lower off-seam change rates in 2004 compared to 2001 also deserve comment. We suspect that this effect is mostly due to processing and programming decisions, not respondents' reporting behaviours. The particular design of the 2004 DI procedures obtains a single amount (either the confirmed 'last time' amount or a repaired figure), which is then assigned to each month in the current wave, allowing no chance of variation in an off-seam month-pair.

5.5 CONCLUSIONS AND DISCUSSION

Despite the limitations noted earlier, we find the results quite encouraging with regard to the quality of month-to-month change data from the new SIPP questionnaire. They offer strong and consistent evidence, across many diverse characteristics, of the significant positive impact of improved DI procedures on the measurement of month-to-month transitions. In the earnings amount results we even find evidence of 'byproduct' positive effects, where nonresponse reduction, not improvement in transition data, was the primary intent. The new, more precise and focused DI procedures employed in the 2004 SIPP panel with the specific intent of improving data on transitions appear to have reduced reports of change at the seam, and to have increased reports of off-seam changes. Both trends address what have been shown to be the major error tendencies in the measurement of change in longitudinal surveys – over-reporting of change at the seam and under-reporting of off-seam changes (Moore and Marquis, 1989). As a result, the likelihood of recording a transition at the interview seam in the current SIPP panel is, for virtually every characteristic examined, significantly more in line with what would be expected in the absence of measurement error than is the case with the previous panel. Despite the significant improvements, however, much seam bias still remains.

Fortunately, the results presented here also highlight an additional area in which there is still much untapped potential for further improvements: 'no-to-yes' changes at the seam. DI as it has been introduced in SIPP focuses exclusively on the *presence* of some characteristic – being enrolled in school, receiving Food Stamps, etc. – in the last months of the prior wave's reference period. A previously identified, likely-to-continue spell is carefully addressed in the new postwave-1 questionnaire; the same attention is *not* paid, however, to the onset of a new spell at the seam. We recommended that such procedures be developed and tested. Their general form seems fairly straightforward.

When a respondent reports that a new spell of some characteristic has started – that is, reports a 'yes' for a characteristic that was *not* a 'yes' at the end of the previous wave's reference period – then questioning about the start of that spell should refer to what is known from the previous wave, e.g.: 'When we interviewed you back in early March you weren't receiving Food Stamps. When did you start to receive them?' Addressing, in this or some similar manner, the continuation of the *absence* of some characteristic across the seam is likely to produce additional gains in the overall quality of transition data.

ACKNOWLEDGEMENTS

This report is released to inform interested parties of research and to encourage discussion of work in progress. The views expressed are the authors' and not necessarily those of the US Census Bureau. The authors prepared this chapter as part of their official duties as US government employees. Therefore it is a US government work and does not attract copyright. Many people provided assistance with various aspects of the work presented in this chapter. At the risk of inadvertently omitting key names, we specifically note the following: our immediate colleagues for many years on the SIPP Methods Panel project – Anna Chan, the late, great Pat Doyle, Tim Gilbert, Julia Klein Griffiths, Elaine Hock and Heather Holbert. We also thank our Census Bureau colleagues Aref Dajani, for assistance with data file preparation, and John Boies, Kathy Creighton and Chuck Nelson, who provided many useful comments on early drafts of this chapter. The final version also benefited greatly from the comments of the editor of this volume, Peter Lynn, and two anonymous reviewers.

REFERENCES

Brown, A., Hale, A. and Michaud, S. (1998). Use of computer assisted interviewing in longitudinal surveys. In M. Couper, R. Baker, J. Bethlehem, C. Clark, J. Martin, W. Nicholls and J. O'Reilly (Eds), *Computer Assisted Survey Information Collection* (pp. 185–200). New York: John Wiley & Sons, Inc.

Burkhead, D. and Coder, J. (1985). Gross changes in income recipiency from the Survey of Income and Program Participation. *Proceedings of the American Statistical Association, Social Statistics Section*, pp. 351–356.

Cantor, D. and Levin, K. (1991). Summary of Activities to Evaluate the Dependent Interviewing Procedure of the Current Population Survey. Westat, Inc. Report submitted to the Bureau of Labor Statistics (Contract No. J-9-J-8-0083).

Coder, J., Burkhead, D., Feldman-Harkins, A. and McNeil, J. (1987). Preliminary data from the SIPP 1983–84 Longitudinal Research File. *SIPP Working Paper*, No. 8702. Washington, DC: US Census Bureau.

Collins, C. (1975). Comparison of month-to-month changes in industry and occupation codes with respondent's report of change: CPS job mobility study. *U.S. Census Bureau Response Research Staff Report*, No. 75-5, May 15, 1975.

Cotton, C. and Giles, P. (1998). The seam effect in the Survey of Labour and Income Dynamics. *Statistics Canada SLID Research Paper Series*, Catalogue No. 98-18.

Corti, L. and Campanelli, P. (1992). The utility of feeding forward earlier wave data for panel studies. In A. Westlake, *et al.* (Eds), *Survey and Statistical Computing* (pp. 109–118). London: Elsevier.

Czajka, J. (1983). Subannual income estimation. In M. David (Ed.), *Technical, Conceptual, and Administrative Lessons of the Income Survey Development Program (ISDP)* (pp. 87–97). New York: Social Science Research Council.

Dibbs, R., Hale, A., Loverock, R. and Michaud, S. (1995). Some effects of computer-assisted interviewing on the data quality of the survey of labour and income dynamics. *Statistics Canada SLID Research Paper Series*, Catalogue No. 95-07.

Doyle, P., Martin, E. and Moore, J. (2000). Methods panel to improve income measurement in the Survey of Income and Program Participation. In *Proceedings of the Section on Survey Research Methods* (pp. 953–958). Washington, DC: American Statistical Association.

Grondin, C. and Michaud, S. (1994). Data quality of income data using computer assisted interviewing: SLID experience. *Statistics Canada SLID Research Paper Series*, Catalogue No. 94-15.

Hale, A. and Michaud, S. (1995). Dependent interviewing: impact on recall and on labour market transitions. *Statistics Canada SLID Research Paper Series*, Catalogue No. 95-06.

Hill, D. (1987). Response errors around the seam: analysis of change in a panel with overlapping reference periods. In *Proceedings of the Section on Survey Research Methods* (pp. 210–215). Washington, DC: American Statistical Association.

Hill, D. (1994). The relative empirical validity of dependent and independent data collection in a panel survey. *Journal of Official Statistics*, 10, 359–380.

Holmberg, A. (2004). Pre-printing effects in official statistics: an experimental study. *Journal of Official Statistics*, 20, 341–355.

Hoogendoorn, A. (2004). A questionnaire design for dependent interviewing that addresses the problem of cognitive satisficing. *Journal of Official Statistics*, 20, 219–232.

Jabine, T., King, K. and Petroni, R. (1990). *Survey of Income and Program Participation Quality Profile*. Washington, DC: US Census Bureau.

Jäckle, A. and Lynn, P. (2007). Dependent interviewing and seam effects in work history data. *Journal of Official Statistics*, 23, 529–551.

Kalton, G. (1998). SIPP quality profile 1998. *SIPP Working Paper No. 230*. Washington, DC: US Census Bureau.

Kalton, G. and Miller, M. (1991). The seam effect with social security income in the Survey of Income and Program Participation, *Journal of Official Statistics*, 7, 235–245.

LeMaître, G. (1992). Dealing with the seam problem for the Survey of Labour and Income Dynamics. *Statistics Canada SLID Research Paper Series*, Catalogue No. 92-05.

Lynn, P., Jäckle, A., Jenkins, S. and Sala, E. (2006). The effects of dependent interviewing on responses to questions on income sources. *Journal of Official Statistics*, 22, 357–384.

Lynn, P. and Sala, E. (2006). Measuring change in employment characteristics: the effects of dependent interviewing. *International Journal of Public Opinion Research*, 18, 500–509.

Marquis, K. and Moore, J. (1990). Measurement errors in the Survey of Income and Program Participation. *Proceedings of the US Census Bureau's Annual Research Conference*, pp. 721–745.

Martini, A. (1989). Seam effect, recall bias, and the estimation of labor force transition rates from SIPP. *Proceedings of the American Statistical Association, Section on Survey Research Methods*, pp. 387–392.

Mathiowetz, N. and McGonagle, K. (2000). An assessment of the current state of dependent interviewing in household surveys. *Journal of Official Statistics*, 16, 401–418.

Michaud, S., Dolson, D., Adams, D. and Renaud, M. (1995). Combining administrative and survey data to reduce respondent burden in longitudinal surveys. *Statistics Canada SLID Research Paper Series*, Catalogue No. 95-19.

Moore, J. (2006a). The effects of questionnaire design changes on asset income amount nonresponse in waves 1 and 2 of the 2004 SIPP Panel. *Research Report Series (Survey Methodology)*, No. 2006-01.Washington, DC: US Census Bureau.

Moore, J. (2006b). The effects of questionnaire design changes on general income amount nonresponse in waves 1 and 2 of the 2004 SIPP Panel. *Research Report Series (Survey Methodology)*, No. 2006-04. Washington, DC: US Census Bureau.

Moore, J. and Griffiths, J. (2003). Asset ownership, program participation, and asset and program income: improving reporting in the Survey of Income and Program Participation. *Proceedings of the American Statistical Association* [CD-ROM]. Alexandria, VA: American Statistical Association.

Moore, J. and Kasprzyk, D. (1984). Month-to-month recipiency turnover in the ISDP. *Proceedings of the American Statistical Association, Section on Survey Research Methods*, pp. 726–731.

Moore, J. and Marquis, K. (1989). Using administrative record data to evaluate the quality of survey estimates. *Survey Methodology*, 15, 129–143.

Moore, J. Pascale,J., Doyle, P., Chan, A. and Griffiths, J. (2004). Using field experiments to improve instrument design: the SIPP Methods Panel project. In S. Presser, J. Rothgeb, M. Couper, J. Lessler, E. Martin, J. Martin and E. Singer (Eds), *Methods for Testing and Evaluating Survey Questionnaires* (pp. 189–207). New York: John Wiley & Sons, Inc.

Moore, J., Stinson, L. and Welniak, E. (2000). Income measurement error in surveys: a review. *Journal of Official Statistics*, 16, 331–361.

Murray, T., Michaud, S., Egan, M. and LeMaître, G. (1991). Invisible seams? The experiences with the Canadian Labor Market Activity Survey. *Proceedings of the US Census Bureau's Annual Research Conference*, pp. 715–729.

Polivka, A. and Rothgeb, J. (1993). Redesigning the CPS Questionnaire. *Monthly Labor Review*, September, 10–28.

Rips, L., Conrad, F. and Fricker, S. (2003). Straightening the seam effect in panel surveys. *Public Opinion Quarterly*, 67, 522–554.

Ryscavage, P. (1993). The seam effect in SIPP's labor force data: did the recession make it worse?' *SIPP Working Paper*, No. 180. Washington, DC: US Census Bureau.

Stanley, J. and Safer, M. (1997). 'Last time you had 78, how many do you have now?' The effect of providing previous reports on current reports of cattle inventories. *Proceedings of the American Statistical Association, Section on Survey Research Methods*, pp. 875–880.

Weidman, L. (1986). Investigation of gross changes in income recipiency from the Survey of Income and Program Participation. *Proceedings of the American Statistical Association, Section on Survey Research Methods*, pp. 231–236.

Weinberg, D. (2002). The Survey of Income and Program Participation – recent history and future developments. *SIPP Working Paper Series*, No. 232, Washington, DC: US Census Bureau.

US Census Bureau (2001). *SIPP Users' Guide*. Washington, DC: US Census Bureau.

Young, N. (1989). Wave-seam effects in the SIPP. *Proceedings of the American Statistical Association, Section on Survey Research Methods*, pp. 393–398.

CHAPTER 6

Dependent Interviewing: A Framework and Application to Current Research

Annette Jäckle
University of Essex, UK

6.1 INTRODUCTION

Panel surveys increasingly feed forward substantive information from previous waves of data collection, to tailor the wording or routing of questions to the respondent's situation or to include automatic edit checks during the interview. Personalising the questionnaire through 'dependent interviewing' (DI) can reduce respondent burden, increase the efficiency of data collection and improve data quality (for reviews, see Mathiowetz and McGonagle, 2000; Lynn *et al.*, 2006). The main motivation for introducing DI, however, varies across survey organisations, surveys and items, resulting in a variety of designs and applications. As a consequence it is not straightforward to evaluate the effects of different design features or compare the use of DI in different surveys.

Based on a review of current practices, this chapter develops a conceptual framework of DI in an attempt to disentangle design features, channels through which they take effect, and implications for burden, efficiency and data quality. The aim is to aid the understanding of the effects of DI, about which empirical evidence has been very limited until recently. Section 6.2 describes situations in which DI can be beneficial and discusses the causes of 'seam effects', a major type of error in panel surveys that can be reduced with DI. Section 6.3 conceptualises the different design options and their effects and Section 6.4 evaluates different (proactive and reactive) DI designs, based on findings from the 'Improving Survey Data on Income and Employment' study. Section 6.5 discusses effects of DI on data quality, highlighting limitations and pointing out other factors that need to be taken into account when assessing the effect of DI across surveys

Methodology of Longitudinal Surveys P. Lynn

and items. Section 6.6 concludes with issues that remain unresolved concerning the practical implementation of DI and effects on data quality, comparability and analysis.

6.2 DEPENDENT INTERVIEWING – WHAT AND WHY?

Dependent interviewing refers to a method of designing questions in panel surveys, where substantive answers from previous interviews are fed forward and used to improve data quality, in terms of longitudinal consistency and item non-response, and survey processes, in terms of efficiency of data collection and respondent burden. This differs from traditional independent interviewing, where respondents are typically asked the same questions about their situation at different points in time, without reference to previous answers.

6.2.1 Data Quality

When information from successive panel interviews is combined to create event histories, one typically finds large concentrations of transitions at the 'seam' between reference periods. Such seam effects are well documented for data on benefit income (e.g. Czajka, 1983; Moore and Kasprzyk, 1984; Burkhead and Coder, 1985) and labour market activities (e.g. Martini, 1989; Murray *et al.*, 1991; Ryscavage, 1993; Hill, 1994; Jäckle and Lynn, 2007). A common misconception is that the concentration at the seam is the result of misdating of changes, and that respondents have a tendency to *report* more transitions at the seam. In reality seam effects are the result of combining data from panel interviews in the presence of measurement and data processing errors. These errors lead to a combination of *under-reporting* of change within a wave, and *spurious change* between waves, causing the observed concentration of transitions at the seam (see Young, 1989; Rips *et al.*, 2003).

Dated history information is mainly collected in one of two ways in surveys: by asking respondents to report spells and transition dates (sequential spell questions) or by asking whether the respondent experienced a particular event or status during each subperiod, e.g. week, of the reference period (period status questions). In sequential spell questions, changes within a wave are given by the reported transition dates. (These correspond to 'time-of-occurrence' questions in the typology of temporal questions by Tourangeau *et al.* (2000).) In period status questions, changes are inferred from changes in yes/no answers. In both cases, analysts may infer a change at the seam if the status report for the last month from the first interview does not match the retrospective report for the first month from the second interview. Such mismatches may reflect a real change, but could occur due to keying errors, coding variability or misclassification by the respondent. Thus, some observed changes at the seam will be spurious and not correspond to any true changes (Figure 6.1). On the other hand, omissions of events or errors in the dating of events lead to under-reporting of change during the recall period, and may lead to dates being misplaced to the seam. In this case, transitions observed at the seam might correspond to true changes that in fact took place during the period between seams.

Dependent interviewing can reduce the occurrence of spurious changes at the seam and of constant wave responses caused by under-reporting events. The increase in longitudinal consistency of responses is, however, not necessarily achieved by reducing

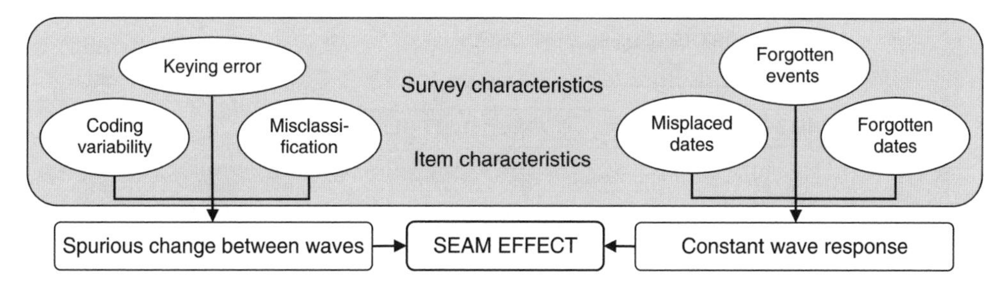

Figure 6.1 Determinants of seam effects.

errors. For example, if erroneous reports are fed forward, cross-sectional estimates of *prevalence* may not be improved by dependent interviewing. Longitudinal estimates of *change* would, however, be improved if spurious changes are reduced.

The second aspect of data quality that can be improved with DI is item nonresponse. This problem affects all surveys, but panels offer an opportunity that one-off surveys do not: respondents who do not give a legitimate answer can be reminded of previous reports to jog their memory and questions that remained unanswered can be fed forward and repeated in the subsequent interview.

6.2.2 Survey Processes

Panel respondents often complain about having to answer the same questions repeatedly even when their circumstances have not changed (Phillips *et al.*, 2002). This is especially problematic for surveys with short intervals between interviews and for inherently stable items. In so far as there is genuine stability, dependent interviewing can be used to identify and route around redundant questions and thereby reduce respondent burden, interview durations and possibly the number of open-ended answers requiring coding. In addition, tailoring questions to the respondent's situation improves the flow of the interview and reminding respondents of previous answers simplifies the response task, for example by replacing recall by recognition (Hoogendoorn, 2004) or requiring yes/no instead of open-ended answers (Jäckle, 2008). (See also Holmberg (2004) for a discussion of reasons for the use of DI by Statistics Sweden.)

6.3 DESIGN OPTIONS AND THEIR EFFECTS

Dependent interviewing questions are typically classified as either proactive or reactive (Brown *et al.*, 1998). With reactive DI, information fed forward from the previous interview is used to carry out edit checks during the interview; with proactive DI, previous information is used to determine question routing or wording. This broad classification encompasses a number of design features. The following conceptualisation of the different design options and their effects is based on a review of current practices and research, where the classification of designs was inspired by Pascale and Bates' (2004) description of DI questions in the US Survey of Income and Program Participation (SIPP). Unless stated otherwise, examples are taken from the 'Improving Survey Measurement of Income and Employment' study, described in Section 6.4.

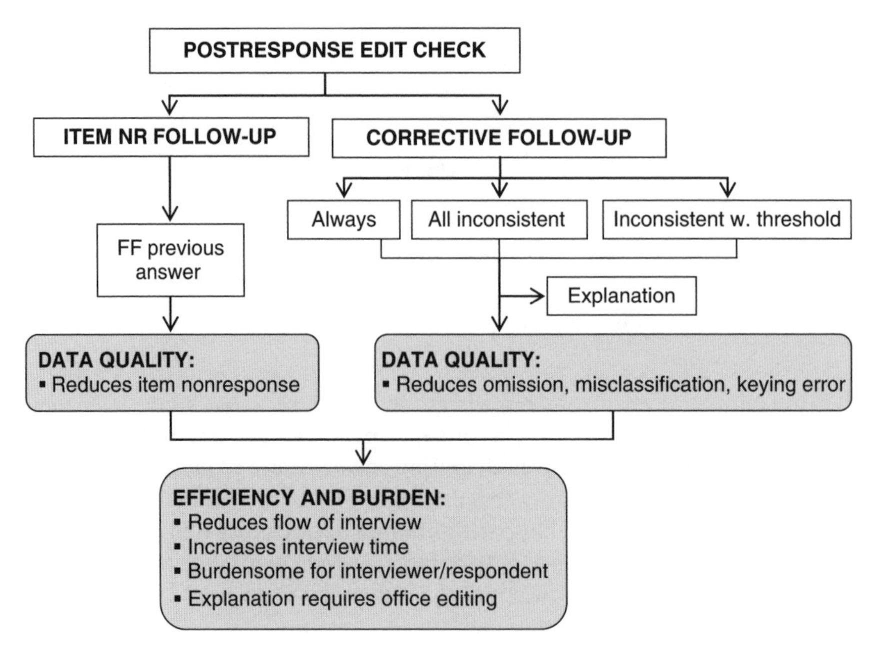

Figure 6.2 Reactive dependent interviewing: design options and effects. NR = 'nonresponse', FF = 'feed forward'. White boxes indicate design options, grey boxes indicate effects and arrows indicate conceptual associations.

6.3.1 Reactive Dependent Interviewing

Edit checks based on information from previous interviews can be built in to either follow up on item nonresponse or check consistency with previous reports. Edit checks are used to improve data quality, but because they imply additional and potentially difficult questions the data improvements may come at the cost of respondent burden and efficiency of data collection. Figure 6.2 illustrates design options (white boxes) for edit check questions and their effects (grey boxes) on data quality, efficiency of data collection and respondent burden, where the arrows indicate conceptual associations.

Item Nonresponse Follow-up

The SIPP, for example, does not accept 'don't know' or refusal as a response to questions on income amounts from earnings, unearned income and assets. If the respondent does not volunteer an amount, he is reminded of his report in the previous interview and asked if this still sounds about right: '. . . last time you received *<$$$>* in food stamps. Does that still sound about right?' (Moore *et al.*, 2004, p. 193).

Corrective Follow-up

An edit check can also be designed to check consistency with previous reports and be prompted:

(1) always, for example to check consistency of verbatim answers with previous reports (follow-up, always): 'Can I just check, is that the same employer that you were

working for last time we interviewed you, on <*DATE OF INTERVIEW*>, when we recorded your employer as <*EMPLOYER*>?';

(2) to clarify reports that are inconsistent with previous reports (follow-up, all inconsistent): 'Can I just check, according to our records you have in the past received <*INCOME SOURCE*>. Have you received <*INCOME SOURCE*> at any time since <*DATE OF INTERVIEW*>?';

(3) selectively, if reports differ from previous reports beyond a defined threshold, for example if usual earnings for a standardised period differ by more than ±10 % (follow-up, inconsistent with threshold): 'So, your net/gross pay has gone <*UP/DOWN*> since last time we interviewed you, from <*AMOUNT*$_1$> per <*PERIOD*$_1$> to <*AMOUNT*$_2$> per <*PERIOD*$_2$>, is that correct?' The respondent may also be asked to clarify reasons for the discrepancy and the explanation may be recorded either as verbatim text or as a pre-coded answer.

Effects on Data Quality

The nonresponse follow-up can significantly reduce item nonresponse (see Moore *et al.* (2004) for evidence from the SIPP), while corrective follow-ups can significantly reduce recall and keying errors leading to spurious changes, and omissions of spells leading to constant wave responses. Evidence is discussed in Section 6.4.

Effects on Survey Processes

On the other hand, reactive follow-ups are likely to have adverse effects on respondent burden and efficiency of data collection. Edit checks interrupt the flow of the interview and may be ignored by interviewers if there are too many (Dibbs *et al.*, 1995); asking the respondent to clarify inconsistencies and possibly provide explanations may be a difficult task; and additional (difficult) questions are likely to increase the duration of the interview. Attention also needs to be given to the way in which the respondent's answers are queried (see Mathiowetz and McGonagle, 2000, p. 409) and how often they are queried, in order to prevent spoiling the rapport between interviewer and respondent. Finally, explanations of discrepancies require office coding and decisions about how to incorporate corrections of previous data.

6.3.2 Proactive Dependent Interviewing

With proactive DI respondents can be reminded of previous answers or asked questions they did not answer previously. Reminders primarily simplify the response task and can also be used to reduce redundancies of questions, both of which are hoped to improve data quality. Figure 6.3 illustrates the design options (white boxes) for proactive reminders and their effects (grey boxes) on data quality, efficiency of data collection and respondent burden, where the arrows indicate conceptual associations (the use of proactive DI to feed forward previously unanswered questions is not shown).

Feeding Forward Item Nonresponse

The Canadian National Longitudinal Survey of Children and Youth (NLSCY), for example, asks respondents about the number of times they have moved since the previous interview. Respondents who have not reported on moves previously (because of item nonresponse or because they were not previously eligible for the question) are asked about

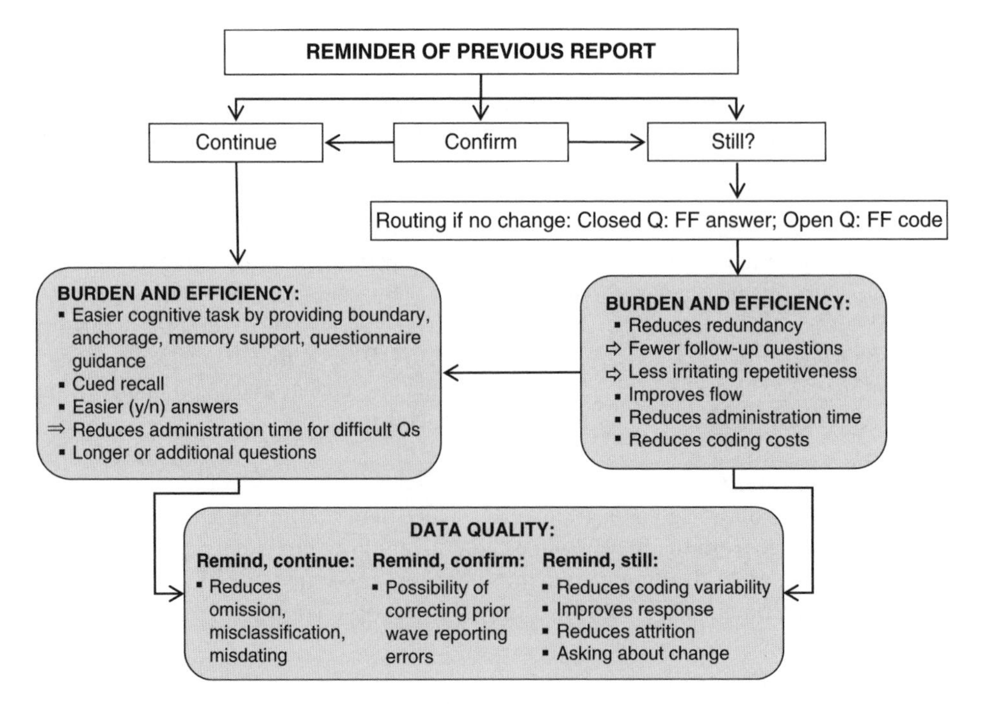

Figure 6.3 Proactive dependent interviewing: design options and effects. Q = 'question', FF = 'feed forward'. White boxes indicate design options, grey boxes indicate effects and arrows indicate conceptual associations.

the total number of moves in their lifetime. 'Thus the survey ensures the same data for all respondents, regardless of what happened in the previous interview' (Brown *et al.*, 1998, p. 195). Feeding forward nonresponse to previous questions could be particularly useful for baseline demographic information collected only in the first interview.

Reminder of Previous Reports

Previous information can be used as a reminder:

(1) to aid the respondent's memory and provide a boundary before asking the standard independent question (remind, continue): 'According to our records, when we last interviewed you, on *<DATE OF INTERVIEW>*, you were receiving *<INCOME SOURCE>*, either yourself or jointly. For which months since *<MONTH OF INTERVIEW>* have you received *<INCOME SOURCE>*?';

(2) to ask respondents to check and confirm previously recorded answers (remind, confirm): 'When we last interviewed you, on *<DATE OF INTERVIEW>*, our records show that you were *<LABOUR MARKET ACTIVITY*$_1$*>*. Is that correct?';

(3) explicitly to ask about changes (remind, still): 'Last time we interviewed you, on *<DATE OF INTERVIEW>*, you said your occupation was *<OCCUPATION>*. Are you still in that same occupation?'.

'Remind, continue' questions are equivalent to the methods used in bounded interviews to reduce telescoping (see, Neter and Waksberg, 1964). 'Remind, still' type questions are often combined with routing if circumstances have not changed (possibly with subsequent imputation of previous data), for example routing around questions about the characteristics of a job. 'Remind, confirm' questions are usually either followed by the standard independent question (as are 'remind, continue' questions) or followed by questions about change (similar to 'remind, still' questions) and sometimes combined with routing.

Effects on Survey Processes

'Remind, continue' designs increase the length of question wording; 'remind, confirm' designs imply at least one additional question for all sample members. Nonetheless, reminders primarily reduce respondent burden and as a result lead to efficiency gains and improvements in data quality. 'Remind, continue' questions simplify the response task by providing memory support, guidance on the type of information required and temporal boundaries. The cognitive task of recall is replaced by the less demanding task of cued recall (Mathiowetz and McGonagle, 2000). As a result, questions that are typically difficult, such as long list or retrospective questions, may be easier and quicker to administer when combined with reminders. 'Remind, still' designs with routing remove redundancies and can reduce respondent frustration at seeming to have to answer the same questions at every wave (Hoogendoorn, 2004). In fact, respondents expect interviewers to be able to use their responses from previous interviews and are not concerned by privacy issues (Pascale and Mayer, 2004). Reducing redundancies can also improve the flow of interviews and reduce administration times. In addition, 'remind, still' questions are often phrased as yes/no questions, which are easier and quicker to answer, especially for open-ended questions (Jäckle, 2008). Finally, the reduction of redundancies can lead to significant savings in coding costs. The potential for efficiency gains and burden reduction depends on the degree of stability experienced by respondents, which is determined by characteristics of the survey (such as the length of the reference period) and the inherent stability of the item (see Jäckle (2008) for a discussion).

Effects on Data Quality

By simplifying the response task, reminders can improve data quality significantly, by reducing omissions, misdating of changes and misclassifications and thereby reducing both spurious change and spurious stability. Reminders could, however, also have negative effects if they prompt respondents to use least effort strategies such as constant wave responding. 'Remind, confirm' questions can in addition be used to verify previous records and 'remind, still' questions can be used to explicitly ask about change instead of inferring change from differences in reports. Coupled with routing and subsequent imputation of previous data, 'remind, still' questions can further reduce coding variation and are likely to improve response and reduce attrition by removing detrimental impact on respondent motivation.

6.4 EMPIRICAL EVIDENCE

The 'Improving Survey Measurement of Income and Employment (ISMIE)' study presents a unique opportunity to compare the effects of alternative DI designs. This experimental study was funded by the Research Methods Programme of the

UK Economic and Social Research Council. The study compared independent interviewing, reactive DI and proactive DI for sets of questions on income sources, current employment, earnings, activity histories and school-based qualifications, based on the 2002 wave of the British Household Panel Survey and covering a reference period of, on average, 18 months. Respondents were randomly allocated to one of the three treatment groups. In addition, the survey responses on unearned income sources were linked to individual-level validation data from administrative records. For a detailed description, see Jäckle *et al.* (2004). The DI questions were primarily designed to improve data quality, although questions on occupation and industry were also designed to reduce redundancies. Table 6.1 provides a summary of the experimental ISMIE questions.

6.4.1 Income Sources

The ISMIE DI questions were designed to reduce omission among respondents who had previously reported receipt of a source and thereby to reduce spurious transitions. Respondents were first asked which of a list of income sources they had received since the previous interview and then asked period status questions about the timing of receipt. With *reactive DI* this led to a 'follow-up, all inconsistent' design: 'Can I just check, according to our records you have in the past received <*INCOME SOURCE*>. Have you received <*INCOME SOURCE*> at any time since <*DATE OF INTERVIEW*>?' This is similar to the approach used in the Canadian Survey of Labour and Income Dynamics (SLID) for questions on unemployment insurance, social assistance and workers' compensation (see Dibbs *et al.*, 1995). For *proactive DI*, a 'remind, continue' design was used: 'According to our records, when we last interviewed you, on <*DATE OF INTERVIEW*>, you were receiving <*INCOME SOURCE*>, either yourself or jointly. For which months since <*MONTH OF INTERVIEW*> have you received <*INCOME SOURCE*>?' This is similar to questions used in the SIPP until the redesign in 2004 (see, Burkhead and Coder, 1985).

Compared to administrative benefit records, both DI designs reduced under-reporting but did not eliminate it, confirming the findings reported for the SLID by Dibbs *et al.* (1995). There was evidence that cessation of receipt during the reference period was associated with an increased risk of under-reporting and DI was particularly successful in these cases. There was no evidence that reminding respondents of previous income sources led to any increase in over-reporting (Lynn *et al.*, 2004, 2006). Although DI reduced under-reporting, the administration time for the corresponding section of the questionnaire did not increase significantly (Jäckle, 2008).

Comparison of DI Designs

Proactive and reactive DI were similarly effective at reducing under-reporting when compared to administrative records. Reported receipt with reactive DI, when answers to the follow-up questions were not considered, was similar to that with independent questions, suggesting that the experience of reactive follow-ups earlier in the interview did not alter response to later independent questions. Reactive DI might therefore be an attractive design to introduce in ongoing panel studies, since longitudinal consistency is maintained if analysts can identify answers given in response to the edit check (Lynn *et al.*, 2006). On the other hand, initial evidence suggests that proactive DI might be better at reducing seam effects (see Section 6.5).

Table 6.1 ISMIE experimental questions and rationale for DI designs.

Income sources	
Reduce omission among respondents who had previously reported receipt of a source	*INDI* Please look at this card and tell me if, since September 1st 2001, you have received any of the types of income or payments shown, either just yourself or jointly? And for which months since September 1st 2001 have you received . . . ? *RDI (follow-up, inconsistent) – for each income source reported in previous wave but not in the ISMIE survey* Can I just check, according to our records you have in the past received <INCOME SOURCE>. Have you received <INCOME SOURCE> at any time since <DATE OF INTERVIEW>? For which months since <MONTH OF INTERVIEW> have you received <INCOME SOURCE>? *PDI (remind, continue) – for each income source reported in previous interview* According to our records, when we last interviewed you, on <DATE OF INTERVIEW>, you were receiving <INCOME SOURCE>, either yourself or jointly. For which months since <MONTH OF INTERVIEW> have you received <INCOME SOURCE>? Then INDI question to catch new sources.
Current earnings	
Catch keying errors in earnings amounts and periods that would lead to spurious change	*INDI* The last time you were paid, what was your gross pay – that is including any overtime, bonuses, commission, tips or tax refund, but before any deductions for tax, national insurance, or pension contributions, union dues and so on? *RDI (follow-up, selective) – if net/gross pay has gone up or down by more than 10 %* So, your net/gross pay has gone <UP/DOWN> since last time we interviewed you, from <AMOUNT$_1$> per <PERIOD$_1$> to <AMOUNT$_2$> per <PERIOD$_2$>, is that correct? *PDI (remind, still)* Last time we interviewed you, on <DATE OF INTERVIEW>, our records show that your pay was <AMOUNT$_1$> per <PERIOD$_1$> <GROSS/NET>. Is that still the case now, or has your pay changed?

Table 6.1 (Continued).

Current employment	
Reduce spurious changes compared to previous wave report	*INDI* What was your (main) job last week? Please tell me the exact job title and describe fully the sort of work you do. What does the firm/organisation you work for actually make or do (at the place where you work)? What is the exact name of your employer or the trading name if one is used? Are you an employee or self-employed? If employee: Do you have any managerial duties or do you supervise any other employees? If employee: How many people are employed at the place where you work? *RDI (follow-up, always) – for open-ended occupation and industry questions* Can I just check, is that the same employer that you were working for last time we interviewed you, on <DATE OF INTERVIEW>, when we recorded your employer as <EMPLOYER>? *RDI (follow-up, inconsistent) – for other employment characteristics* So, since last time we interviewed you, on <DATE OF INTERVIEW>, you've changed from being <EMPLOYMENT STATUS$_1$> to <EMPLOYMENT STATUS$_2$>: is that correct? *PDI (remind, still)* Last time we interviewed you, on <DATE OF INTERVIEW>, you said your occupation was <OCCUPATION>. Are you still in that same occupation?
Activity history	
Reduce spurious transitions at seam in reports of activities since previous interview	*INDI* What were you doing immediately before [your current job / spell of <ACTIVITY> which started on <DATE>]? *RDI (follow-up, inconsistent) – if wave 11 current activity different from ISMIE retrospective report of wave 11 activity* May I just check something? According to our records, when we last interviewed you, on <DATE OF INTERVIEW>, you were <ACTIVITY$_1$>. That spell had started in <MONTH>. But the information that you have just given me implies that you were <ACTIVITY$_2$> at that time. It

	may be that I have recorded something wrongly and it is important to us that our information is accurate, so can you just clarify that for me? *If activity reports correspond but start month does not* May I just check something? According to our records, when we last interviewed you, on <DATE OF INTERVIEW>, your current spell of <ACTIVITY> had started in <MONTH$_1$>. But the information that you have just given me implies that this spell started in <MONTH$_2$>. It may be that I have recorded something wrongly and it is important to us that our information is accurate, so can you just clarify that for me? *PDI (remind, confirm) – ask all* When we last interviewed you, on <DATE OF INTERVIEW>, our records show that you were <ACTIVITY>. Is that correct?
Qualifications	
Improve reliability of highest qualification measures derived from annual questions about education and training during reference period	*INDI* Have you attended any education institution full-time since September 1st 2001? If yes: Any qualifications obtained? *RDI (follow-up, always) – for each new qualification reported* You have told me that you have gained <N$_2$> <QUALIFICATIONS TYPE X> since last time we interviewed you, and my records show that you previously had <N$_1$> <QUALIFICATIONS TYPE X>, so, you now have a total of < N$_1$ + N$_2$> <QUALIFICATIONS TYPE X>: is that correct? *PDI (remind, confirm) – ask all* We are particularly interested in checking the accuracy of the information we hold about school-based qualifications. According to our records from previous interviews, you have <HIGHEST QUALIFICATION>. Is that correct?

INDI, independent interviewing; RDI, reactive dependent interviewing; PDI, proactive dependent interviewing.

6.4.2 Current Earnings

The ISMIE DI questions were designed to catch keying errors in earnings amounts and periods that would lead to spurious change. With *reactive DI*, a 'follow-up, inconsistent with threshold' design was chosen, prompted by a change in earnings by more than ± 10 %: 'So, your net/gross pay has gone <*UP/DOWN*> since last time we interviewed you, from <*AMOUNT*$_1$> per <*PERIOD*$_1$> to <*AMOUNT*$_2$> per <*PERIOD*$_2$>, is that correct?' Respondents who did not confirm the change were asked to explain the reason for the discrepancy and to correct the amount or period. Similar designs are used in the SLID (Hale and Michaud, 1995) and the SIPP (Moore and Griffiths, 2003). With *proactive DI*, a 'remind, still' design was used: 'Last time we interviewed you, on <*DATE OF INTERVIEW*>, our records show that your pay was <*AMOUNT*$_1$> per <*PERIOD*$_1$> <*GROSS/NET*>. Is that still the case now, or has your pay changed?' To test whether proactive DI would capture changes, respondents were in addition asked the independent earnings questions.

Although 59 % of respondents were asked the reactive edit check question, all but one confirmed the change. In comparison, Hale and Michaud (1995) reported that only 8.3 % of SLID respondents reported earnings that differed by more than ± 10 % and under a third confirmed an error. This suggests that there may not be an optimal band width to query changes in earnings across surveys, but that the bands should be determined from the actual distributions of change observed in the data.

Proactive DI did not lead to aggregate under-reporting of change, which would be expected if the reminder led respondents to give least effort responses by saying their earnings have not changed. All proactive DI respondents were also asked the independent interviewing questions about their earnings, to test whether proactive DI would lead to an underestimation of change. At the individual level the answers to the proactive DI 'remind, still' question were inconsistent with the independent reports of earnings in some cases: for 12 % of respondents who claimed their earnings had not changed, the independent reports varied by more than ± 10 %. Unfortunately, it is impossible to distinguish whether the reminder led to false reports of stability, or whether the independent amounts contained errors leading to false rates of change. At the aggregate level proactive DI did not lead to different results. Substituting the previous wave information for respondents who reported that their earnings had not changed led to similar estimates of mean earnings, mean changes in earnings and proportions of respondents who experienced a change by more than ± 10 %.

Comparison of DI Designs

Proactive DI is not recommended because of the potential under-reporting of change over time, although this was not evident in the ISMIE survey. Ideally one would want to know how well proactive DI captures change after a number of successive interviews, since an initial amount of earnings may be fed forward several times if the respondent repeatedly answers the 'remind, still' question saying he has not experienced a change. Reactive DI designs appear most appropriate, with bands chosen to target potential errors and minimise the additional burden for respondents. For example, one might specify that only those 5 % of the sample with the largest apparent changes in earnings should be asked the follow-up question. The distribution of earnings changes between previous waves could then be used to identify the corresponding threshold value. This would be a value of change in earnings that is only exceeded by 5 % of the sample.

6.4.3 Current Employment

Dependent interviewing was implemented for a set of questions about the respondent's job and employer, with the objective of reducing spurious changes compared to the previous wave report. With *reactive DI*, 'follow-up, always' questions were used for the open-ended occupation and industry questions: 'Can I just check, is that the same employer that you were working for last time we interviewed you, on *<DATE OF INTERVIEW>*, when we recorded your employer as *<EMPLOYER>*?' For the other employment questions, 'follow-up, all inconsistent' questions were used: 'So, since last time we interviewed you, on *<DATE OF INTERVIEW>*, you've changed from being *<EMPLOYMENT STATUS*$_1$*>* to *<EMPLOYMENT STATUS*$_2$*>*: is that correct?' With *proactive DI*, 'remind, still' questions with routing were used: 'Last time we interviewed you, on *<DATE OF INTERVIEW>*, you said your occupation was *<OCCUPATION>*. Are you still in that same occupation?' This was similar to designs in the SLID (Hiltz and Cléroux, 2004), SIPP (Moore and Griffiths, 2003) and the US Current Population Survey (CPS; see Kostanich and Dippo, 2002), although these surveys make more use of routing around questions on employment characteristics if the respondent is in the same occupation and working for the same employer. For respondents in the same job since the previous interview, levels of change in employment characteristics were implausibly high with independent interviewing. Proactive DI resulted in lower levels of observed change for occupation, industry, managerial duties and size of workforce; reactive DI did not significantly reduce rates of change (Lynn and Sala, 2006). This suggests that proactive DI led to reductions in measurement error and confirms reports from the CPS (Polivka and Rothgeb, 1993, p. 19), about which Hill concluded that 'most of the observed 'change' [in industry and occupation codes] with independent data collection methods is a result of variability in the response/coding process' (Hill, 1994, p. 366). Proactive DI also achieved substantial savings in coding costs: routing around open-ended industry and occupation questions reduced the number of items to be coded by two-thirds (Jäckle, 2008). Similar gains were reported by Kostanich and Dippo (2002, p. 9-1) for the CPS.

Comparison of DI Designs

Reactive DI does not appear effective at reducing measurement error for questions on current employment. In addition, proactive designs offer large scope for efficiency gains, especially for open-ended occupation and industry questions, if combined with routing and feeding forward of previous codes. There are, however, open issues as to which employment characteristics can be assumed unchanged if the respondent is still in the same occupation and working for the same employer.

6.4.4 Labour Market Activity Histories

The ISMIE DI questions were designed to reduce spurious transitions at the seam in reports of labour market activities since the previous interview. With *reactive DI*, a 'follow-up, all inconsistent' design was used to query retrospective reports that were inconsistent with current activity reports from the previous interview: 'May I just check something? According to our records, when we last interviewed you, on *<DATE OF INTERVIEW>*, you were *<ACTIVITY*$_1$*>*. That spell had started in *<MONTH>*. But the information that you have just given me implies that you were *<ACTIVITY*$_2$*>* at that time. It may be that I have recorded something wrongly and it is important to us that our information is accurate, so can you just clarify that for me?' With *proactive DI*, a 'remind, confirm' design was used, reminding respondents of the previous wave current activity as an entry into the history: 'When we last interviewed you, on <*DATE OF INTERVIEW*>, our records show that you were <*ACTIVITY*>. Is that correct?'

With reactive DI, inconsistencies arose in 14 % of cases. Half the respondents said the earlier report was correct: the respondents had forgotten about spells and confirmed them when they were presented to them. The remainder of respondents either said both reports were correct (for example, retired and in part-time employment) or that they no longer remembered (Jäckle and Lynn, 2007). When reminded of their previous current activity, nearly 99 % of respondents confirmed this (Hale and Michaud (1995) reported similar confirmation rates for the SLID). Proactive DI led to a significant reduction of seam transitions, especially for job-to-job changes, confirming findings from the Canadian Labour Market Activity Survey (LMAS) (Murray *et al.*, 1991). Proactive DI did not appear to lead to under-reporting of change, since transition rates at the seam still tended to be higher than in nonseam months. Eliminating the redundancy in reporting the previous wave current activity reduced the number of questions answered in this section by 55 % and reduced coding time for open-ended industry and occupation descriptions by 81 % (Jäckle, 2008).

Comparison of DI Designs

Reactive DI is not appropriate to collect activity histories with sequential spell questions: if the respondent corrects a report in reaction to an 'inconsistent, follow-up', the entire sequence of spells may in fact be erroneous and would probably have to be asked again. Proactive DI is effective both at reducing redundancies of reports across waves and at reducing seam effects. Errors are not completely eliminated, however, because proactive DI remains sensitive to errors in reporting of transition dates.

6.4.5 School-Based Qualifications

The DI questions were designed to improve the reliability of highest qualification measures, derived from annual questions about education and training during the reference period. With *reactive DI*, respondents were asked 'follow-up, always' questions for any new qualifications reported: 'You have told me that you have gained <N_2> <*QUALIFICATIONS TYPE X*> since last time we interviewed you, and my records show that you previously had <N_1> <*QUALIFICATIONS TYPE X*>, so, you now have a total of <$N_1 + N_2$> <*QUALIFICATIONS TYPE X*>: is that correct?' With *proactive DI*, respondents were asked 'remind, confirm' questions about their qualifications recorded previously: 'We are particularly interested in checking the accuracy of the information we hold about school-based qualifications. According to our records from previous interviews, you have <*HIGHEST QUALIFICATION*>. Is that correct?' A similar design is used in the National Longitudinal Survey of Youth 1979, which starts by asking respondents to confirm or correct the highest educational attainment recorded in the previous interview, before asking about changes since (Hering and McClain, 2003 p. 98).

Sample sizes were extremely small for the reactive follow-up, since only a few respondents reported obtaining new qualifications during the reference period ($N = 6$). Only in one of these cases were the records reported to be incorrect. With proactive DI, 6 % ($N = 22$) of the treatment group corrected their record (not taking account of qualifications gained since the previous interview), but only in four cases did this affect the derived highest qualification measures. Respondents who had not been interviewed in all waves since the start of the panel were more likely to have incorrect records, although wave nonresponse did not explain all cases of erroneous records.

Comparison of DI Designs

The 'remind, always' design can be useful to verify information that is updated annually. If wave nonresponse is related to changes, for example if students in their final year are too busy to take part in the survey but join the survey again later, then their qualification will not be picked up since the following survey will only ask about changes in the current reference period. The reminder may also be used for occasional verification purposes, or only asked after a wave nonresponse, rather than at every wave. The reactive design appears less suitable, since it will only detect errors for respondents who have obtained and reported a new qualification.

6.5 EFFECTS OF DEPENDENT INTERVIEWING ON DATA QUALITY ACROSS SURVEYS

The evidence suggests that DI improves cross-wave continuity: both reactive and proactive DI reduce misclassification and under-reporting of events; in addition, reactive DI reduces keying error and proactive DI with routing reduces coding variability and consequent spurious change. Returning to Figure 6.1, DI reduces spurious changes at the seam and constant wave responses caused by under-reporting of events or statuses, especially for respondents who experienced a change during the reference period.

There is no evidence to support the concern voiced by Mathiowetz and McGonagle (2000) that proactive DI may lead to bias caused by respondents simply agreeing with previous information: in the ISMIE, transition rates for labour market activities remained higher for seam than nonseam months, even with proactive DI; the reminder did not increase over-reporting of income sources; aggregate measures of change in earnings were not affected; confirmation rates for previous reports of labour market activities in the SLID and ISMIE were slightly less than 100 %, which can be taken as evidence that respondents do not blindly confirm what is presented to them (Hale and Michaud, 1995), and in the LMAS, transitions from employment were not underestimated at the seam compared to administrative records (Murray *et al.*, 1991). The ISMIE findings reinforce Hale and Michaud's (1995, p. 10) conclusion that survey organisations 'should be using feedback to help [. . .] collect data that otherwise might be forgotten rather than worry about feedback preventing reporting of change'.

Several problems remain, however. First, some under-reporting remains, because DI can only have an impact on respondents for whom feed forward information is available from the previous wave. Given that the propensity to under-report is likely to be associated with some fixed characteristics of the survey respondent, those who under-report at the current wave could be expected to have an increased propensity to have under-reported also at the previous wave. Under-reporting could further be reduced if the DI questions could be extended to other sample members. Lynn *et al.* (2006), for example, suggested extending the DI questions for income sources to include all who reported receipt at any of the previous three waves (three appearing optimal in terms of coverage of under-reporters versus additional burden for nonrecipients), plus groups with a high likelihood of receipt, by filtering DI questions on predictors of receipt (e.g. mothers for child benefit) to increase coverage of under-reporters.

Second, the current DI designs do not improve the dating of events. In sequential spell questions, DI does not reduce nonresponse to date questions and it is not clear to what extent misdating of changes is reduced. As a result, the capacity of DI to reduce seam effects is limited (see Jäckle and Lynn, 2007). In period status questions respondents still tend to provide constant wave responses. In ISMIE around 86 % of income sources were reported as having been received for all months in the reference period, regardless of treatment group. By reducing under-reporting, DI did reduce spurious changes but could not eliminate the concentration of transitions at the seam: with independent interviewing 76 % of all transitions observed in the reference period occurred at the seam, compared to 64 % with reactive DI and 58 % with proactive DI. The comparable figure from the administrative records was 8 %.

This leads to a puzzle regarding the impact of DI on errors in measures of change. While proactive DI has been shown to successfully reduce seam effects in transitions from employment (Murray *et al.*, 1991) and to a lesser extent transitions in labour market activity status in general (Jäckle and Lynn, 2007), proactive DI is not effective at reducing seam problems for income data (Burkhead and Coder, 1985). LeMaître's (1992) discussion suggested that the differences are due to differences in design of the DI questions: the LMAS used 'remind, confirm' questions where information about employment at the time of the previous interview served as the starting point for the following interview. In contrast the SIPP and ISMIE income questions used 'remind, continue' designs. Apparently inspired by this critique, the SIPP questions were revised in 2004 as 'remind, confirm' questions, such that any apparent changes at the seam were queried. This new design was, however, disappointing in that it did not eliminate seam effects for income from welfare programmes: the percentage of transitions in receipt observed at the seam was reduced from 66 % with the original DI design to 52 %. The expected value in the absence of seam effects would have been 17 % (Moore *et al.*, 2004, table 10.6).

This suggests that factors that determine the nature of errors need to be examined when evaluating the effect of DI for different items or in different surveys. The nature of errors is likely to be influenced by characteristics of the *survey*, for example the length of the reference period (since recall errors increase with time), following rules (since alternations between proxy and self-response will lead to response variation) or changes in interviewers or coders. The nature of errors is also determined by characteristics of the *item* of interest. Labour market activities are likely to be more salient, not least because they may form an important part of the respondent's identity, and therefore be easier to recall correctly than income sources. In addition, income sources are multidimensional, since respondents may have received any number during the reference period, whereas labour market activities are mutually exclusive sequences of spells (at least if the survey asks about a well-defined 'main' activity). In general, the extent to which the item is clearly defined and response categories are unambiguous and mutually exclusive can determine the effectiveness of DI. The puzzle also suggests that the *temporal structure* of a question may have an impact on seam effects, regardless of the features of the DI design. As Burkhead and Coder (1985) pointed out, period status questions do not require the respondent to explicitly think about dates of changes. This may also explain Rips, Conrad and Fricker's (2003) low expectations as to the effectiveness of DI at reducing seam effects, since their laboratory study of the causes of seam effects was based on period status questions.

6.6 OPEN ISSUES

Much has been learned since Mathiowetz and McGonagle (2000, p. 416) concluded that 'With respect to dependent interviewing, the empirical literature is virtually nonexistent'. Nonetheless, many questions remain unanswered and many new questions have arisen:

- *Practical issues of implementation*. Little is still known about the effects of alternative question wording for given DI designs or about the impact of DI on the interviewer–respondent interaction, e.g. about how respondents react to reminders and follow-up queries and how these affect the cognitive response process and rapport with interviewers.
- *Data quality*. Although extensions to further reduce under-reporting have been proposed (see Lynn *et al.*, 2006), their effectiveness has not been tested. Improving the reporting of dates remains an unsolved problem, as does the question to what extent DI really reduces errors or merely leads to correlations in errors. Also, there is only anecdotal evidence about items on which DI has had detrimental impact and no clear evidence whether DI has any positive effect on attrition (see Moore *et al.*, 2004).
- *Data comparability*. Within one wave of a survey, different respondents in fact answer different questions, depending on whether feed forward information is available for them. Comparability over time in an ongoing panel survey can be hampered, since the introduction of DI is likely to introduce major breaks. Comparability across surveys is also affected if different surveys use different designs. This is especially problematic in cross-national data collection exercises, such as the EU Survey of Income and Living Conditions, where no central guidance is given on the use of dependent methods of data collection. This suggests that the standardisation of DI approaches, at least for the surveys run by one agency, or as part of a single statistical programme, would be an important future development.
- *Impact on analysis*. Most studies of the effect of DI have been carried out by government statistical offices, which tend to be interested in univariate statistics rather than relationships between variables. Not much evidence exists about the impact of different data collection methods on other types of analysis for which panel data are used (a notable exception is Hill, 1994). Jäckle and Lynn (2007), for example, suggested that although DI significantly reduces errors it may not lead to different results depending on the focus of analysis.

ACKNOWLEDGEMENTS

This paper derives from a project on 'Improving Survey Measurement of Income and Employment' (ISMIE), funded by the UK Economic and Social Research Council Research Methods Programme (H333250031), and developed following a workshop on Dependent Interviewing, held at the University of Essex (16–17 September 2004). I am grateful to my ISMIE colleagues, Peter Lynn, Stephen P. Jenkins and Emanuela Sala, to ISER colleagues for their assistance in producing the ISMIE dataset, especially Nick Buck, Jon Burton, John Fildes, Heather Laurie, Mike Merrett and Fran Williams, and to Mario Callegaro and two anonymous referees for helpful comments.

REFERENCES

Brown, A., Hale, A. and Michaud, S. (1998). Use of computer assisted interviewing in longitudinal surveys. In M. P. Couper *et al.* (Eds), *Computer Assisted Survey Information Collection* (pp. 185–200). New York: John Wiley & Sons, Inc.

Burkhead, D. and Coder, J. (1985). Gross changes in income recipiency from the Survey of Income and Program Participation. In *Proceedings of the Social Statistics Section* (pp. 351–356). Washington, DC: American Statistical Association.

Czajka, J. (1983). Subannual income estimation. In M. David (Ed.), *Technical, Conceptual and Administrative Lessons of the Income Survey Development Program (ISDP)* (pp. 87–97). New York: Social Science Research Council.

Dibbs, R., Hale, A., Loverock, R. and Michaud, S. (1995). Some effects of computer assisted interviewing on the data quality of the Survey of Labour and Income Dynamics. *SLID Research Paper Series*, No. 95-07. Ottawa: Statistics Canada.

Hale, A. and Michaud, S. (1995). Dependent interviewing: impact on recall and on labour market transitions. *SLID Research Paper Series*, No. 95-06. Ottawa: Statistics Canada.

Hering, J. and McClain, A. (2003). *NLSY97 User's Guide*. Washington, DC: US Department of Labor. Retrieved March 6, 2008, from http://www.bls.gov/nls/97guide/rd5/nls97ugall.pdf

Hill, D. H. (1994). The relative empirical validity of dependent and independent data collection in a panel survey. *Journal of Official Statistics*, 10, 359–380.

Hiltz, A. and Cléroux, C. (2004). SLID Labour Interview Questionnaire, January 2003: survey of labour and income dynamics. *Income Research Paper Series*, No. 75F0002MIE2004007. Ottawa: Statistics Canada. Retrieved March 6, 2008, from http://dissemination.statcan.ca/cgi-bin/downpub/listpub.cgi?catno = 75F0002MIE2004007

Holmberg, A. (2004). Pre-printing effects in official statistics: an experimental study. *Journal of Official Statistics*, 20, 341–355.

Hoogendoorn, A. W. (2004). A questionnaire design for dependent interviewing that addresses the problem of cognitive satisficing. *Journal of Official Statistics*, 20, 219–232.

Jäckle, A. (2008). Dependent interviewing: effects on respondent burden and efficiency. *Journal of Official Statistics*, 24, 411–430.

Jäckle, A. and Lynn, P. (2007). Dependent interviewing and seam effects in work history data. *Journal of Official Statistics*, 23, 529–551.

Jäckle, A., Sala, E., Jenkins, S. P. and Lynn, P. (2004). Validation of survey data on income and employment: the ISMIE experience. *ISER Working Paper*, No. 2004-14. Colchester: University of Essex. Retrieved March 6, 2008, fromhttp://www.iser.essex.ac.uk/pubs/workpaps/pdf/2004-14.pdf

Kostanich, D. L. and Dippo, C. S. (2002). Current population survey – design and methodology. *Technical Paper*, No.63RV. Washington, DC: US Census Bureau. Retrieved March 6, 2008, from http://www.census.gov/prod/2002pubs/tp63rv.pdf

LeMaître, G. (1992). Dealing with the seam problem for the Survey of Labour and Income Dynamics. *SLID Research Paper Series*, No. 92-05. Ottawa: Statistics Canada.

Lynn, P., Jäckle, A., Jenkins, S. P. and Sala, E. (2004). The impact of interviewing method on measurement error in panel survey measures of benefit receipt: evidence from a validation study. *ISER Working Paper*, No. 2004-28. Colchester: University of Essex. Retrieved March 6, 2008, from http://www.iser.essex.ac.uk/pubs/workpaps/pdf/2004-28.pdf

Lynn, P., Jäckle, A., Jenkins, S. P. and Sala, E. (2006) The effects of dependent interviewing on responses to questions on income sources. *Journal of Official Statistics*, 22, 357–384.

Lynn, P. and Sala, E. (2006). Measuring change in employment characteristics: the effects of dependent interviewing. *International Journal of Public Opinion Research*, 18, 500–509.

Martini, A. (1989). Seam effect, recall bias, and the estimation of labor force transition rates from SIPP. In *Proceedings of the Survey Research Methods Section* (pp. 387–392). Washington, DC: American Statistical Association.

Mathiowetz, N. A. and McGonagle, K. A. (2000). An assessment of the current state of dependent interviewing in household surveys. *Journal of Official Statistics*, 16, 401–418.

Moore, J. and Griffiths, J. K. (2003). Asset ownership, program participation, and asset and program income: improving reporting in the Survey of Income and Program Participation. In *Proceedings of the American Statistical Association* (pp. 2896–2903). Alexandria VA: American Statistical Association.

Moore, J. and Kasprzyk, D. (1984). Month-to-month recipiency turnover in the ISDP. In *Proceedings of the Survey Research Methods Section* (pp. 726–731). Washington, DC: American Statistical Association.

Moore, J., Pascale, J., Doyle, P., Chan, A. and Griffiths, J. K. (2004). Using field experiments to improve instrument design: the SIPP Methods Panel Project. In S. Presser *et al.* (Eds), *Methods for Testing and Evaluating Survey Questionnaires* (pp. 189–207). New York: John Wiley & Sons, Inc.

Murray, T. S., Michaud, S., Egan, M. and Lemaitre, G. (1991). Invisible seams? The experience with the Canadian Labour Market Activity Survey. In *Proceedings of the US Bureau of the Census Annual Research Conference* (pp. 715–730). Washington, DC: US Census Bureau.

Neter, J. and Waksberg, J. (1964). A study of response errors in expenditure data from household interviews. *Journal of the American Statistical Association*, 59, 18–55.

Pascale, J. and Bates, N. (2004). Dependent interviewing in the US Census Bureau's Survey of Income and Program Participation (SIPP). *Presentation given at the Workshop on Dependent Interviewing (16–17 September 2004)*, University of Essex, Colchester. Retrieved March 6, 2008, from http://www.iser.essex.ac.uk/home/plynn/ismie/Workshop.htm

Pascale, J. and Mayer, T. S. (2004). Exploring confidentiality issues related to dependent interviewing: preliminary findings. *Journal of Official Statistics*, 20, 357–377.

Phillips, M., Woodward, C., Collins, D. and O'Connor, W. (2002). Encouraging and maintaining participation in the Families and Children Survey: understanding why people take part. *Working Paper*. London: Department of Work and Pensions.

Polivka, A. E. and Rothgeb, J. M. (1993). Redesigning the CPS questionnaire. *Monthly Labor Review*, September, 10–28.

Rips, L. J., Conrad, F. G. and Fricker, S. S. (2003). Straightening the seam effect in panel surveys. *Public Opinion Quarterly*, 67, 522–554.

Ryscavage, P. (1993). The seam effect in SIPP's labor force data: did the recession make it worse? *SIPP Working Paper*, No. 180. Washington DC: US Census Bureau. Retrieved March 6, 2008, from http://www.sipp.census.gov/sipp/wp9308.pdf

Tourangeau, R., Rips, L. J. and Rasinski, K. (2000). *The Psychology of Survey Response*. New York: Cambridge University Press.

Young, N. (1989). Wave-seam effects in the SIPP. In *Proceedings of the Survey Research Methods Section* (pp. 393–398). Washington, DC: American Statistical Association.

CHAPTER 7

Attitudes Over Time: The Psychology of Panel Conditioning

Patrick Sturgis
University of Southampton, UK
Nick Allum
University of Essex, UK
Ian Brunton-Smith
University of Surrey, UK

7.1 INTRODUCTION

Panel studies are of crucial importance to our understanding of the complex, interacting and dynamic nature of causal processes in the social world. A limitation to valid inference from panel studies, however, derives from the reflexive nature of humans as research subjects; the very act of observation can serve to transform the behaviour of those being observed (Kalton and Citro, 2000). In panel studies this source of error falls under the general heading of 'panel conditioning' or 'time in sample bias' and relates to the fact that responses to questions in later rounds of the panel may be influenced by the interviews conducted at previous waves. The majority of empirical investigations of panel conditioning effects have focused on estimating biases in marginal totals of behavioural frequency, such as electoral turnout (Clausen, 1969), consumer spending (Pennell and Lepkowski, 1992) and alcohol and drug use (Johnson *et al.*, 1998). Less attention has been paid, in contrast, to how repeated interviewing might influence attitudinal responses. The cognitive processes underlying responses to attitude questions are, however, quite different in nature to those involved in reporting the frequency of behaviours (Schuman and Presser, 1981). Additionally, much of the research in the area

Methodology of Longitudinal Surveys P. Lynn

of panel conditioning has been largely atheoretical, with a primary focus on estimating the direction and magnitude of possible biases, as opposed to an elucidation of the causal mechanisms underlying conditioning effects (Holt, 1989).

In this chapter, our focus is on panel conditioning with respect to attitude questions. Our methodological approach is different from the majority of previous studies in this area, in that we do not attempt to estimate biases in marginal and associational distributions through comparison with a fresh cross-sectional sample. Rather, our approach is based on testing hypotheses on a single dataset, derived from an explicit theoretical model of the psychological mechanism underlying conditioning effects in repeated measures of the attitude. We refer to this theoretical account as *the cognitive stimulus* (CS) model. Specifically, we use a range of empirical indicators to evaluate the theory that repeatedly administering attitude questions serves to stimulate respondents to reflect and deliberate more closely on the issues to which the questions pertain. This, in turn, results in stronger and more internally consistent attitudes in the later waves of a panel.

The chapter proceeds in the following manner. First, we review the existing literature on panel conditioning effects. Next, we set out in more detail the rationale underlying the CS hypothesis. We then use data from the first eleven waves of the British Household Panel Survey (BHPS) to test four inter-related hypotheses expressed as empirical expectations of the CS model. We conclude with a discussion of the implications of our findings for the validity of attitude measures in panel surveys.

7.2 PANEL CONDITIONING

It is a truth universally acknowledged in discussions of the advantages and disadvantages of repeated measures surveys that panel conditioning is a serious threat to valid inference (Kasprzyk *et al.*, 1989; Rose, 2000; Ruspini, 2002). In one of the first methodological treatments of the repeated measures design, Paul Lazarsfeld, an early pioneer, noted that "the big problem, as yet unsolved is whether repeated interviews are likely, in themselves, to influence a respondent's opinion" (Lazarsfeld, 1940). There are, to be sure, strong reasons to believe that by repeatedly administering the same question to the same respondent over a period of time we might alter the responses in later rounds from those we would have obtained in the absence of the earlier interviews. Among the most frequently cited reasons why such an effect might occur are that respondents become more knowledgeable as a function of information imparted through the interview, become 'committed to' a particular position by stating it explicitly at first interview, reflect and deliberate more closely on an issue as a result of being asked about it and become 'better' respondents through learning the requirements of the survey interview procedure. However, despite Lazarsfeld's explicit focus on the conditioning of *opinions*, the majority of studies that have addressed the problem have focused instead on survey estimates of behaviour.

Typical of this focus on behavioural outcomes is Clausen's (1969) investigation of the effect of earlier interviews on subsequent voting behaviour. Noting the tendency of post-election surveys to overestimate voter turnout relative to official records, Clausen adjusted the sample of the 1964 National Election Study (NES) to better match the population of registered voters. He found that the NES sample overestimated turnout, even after appropriate matching adjustments had been made, leading him to conclude that the pre-election survey had 'stimulated' respondents to become more interested and engaged in the election and, therefore, more likely to vote. A later study by Kraut and McConahay replicated Clausen's finding using individual-level turnout data across randomly allocated interviewed and non-interviewed samples. They concluded, however, that rather than stimulating respondents to political action, the process of being interviewed reduces 'alienation' by forming a connection with society through the social interview and also changes a respondent's 'self-concept' so that the respondent thinks of him/herself as "more politically inclined than previously" (Kraut and McConahay, 1973). This, however, was purely conjectural as their analysis was based on a sample of only 104 voters and contained no measures of either of their key concepts of alienation and self-concept.

Yalch (1976), too, employed a design in which a sample of registered voters was randomly assigned to interview and noninterview conditions prior to a series of local elections. He again found significantly higher turnout amongst the interviewed sample, although this effect was apparent only in the first election after the interview and had disappeared by the time of the election occurring eight months after the interview. Yalch concluded, again on a rather thin evidential base, that his results were more consistent with Clausen's stimulus interpretation than the alienation and self-concept model of Kraut and McConahy. Traugott and Katosh (1979) were able to gain better leverage on the effect of previous interviews on turnout by comparing self-reported voter registration and turnout amongst members of a panel study to official turnout figures for the 1976 US presidential election. They found significantly higher levels of both registration and vote amongst the interviewed sample than in the general population. Additionally, they found this effect to be cumulative, with the registration and turnout rates increasing significantly with each additional interview. They found no evidence to suggest that repeated interviewing changed respondents' feelings of alienation or political self-concept. Thus, although it is not possible to indisputably attribute the effect to conditioning rather than attrition, they concluded that the very act of being interviewed can stimulate respondents to higher rates of actual, rather than merely reported, political participation.

A second strand of research on panel conditioning has utilized serendipitous or planned co-occurrences of panel and cross-sectional surveys and 'rotating panel' studies to estimate conditioning effects on marginal totals for a range of behavioural outcomes. Generally, these studies have shied away from identifying theoretical mechanisms underlying conditioning effects, focusing instead on estimating the direction and magnitude of time in sample bias on specific variables. This tradition of research has identified significant but small conditioning effects (Bailar, 1975, 1989; Pearl, 1979; Ghangurde, 1982; Wang *et al.*, 2000), although others have found no real evidence of conditioning effects at all (Pennell and Lepkowski, 1992). One of the few studies to concentrate specifically on panel conditioning of attitude responses using this type of design is that conduced by Waterton and Lievesley (1989). They tested six different hypotheses about the effects of repeated interviewing by comparing distributions on a range of variables on a three-wave panel to fresh annual cross-sections containing the same variables and carried out to near-identical designs. They found support for three of these six hypotheses: respondents became more 'politicized', reported more honestly and became less likely to answer 'don't know' as a result of participation in earlier waves. Although this study was unusual in its explicit focus on testing possible theoretical mechanisms underlying conditioning, no attempt is made to integrate the six hypotheses into a coherent causal narrative. It is not, therefore, entirely clear *why* we should expect to observe the pattern of empirical regularities they demonstrate, although, as we shall see, they are all consistent with the CS model we set out in the following section.

All the studies reviewed so far suffer, furthermore, from an inferential limitation deriving from what, at first sight, appears to be their primary analytical advantage; they estimate the effect of panel conditioning by comparing marginal totals from the $1 + n$th wave of the panel to a fresh cross-sectional, or earlier panel wave sample drawn at the same time from the same target population. However, while this approach allows an estimate of the more generic 'time in sample bias', the gross difference between samples confounds a number of potential causes, including nonresponse bias, interviewer conditioning, questionnaire content, questionnaire mode and variable measurement errors (see Holt (1989) for a detailed treatment of the limitations of this design; Kalton *et al.*, 1989; O'Muircheartaigh, 1989). Our approach in this paper, then, diverts from this standard design. Before turning to a presentation of our empirical results, we set out in detail below the proposed causal mechanism underlying panel conditioning of attitudes.

7.3 THE COGNITIVE STIMULUS HYPOTHESIS

Perhaps the principal finding from decades of empirical research into the social and political attitudes of the general public is that opinions are often weakly held, easily influenced and founded on a rather thin informational base. Survey researchers have repeatedly demonstrated that respondents: willingly offer opinions on nonexistent issues (Bishop *et al.*, 1986); demonstrate only weak consistency between issues, which elites routinely parcel together (Converse, 1964); switch from one side of an issue to the other in a quasi-random manner over time (Iyengar,1973; Sturgis, 2002); and are strongly influenced by the context in which a question is asked (Schuman and Presser, 1981). This pattern of empirical findings famously led Philip Converse to characterize survey measures of public opinion as "nonattitudes", uncognized responses constructed on the spot, in order to conform to the behavioural protocol of the survey interview. Importantly, for our purposes here, nonattitudes appear to result from a lack of information about and involvement in the issues to which attitude statements pertain. Thus, respondents with lower levels of political knowledge and engagement generally exhibit lower temporal and inter-item consistency, higher levels of nonsubstantive responding and lower resistance to persuasive communications (Zaller, 1992; Delli Carpini and Keeter, 1996). Why is this body of work from political science germane to an understanding of panel conditioning effects? Well, to the extent that many, possibly most, respondents will begin their first interview without having given much thought to their positions on important social and political matters, being asked questions about them may stimulate such respondents to reflect more closely on their views once the interview has ended. This may amount to little more than a cursory moment of private deliberation. Alternatively, it may in some instances extend to discussions with friends and family, or even to acquiring information through paying closer attention to relevant news in the media and elsewhere. Waterton and Lievesley (1989) report evidence for just such an effect from qualitative interviews with respondents to the 1983–1986 British Social Attitudes Panel.

These processes of public and private deliberation may, in due course, lead individuals to adopt a different attitudinal position than would have been the case in the absence of the initial interview. Fishkin's experiments with Deliberative Polling methodology convincingly demonstrate that information and deliberation can result in sizeable net and gross opinion change amongst random samples of the general public (Fishkin, 1997). Goodin and Niemeyer (2003) show, furthermore, that much of the opinion change induced via deliberation can come through private internal reflection, without any subsequent social or interactional element. Thus, the very process of administering attitude questions may be sufficient to engender some degree of attitude change, through stimulating respondents to reflect more closely on the subject matter of the questions (Jagodzinski *et al.*, 1987). Yet, it is not only shifts in marginal distributions that may result from this stimulus effect. Even in the absence of attitude change, respondents may come to hold their attitudinal positions more strongly and with greater consistency in relation to other idea elements in their belief systems as a whole (Sturgis *et al.*, 2005). So, increases in information, reflection and deliberation should lead to more considered, stable attitudes that have been arrived at through rational, preference-based judgement, as opposed to being constructed, more or less randomly, on the spot. An empirical expectation resulting from the CS model is, therefore, that indicators of attitude strength or crystallization should increase between the first and subsequent waves of the panel.

7.4 DATA AND MEASURES

Data come from the first 11 waves of the British Household Panel Survey (BHPS), 1991–2001. Our analytical sample comprises only those respondents who responded at all 11 waves, yielding a sample size of 5122, 60 % of the eligible wave 1 sample (see Taylor (2005) for full details of the sample design and response rate data for the study). Although taking only the completers reduces the inferential value of the sample with respect to the target population of all adults in Britain, it has the benefit of discounting differences in sample composition over time due to differential nonresponse as a cause of observed change over time. Conversely, it also means that our estimates of conditioning may be biased if attrition is correlated with change over time in our outcome measures, though this seems an unlikely contingency. Our analysis focuses on four different multi-item attitude scales: gender role attitude (administered every other year between 1991 and 2001), work and family attitude (administered every other year between 1991 and 2001), left–right attitude (administered 1991, 1993, 1995, 1997 and 2000) and life satisfaction (administered every year between 1997 and 2001), and a single item measure of interest in politics (administered every year between 1991 and 1996). Exact wordings for these items are provided in Table 7.1.

Table 7.1 Question wordings for the attitude scales and a measure of political interest.

Question Wordings for Work and Family Life Scale
1 = strongly agree, 5 = strongly disagree
A preschool child is likely to suffer if his or her mother works
All in all, family life suffers when the woman has a full-time job
A husband's job is to earn money; a wife's job is to look after the home and family

Question Wordings for Left–Right Scale
1 = strongly agree, 5 = strongly disagree
Ordinary people get their fair share of the nation's wealth
There is one law for the rich and one for the poor
Private enterprise is the best way to solve Britain's economic problems
Major public services and industries ought to be in state ownership
It is government's responsibility to provide a job for everyone who wants one
There is no need for strong trade unions to protect employee's working conditions and wages

Question Wordings for Gender Role Scale
1 = strongly agree, 5 = strongly disagree
A woman and her family would all be happier if she goes out to work
Both the husband and wife should contribute to the household income
Having a full-time job is the best way for a woman to be an independent person

Question Wordings for Life Satisfaction Scale
1 = not at all satisfied, 7 = completely satisfied
Here are some questions about how you feel about your life. Please tick the number that you feel best describes how dissatisfied or satisfied you are with the following aspects of your current situation:
Your health
The income of your household
Your house/flat
Your life overall

Question Wordings for Political Interest
1 = very interested, 4 = not at all interested
How interested would you say you are in politics? Would you say you are ...

7.5 ANALYSIS

As we noted earlier, estimating the effect of panel conditioning through comparison to a fresh cross-section is confounded with so many other factors that little confidence can be placed in the ability of this analytical strategy to isolate conditioning effects. Instead, our approach here is based on testing hypotheses expressed in terms of empirical expectations from a theoretical model on a single panel dataset over time. Although this approach is subject to its own inferential limitations, it does have the benefit of providing an explicit theoretical explanation of the conditioning effect, as opposed to "simply presenting additional evidence from a new survey that we, the survey community, must integrate with previous results" (Holt, 1989). Our first hypothesis is derived from the notion that being administered attitude questions stimulates respondents to reflect more carefully on the subject matter of the questions, resulting in a 'crystallization' or 'strengthening' of attitudinal positions. In the language of signal detection theory, we expect the ratio of signal to noise for individual items to increase as a result of this process. In the aggregate,

Table 7.2 Average item reliabilities 1991–2001.

Attitude scale	t_1	t_2	t_3	t_4	t_5	t_6
Left–right	0.17	0.19*	0.19	0.15*	0.14	
Work and family life	0.54	0.55*	0.57*	0.58*	0.59	0.59
Gender role	0.34	0.35*	0.35	0.36*	0.36	0.34
Life satisfaction (annual, 1997–2000)	0.38*	0.40*	0.41*	0.42*	—	—

* Significantly different from previous wave ($P < 0.05$).

this should serve to reduce the random component of variance in responses to the respective survey items, increasing the magnitude of correlation between related items at a single time point (Bollen, 1989). Thus our first hypothesis becomes:

H1 The reliability of attitude items will increase between the first and subsequent waves of the panel

Our measure of reliability is the mean of the communalities for the items in each scale from a confirmatory factor model, where the items are specified as measuring a single factor. In essence, this is a measure of the average percentage variance explained in the items by the factor they are intended to measure. Table 7.2 shows the reliabilities for each of the four scales at each wave they were administered between 1991 and 2001.

On all four scales, the prediction of the CS model is confirmed; the average item reliabilities increase from the first to subsequent waves. The pattern is seen most consistently in the 'work and family life' and 'life satisfaction' items, where there is a clear upward gradient from each wave to the next. The reliabilities for these items increase significantly from each wave to the next. Our test of significance is the difference in the likelihood ratios for a model in which the standardized factor loadings are constrained to be the same across waves and a model in which the standardized factor loadings are freely estimated (Agresti, 1986).

The gender role and left–right items also show an increase in reliabilities between the first and subsequent waves, although the trend is not as strong and tails off in later waves, particularly for the left–right items where there is a marked drop-off in the reliability of the scale items at the fourth measurement. It is interesting to note that the life satisfaction scale shows a consistent increase in reliability from each wave to the next, despite being administered for the first time at the seventh wave of the panel. This suggests that the mechanism underlying the effect is more likely a function of the *subject matter* of the questions than respondents simply becoming more familiar with survey procedures and standard item formats. If conditioning effects are caused by growing familiarity with interview procedures, we should expect such effects to be significantly smaller, or nonexistent, for questions introduced for the first time after six waves of interview. So the results of the test of H1 provide support for the CS model. An additional implication of these results is that the Wiley correction for measurement error in a single item (Wiley and Wiley, 1970) may often be inappropriate for attitudinal data. This is because, to identify the relevant parameter, the model assumes the error variance of the item to be constant over time. These results suggest that this assumption is unlikely to be met in practice.

Our second hypothesis is also derived from the idea that administering attitude questions serves to strengthen, or 'crystallize', opinions. Studies of the stability of social and political attitudes over time have revealed that there is usually a good deal of individual-level 'churn' in the response alternative selected from one wave to the next, without any noticeable change in the aggregate marginals (Johnston and Pattie, 2000). The primary explanation for this effect is that, because the majority of opinions are generally weak, respondents tend to show little consistency in the response alternative selected between each wave and the next (Converse, 1964). Furthermore, because better informed respondents have been shown to exhibit significantly greater stability over time (Delli Carpini and Keeter, 1996), we should expect, *ceteris paribus*, stability coefficients to increase over repeated administrations of the same items, as respondents develop stronger and more well-founded preferences. Our second hypothesis is, therefore:

H2 The stability of attitude items will increase from the first to subsequent waves of the panel

where stability is operationalized as the bivariate Pearson's correlation between the summed scale at adjacent pairs of waves of the panel. Figure 7.1 plots the stability coefficients for each of the four scales across adjacent waves. Although only the 'work and family life' scale shows a statistically significant upward gradient ($P < 0.05$), all four scales show the same general pattern of increasing stability over time. The 'left–right' scale shows a statistically significant increase in stability across the first three administrations but this tails off between the third and fourth administrations, possibly as a result of the additional year between the third and fourth administrations of these items. Despite this discrepancy, H2 also provides support for the CS model of panel conditioning; the stability of attitudinal responses increases over repeated administrations.

In addition to the effect on inter- and intra-item consistency induced through public and private deliberation, it is likely that administering attitude questions *per se* will lead respondents to be more likely to declare an opinion at all, as opposed to saying 'Don't Know' (DK). Such an increase in 'opinionation' has been observed between the pre- and

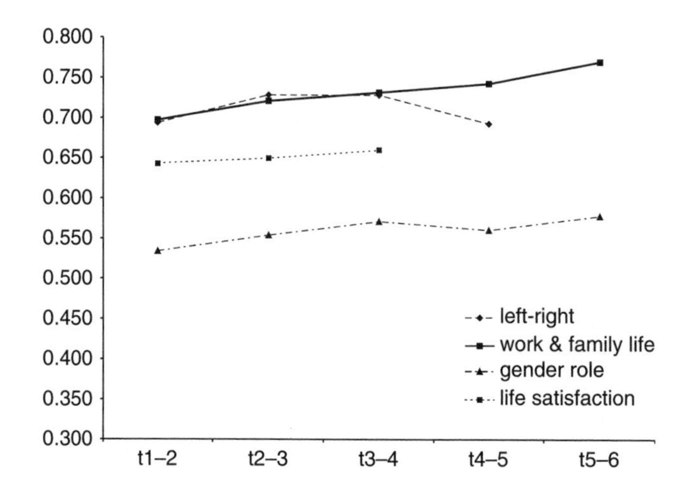

Figure 7.1 Interwave stabilities, 1991–2001.

postsurveys of deliberative polls (Moller Hansen, 2004) and was also noted by Waterton and Lievesley in their analysis of the British Social Attitudes Panel (Waterton and Lievesley,1989). Our third hypothesis is therefore:

H3 The degree of opinionation will increase between the first and subsequent waves of the panel

Figure 7.2 plots our measure of opinionation, the proportion of responses across the six items of the 'left–right' attitude scale recorded as DK, or 'can't choose', at each wave. Only the left–right scale was suitable for testing the opinionation hypothesis because the other three scales were all administered via a self-completion questionnaire, which offered no explicit DK option. The degree of opinionation on the left–right scale increases steadily and significantly from the first to subsequent waves of the panel ($P < 0.05$), providing further support for the CS model. There is a very small increase in DK responses between the fourth and fifth time points, again possibly resulting from the longer time lag between administration of these waves. The general decline in DK responses replicates the results of Waterton and Lievesley (1989), who also found a significant decline in DK responses over time, although they interpreted the finding as indicating that respondents acquired a better understanding of survey procedures and requirements over repeated waves of the panel. Although there is no obvious way of determining which of these interpretations is correct, there seems to us no obvious reason why respondents should come to believe that DK responses are somehow inappropriate. A DK option is, after all, explicitly offered along with the substantive response alternatives at each wave, so it is hard to see how repeated interviewing might lead to their developing such a view. An increase in the proportion of substantive responses is, on the other hand, a direct expectation of the CS model, has been empirically demonstrated in comparable studies, and is easily integrated with the other empirical regularities we present here.

This brings us to the last of our empirical tests. We have shown that administering attitude items to respondents on a number of occasions leads to increased inter- and intra-item consistency and a reduction in the proportion of nonsubstantive responses.

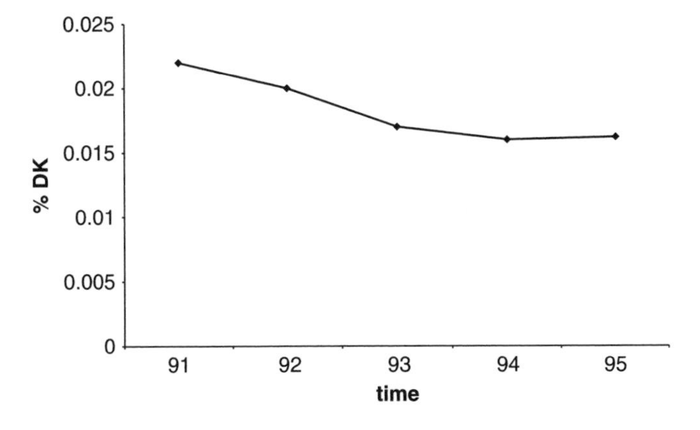

Figure 7.2 Proportion of 'Don't know' (ok) responses on the 'left–right' scale, 1991–2001.

We have argued that these effects occur as a result of respondents paying greater attention to and reflecting more closely on the issues to which the attitude questions pertain. Another way of putting this is that respondents become more interested in these issues. Indeed, this is what the term 'interest' really denotes: an inclination to pay attention to, reflect on and discuss a particular subject or topic domain (Smith, 1989). Our fourth hypothesis is therefore:

H4 The expressed level of interest in politics will increase between the first and subsequent waves of the panel

Figure 7.3 shows the mean self-reported interest in politics amongst the sample of completers in the BHPS between 1991 and 1996, where higher scores denote greater interest. There is an initial upturn in interest between the first two waves ($P < 0.001$) but thereafter interest declines significantly over the following four waves. Fitting a fixed effect cross-sectional time series model to the individual level scores shows that interest decreases significantly over time, controlling for all time-invariant respondent characteristics ($P < 0.001$).

Thus, our results lead us to reject H4, while noting that this is complicated somewhat by the initial increase in interest between waves 1 and 2. It would seem that, counter to our expectation, participation in the panel resulted in respondents becoming *less* interested in politics over time. It is not clear from the data what underlies this trajectory of political interest amongst panel members over time, though we speculate that it might be an influence of the external political environment. That is to say, what we may be seeing here is an upturn in interest in 1992, when a General Election was held, followed by a cyclical inter-election decline. Andersen, Tilley and Heath (2005) report such an effect for political knowledge in the 1992–1997 and 1997–2001 electoral cycles in Britain. Unfortunately, however, the political interest question was not administered in the 1997 wave of the BHPS, when the next election was held, meaning that it is not possible for us to assess this hypothesis empirically.

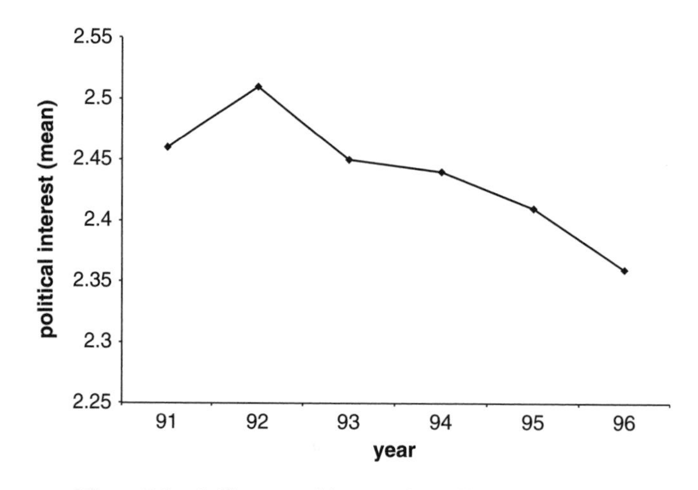

Figure 7.3 Self-reported interest in politics, 1991–1996.

7.6 DISCUSSION

The existing literature on panel conditioning suffers from two primary flaws. First, it has lacked coherence and clarity with regard to the proposed mechanisms underlying conditioning effects. Second, it has relied almost entirely on research designs that confound the effect of conditioning with a range of extraneous factors, such as differential nonresponse and attrition, variable measurement errors, context effects and interviewer conditioning. In this paper we have employed a different analytical approach, testing a number of inter-related hypotheses derived from a clearly articulated theoretical model on a single panel dataset. Overall, our results provide support for the proposed CS model of panel conditioning. This model stipulates that being administered attitude questions leads to public and/or private deliberation on the subject matter of the questions. This process, in turn, leads to a strengthening and crystallization of preferences on the underlying attitude. Three empirical expectations from this model – that attitude items will become more reliable and more stable over time and that opinionation on social and political issues will increase – were all borne out in our empirical analyses. A fourth hypothesis, that the stated level of interest in politics would increase over time, was not supported; although an initial increase in interest was observed between waves 1 and 2 of the panel, subsequent waves showed a consistent and significant decrease in interest.

That the internal consistency and consistency over time of the life satisfaction scale, which was first administered at wave 7, also conformed to the expected pattern suggests that the CS is specific to the particular issue in the items, rather than a more general learning of how to be a 'good respondent'. Note, in addition, that the findings of Waterton and Lievesley (1989) are consistent with the CS model. Although they place a somewhat different interpretation on their results, in their study panel respondents became more likely to identify with political parties and less likely to say 'don't know' with each additional wave of the panel, a pattern of results that is wholly consistent with the CS model we have set out here.

In terms of the implications of our results for the validity and reliability of attitudinal panel data, one could interpret these findings in either of two ways. First, repeated measurements on the same individuals make the responses provided by those individuals differ from the responses that would be obtained from a fresh sample of the same population. From this perspective, panel conditioning effects lead to biased estimates of population parameters. The direction of this bias is toward overestimating the political sophistication and opinion strength of the general public. This is likely to compound the biasing effects of panel attrition, which is also disproportionately situated amongst the less well informed and unengaged members of the public (Brehm, 1993). However, in terms of coming to understand 'true' social and political preferences, such a bias may, paradoxically, be seen as rather beneficial; in a world where citizens are only dimly aware of important social and political issues, the effect of repeated interviews may be to bring expressed preferences more in line with individual and group interests, serving to level out the information asymmetries commonly found in mass publics (Althaus, 2003).

We have attempted, in this study, to overcome the problems inherent in most extant empirical investigations of panel conditioning by setting out in detail a theoretical mechanism and testing a number of hypotheses derived as direct expectations on a single panel dataset. This approach sidesteps many of the problems of cross-sample comparability in the conventional approach but throws up new barriers to generalization in its place. One such barrier is that our observed effects could be a result of simple maturation of the sample; as people get older, they have more consistent, stable attitudes and become more opinionated. More importantly, perhaps, our reliance on a single panel dataset means that we cannot discount the possibility that, when we observe change over time, we are merely picking up trends occurring in the general population, rather than any kind of conditioning effect. This is, after all, what panel studies are primarily intended to do. Finally, our approach does not allow us to discount the possibility that at least some of the observed regularities we are attributing to respondent conditioning are, at least to some degree, also a result of the conditioning of interviewers, who often interview the same household repeatedly in this type of survey. Thus, it is important that this study acts, itself, as a stimulus to replication and refinement in the future on different datasets, containing different measures, administered in different contexts and using different modelling techniques. Only with replication, confirmation and refinement will our analytical approach begin to bear fruit.

REFERENCES

Agresti, A. (1986). *Introduction to Categorical Data Analysis.* New York: John Wiley & Sons, Inc.

Althaus, S. (2003). *Collective Preferences in Democratic Politics: Opinion Surveys and the Will of the People.* New York: Cambridge Press.

Andersen, R., Tilley, J. and Heath, A. (2005). Political knowledge and enlightened preferences: party choice through the electoral cycle. *British Journal of Political Science*, 35, 285–302.

Bailar, B. A. (1975). The effects of rotation group bias on estimation from panel surveys. *Journal of the American Statistical Association*, 70, 23–30.

Bailar, B. A. (1989). Information needs, surveys and measurement errors. In D. Kasprzyk, G. J. Duncan, G. Kalton and M. P. Singh (Eds), *Panel Surveys* (pp. 1–25). New York: John Wiley & Sons, Inc.

Bishop, G., Tuchfarber, A. and Oldendick, R. (1986). Opinions on fictitious issues: the pressure to answer survey questions. *Public Opinion Quarterly*, 50, 240–250.

Bollen, K. (1989). *Structural Equations with Latent Variables.* New York: John Wiley & Sons, Inc.

Brehm, J. (1993). *The Phantom Respondents: Opinion Surveys and Mass Representation.* Ann Arbor: University of Michigan Press.

Clausen, A. R. (1969). Response validity: vote report. *Public Opinion Quarterly*, 37, 398–406.

Converse, P. (1964). The nature of belief systems in mass publics. In D. E. Apter (Ed.), *Ideology and Discontent* (pp. 206–261). New York: Free Press.

Delli Carpini, M. and Keeter, S. (1996). *What Americans Know about Politics and Why it Matters.* New Haven, CT: Yale University Press.

Fishkin, J. (1997). *The Voice of the People: Public Opinion and Democracy* (2nd edn). New Haven, CT: Yale University Press.

Ghangurde, P. D. (1982). Rotation group bias in the LFS estimates. *Survey Methodology*, 8, 86–101.

Goodin, R. and Niemeyer, S. (2003). When does deliberation begin? internal reflection versus public discussion in deliberative democracy. *Political Studies*, 51, 627–649.

Holt, T. (1989). Panel conditioning: discussion. In D. Kasprzyk, G. J. Duncan, G. Kalton and M. P. Singh (Eds), *Panel Surveys* (pp. 340–347). New York: John Wiley & Sons, Inc.

Iyengar, S. (1973). The problem of response stability: some correlates and consequences. *American Journal of Political Science*, 17, 797–808.

Jagodzinski, W., Kuhnel, S. M. and Schmidt, P. (1987). Is there a "socratic effect" in non-experimental panel studies? consistency of an attitude toward guest-workers. *Sociological Methods and Research*, 15, 259–302.

Johnson, R., Gerstein, D. R. and Rasinski, K. (1998). Adjusting survey estimates for response bias: an application to trends in alcohol and marijuana use. *Public Opinion Quarterly*, 62, 354–377.

Johnston, C. and Pattie, G. (2000). Inconsistent individual attitudes within consistent attitudinal structures: comments on an important issue raised by John Bartle's paper on causal modelling of voting in Britain. *Political Studies*, 30, 361–374.

Kalton, G. and Citro, C. (2000). Panel surveys: adding the fourth dimension. In D. Rose (Ed.), *Researching Social and Economic Change* (pp. 36–51). London: Routledge.

Kalton, G., Kasprzyk, D. and McMillen, D. (1989). Nonsampling errors in panel surveys. In D. Kasprzyk, G. J. Duncan, G. Kalton and M. P. Singh (Eds), *Panel Surveys*. New York: John Wiley & Sons, Inc.

Kasprzyk, D., Duncan, G., Kalton, G. and Singh, M. P. (1989). *Panel Surveys*. New York: John Wiley & Sons, Inc.

Kraut, R. E. and McConahay, J. B. (1973). How being interviewed affects voting: an experiment. *Public Opinion Quarterly*, 16, 381–398.

Lazarsfeld, P. (1940). 'Panel' studies. *Public Opinion Quarterly*, 4, 122–128.

Moller Hansen, K. (2004). *Deliberative Democracy and Opinion Formation*. Department of Political Science, University of Southern Denmark.

O'Muircheartaigh, C. (1989). Sources of nonsampling error: discussion. In D. Kasprzyk, G. J. Duncan, G. Kalton and M. P. Singh (Eds), *Panel Surveys* (pp. 271–288). New York: John Wiley & Sons, Inc.

Pearl, R. (1979). *Reevaluation of the 1972–3 US Consumer Expenditure Data by Household Interviews: An Experimental Study*. Washington, DC: US Census Bureau.

Pennell, S. G. and Lepkowski, J. M. (1992). Panel conditioning effects in the Survey of Income and Program Participation. In *Proceedings of the Survey Research Methods Section of the American Statistical Association* (pp. 566–571). Washington, DC: American Statistical Association.

Rose, D. (2000). *Researching Social Change: Household Panel Studies: Methods and Substance*. London: UCL Press.

Ruspini, E. (2002). *Introduction to Longitudinal Research*. London: Routledge.

Schuman, H. and Presser, S. (1981). *Questions and Answers in Attitude Surveys: Experiments on Question Form, Wording and Context*.New York: Academic Press.

Smith, E. (1989). *The Unchanging American Voter*. Berkeley and Los Angeles: University of California Press.

Sturgis, P. (2002). Attitudes and measurement error revisited: a reply to Johnston and Pattie. *British Journal for Political Science*, 32, 691–698.

Sturgis, P., Roberts, C. E. and Allum, N. (2005). A different take on the deliberative poll: information, deliberation and attitude constraint. *Public Opinion Quarterly*, 69, 30–65.

Taylor, M. (2005). *British Household Panel Survey User Manual Volume A: Introduction, Technical Report and Appendices*. Colchester: University of Essex.

Traugott, M. W. and Katosh, J. P. (1979). Response validity in surveys of voter behaviour. *Public Opinion Quarterly*, 43, 359–377.

Wang, K., Cantor, D. and Safir, A. (2000). Panel conditioning in a random digit dial survey. In *Proceedings of the Survey Research Methods Section of the American Statistical Association* (pp. 822–827). Washington, DC: American Statistical Association.

Waterton, J. and Lievesley, D. (1989). Evidence of conditioning effects in the British Attitudes Panel Survey. In D. Kasprzyk, G. J. Duncan, G. Kalton and M. P. Singh (Eds), *Panel Surveys* (pp. 319–339). New York: John Wiley & Sons, Inc.

Wiley, D. and Wiley, E. (1970). The estimation of measurement error in panel data. *American Sociological Review*, 35, 112–117.
Yalch, R. F. (1976). Pre-election interview effects on voter turnout. *Public Opinion Quarterly*, 40, 331–336.
Zaller, J. (1992). *The Nature and Origins of Mass Opinion*. New York: Cambridge University Press.

CHAPTER 8

Some Consequences of Survey Mode Changes in Longitudinal Surveys

Don A. Dillman
Washington State University, USA

8.1 INTRODUCTION

Longitudinal surveys are often undertaken to measure changes in perceptions and behaviours over time. One of the unique challenges in such surveys is measuring the extent to which people's attitudes or behaviours have changed by asking the same questions during multiple periods of data collection and examining the differences (Dillman and Christian, 2005). Examples of questions used to measure change are:

- How satisfied are you with your work: very satisfied, somewhat satisfied, a little satisfied or not at all satisfied?
- How would you rate your health: excellent, good, fair or poor?
- In the last six months how many days have you missed work because of illness?
- Please indicate which of the following personal goals you believe are desirable for employees you supervise? (followed by list of many different goals).
- Which do you consider a higher personal priority—being recognized by your supervisor for doing a good job on a project or obtaining a cash bonus.

In longitudinal surveys, questions such as these are often asked to assess programme interventions or examine how a specific event affected peoples' lives or attitudes. Sometimes they are used simply to measure the maturation of respondents, or to understand how time has affected them. In order to measure change accurately it is therefore critical that respondents be presented with the exact same question stimulus each time a measurement is made.

The accurate measurement of change between surveys for such questions is often threatened, however, by making a switch from one mode of survey data collection to

Methodology of Longitudinal Surveys P. Lynn

another. Use of a different mode may result in unintentional changes of question wording, answers that are affected by whether the communication is aural or written and construction practices that are associated in normal practice with one survey mode but not another. As the need for and ability to administer mixed-mode surveys have evolved, consideration of how survey modes affect the measurement of change has grown from a minor to a major concern (De Leeuw, 1992). These issues require careful contemplation when designing the initial survey and when making procedural decisions for follow-up surveys. The aim of this chapter is to discuss ways in which changes in survey mode produce errors in the measurement of change in longitudinal surveys, and to suggest potential remedies to prevent incorrect inference about such changes.

Achieving accurate measures of change in longitudinal surveys, regardless of mode, is also threatened by a number of other considerations, ranging from changes in question order to the introduction of new questions that provide different contexts, which in turn influence how people think of individual items and formulate their responses. Though important, these issues are not the focus of this chapter. Concern here is limited to the causes and consequences of collecting data by one survey mode initially and then changing to another mode for the follow-up measurement of change.

8.2 WHY CHANGE SURVEY MODES IN LONGITUDINAL SURVEYS?

There are a number of reasons why survey modes sometimes change in the middle of longitudinal studies. Many such studies, for example, were designed to use the most cost effective and accurate mode that was available at the time the study was designed. These might include classroom (or group) administered, postal mail, telephone or face-to-face surveys. Since these studies began though, new modes, such as Internet (Best and Krueger, 2004) and interactive voice response (IVR) surveys (Miller Steiger and Conroy, 2008), have been developed that make it possible to obtain responses with much greater speed and efficiency than previous modes. In addition, both Internet and IVR surveys have significant cost advantages over the traditional modes, making them particularly attractive alternatives. In many cases the potential cost savings are a major factor in the decision to switch modes for future rounds of data collection. For example, the US Current Population Survey, a monthly data collection designed to measure unemployment rates, switches respondents from face-to-face methods for initial household contacts to telephone for follow-up contacts wherever possible. This is done to speed up data collection as well as to reduce costs.

Another reason to choose a different mode is because respondents' situations change between data collection periods. For example, some longitudinal surveys are begun by administering paper questionnaires in classroom settings, e.g. high school or university students. Later, as respondents move to other locales, reassembling them in one location is no longer possible. In such circumstances there has often been a strong push to use the telephone. An example is The Wisconsin Longitudinal Study begun among high school seniors in 1957, which later used telephone for data collection as a means of reducing attrition (Hauser, 2003).

One might also change survey modes between data collection periods in response to limitations in the capability of data collection organizations. Some survey organizations are equipped to do telephone surveys but others are not. Some of the newer organizations have developed web survey capabilities, but have chosen not to use older modes such as mail or telephone. In the USA and certain other countries, such as Sweden and Switzerland, fewer organizations now have the capability to mount large-scale face-to-face interview surveys. When bids are solicited to decide who should conduct an additional round of data collection, organizations that do not have a well-developed capability for conducting the survey by the original mode may place bids using a different data collection strategy.

Changes in the way people manage their personal contact information may also encourage changes in survey modes between data collection periods. Many people now have cellular phones apart from their home phone and a significant portion of Americans have got rid of their landline telephones altogether (Blumberg *et al.*, 2006). Many people also maintain multiple e-mail addresses, some of which they only share selectively. Whereas in the past people might reasonably be asked to provide all of their contact information, including telephone numbers and a mailing address, the situation now is quite different. Many people are hesitant to provide all of their contact information, and they vary in their willingness to provide each of the specific types of information. Under these conditions sometimes the only way to reach the original set of survey respondents for a follow-up is by using a mixed-mode approach.

An issue closely related to the management of personal contact information is people's increasing unwillingness to respond to surveys, resulting in declining response rates for most survey modes. Whereas in the early years of telephone surveying people were likely to answer a ringing telephone, the availability of answering machines and voice mail along with increasing intolerance for unsolicited calls (i.e. telemarketing, campaigning and fund-raising calls) has resulted in some people screening their calls, which they may then ignore, return at their convenience or even respond to by another mode such as e-mail. As a result of this change in behaviour, surveyors increasingly feel a need to switch modes to increase the likelihood of being able to first contact and then complete the survey with potential respondents. Mode changes to increase response rates can occur between data collection periods, but they may also occur in the middle of a single data collection period because research has shown that offering an alternative survey mode can increase response rates (Dillman and Christian, 2005). As a result, sponsors of both initial and follow-up data collection efforts may use a second or third mode to improve overall response and retention in longitudinal surveys. For example, the United States Survey of College Graduates conducted by the National Science Foundation that surveys the same individuals for many years has asked respondents their preferred method of being surveyed, and by using it has been able to speed up data collection (Hoffer *et al.*, 2007).

In some respects then, the decision on survey mode, which used to be primarily under the control of the surveyor, is now switching to being under the control of the respondent. Surveyors are left with little choice but to follow respondent preferences and behaviours if they want respondent cooperation. After taking all these changes into consideration, it should not be surprising that changing modes in longitudinal surveys has switched from being an option to seeming mandatory in some cases. It is against this background that the consequences of switching modes need to be carefully examined.

8.3 WHY CHANGING SURVEY MODE PRESENTS A PROBLEM

8.3.1 Changes in Question Structure

Each survey mode tends to bring with it a preferred way of asking questions. Even when surveyors are consciously trying to maintain a particular way of asking a survey question across two or more modes, changes often inadvertently get made because they fit normal construction procedures for a specific survey mode (Dillman and Christian, 2005).

For example, face-to-face surveyors typically use show cards so respondents can see and read answer categories, or even the entire question. Other reasons typically offered for the use of show cards is that they help the interviewer retain rapport with the respondent, and they give the respondent something to do during the interview. However, as a consequence, respondents to such interviews receive both visual and aural stimuli, while respondents to a telephone survey would receive only an aural stimulus and respondents to a mail survey would receive only a visual stimulus.

Because telephone surveys rely solely on aural information, surveyors often remove words from scales, favouring numerical scales with labelled endpoints over those using labels for every category. This change simplifies the interviewer's task and reduces the likelihood of mistakes. For example, respondents in a face-to-face interview may be presented with the question, 'How satisfied would you say you are with your current position?', followed by five fully labelled options ranging from completely to not at all satisfied. This format, which works perfectly well in the face-to-face interview with a show card, may get changed for a telephone survey to, 'On a five-point scale where 5 is completely satisfied and 1 is not at all satisfied, and you can use any number in between, how satisfied would you say you are with your current position?' This is a significant change in question stimulus that seems 'necessary' given the features of the two modes under consideration but that can have potentially significant effects on responses. In addition, the telephone places pressure on surveyors to keep questionnaires short, which may lead to the removal of certain questions that may either eliminate or introduce order effects.

In another example, surveyors often face significant pressure to change a series of yes/no, forced-choice questions (e.g. Please indicate whether or not each of the following items affected your decision to change jobs) that work perfectly well in aural modes (face-to-face or telephone) to a check-all-that-apply format (e.g., Please check all of the items that affected your decision to change jobs) for visual modes (mail and web). It is assumed that these two formats will produce the same responses. The forced-choice format is often used in aural modes because it is mentally easier to process the response options one at a time than as a whole group when they are only heard and not seen. In visual modes though, the check-all format is thought to be easier and less time consuming for respondents and to reduce the tendency to only mark 'yes' answers. However, research now indicates clearly that the check-all and forced-choice formats produce different results both within and across visual and aural survey modes. In mail and web surveys the forced-choice format encourages deeper processing of response options and endorsement of more items than the check-all format and it discourages satisficing response strategies often used in check-all questions (Smyth *et al.*, 2006c; Thomas and Klein, 2006). Moreover, the forced-choice format produces similar results in web and telephone surveys (Smyth *et al.*, 2006a).

Surveyors also tend to treat open-ended questions differently across modes. Often they avoid these questions in self-administered surveys because they have exhibited a tendency to perform poorly (Sudman and Bradburn, 1982). In contrast, open-ended questions perform much better in telephone and face-to-face surveys because the interviewers can probe for more informative answers. Research is beginning to suggest, however, that open-ended questions produce substantially more information in web compared to mail surveys and even that certain strategies can be used to enhance their performance to the point that web responses are quite comparable to telephone responses (Smyth *et al.*, 2006b).

In addition to the above mode challenges, the Internet introduces new question formats that are not available in other survey modes. One of these formats is drop-down menus, which can be used to avoid open-ended formats and the subsequent coding they require, e.g. asking for birth month or perhaps state of residence. Another example is sliding scales where, instead of marking an answer using a radio button associated with it, respondents click on and drag a marker along a scale to the point they would like to choose and then drop the marker at that point (Funke and Reips, 2006). Internet technology also allows for the use of audio, video, colour and fill-ins from previous answers more so than in other modes. In addition, web survey designers have the option of requiring answers in order to eliminate item nonresponse, but doing so might anger respondents and increase mid-survey terminations (Best and Krueger, 2004).

Interactive voice response surveys also have a number of unique features that differentiate them from the other modes (Miller Steiger and Conroy, 2008). For example, they are characterized by little flexibility as they require responses to all questions and cannot interactively respond to situations as they arise. That is, there is usually no way for respondents to have the questions clarified or to make their own enquiries if they do not understand something. In addition, respondents must coordinate what they are hearing through the receiver with the task of entering numerical answers on the touchtone pad of their telephone. As a consequence, survey designers tend to work towards shortening questions even further in this mode than is typically done with telephone surveys.

Few people would set out deliberately to change question structures across survey modes, yet the pressure to stay with standard organizational practice is powerful, even for routine demographic questions. For example, one organization as standard practice in telephone surveys asked, 'What is your marital status?', as an open-ended question. Interviewers then coded responses into one of five categories: 'single, married, separated, widowed or divorced' (Dillman and Christian, 2005). A change to conducting the same survey over the Internet resulted in asking the same question, except that respondents were provided with the five categories up front and were required to mark one of them. The result of this change was that the proportion of single or married respondents decreased by 4.4 % while the other three categories combined increased by the same amount. This difference could easily be explained by the absence of the last three categories as an explicit stimulus in the telephone survey. Although the simple solution was to offer the same five categories over the telephone as were listed on the web page, doing so required giving up a standard organizational procedure used for nearly all telephone surveys conducted by the sponsoring organization.

This source of differences across survey modes has led to the recommendation that surveyors use unified mode construction as opposed to mode-specific construction when designing mixed-mode surveys, especially for longitudinal surveys (Dillman,

Smyth, and Christian, 2009). As described there, unified mode construction refers to writing questions that minimize differences in answers between modes by finding a common ground for construction. The goal is to present the same mental stimulus, regardless of survey mode and in particular whether it is visual or aural. An example of unified mode construction is to explicitly offer no opinion categories in both visual and aural surveys, rather than the interviewer holding them back unless the desire to use such a category is initiated by the respondent. Another example is the use of forced-choice question formats in both telephone and web surveys, rather than switching in the latter to a check-all format, which would produce different answers (Smyth *et al.*, 2006a). Thus, the adoption of a unified mode philosophy of question construction implies identical wording of questions in all survey modes.

Mode-specific construction, on the other hand, suggests that survey modes may be inherently different in their capabilities for obtaining accurate answers, especially when visual vs. aural modes are used. As a result, certain changes in question structure may sometimes achieve more equivalent answers than if unified mode construction procedures are followed. Several examples of the need for mode-specific construction appear in the paragraphs that follow, including use of technological features available on the web and how people respond to open-ended questions across multiple modes.

One of the barriers to collecting equivalent data across survey modes is that initial benchmark surveys are often conducted without considering the possibility that the survey mode may need to be changed for the follow-up. Thus, question formats that best use the attributes of a particular mode may be chosen with no consideration being accorded to what follow-up data collection in a different survey mode might require in the way of adaptation. The effects of switching between aural and visual communication are especially important in this regard.

8.3.2 Effects of Visual vs. Aural Communication Channels

Aural communication, which is the primary mode of communication in telephone, face-to-face and IVR surveys, differs significantly from visual communication, the primary way in which questions are communicated to respondents in mail and Internet surveys. In both cases paralanguages play a role in how questions are presented to respondents and may influence answers in different ways. For example, Krosnick (1999) has argued that when response category choices are presented visually, weak satisficing is likely to bias respondents towards selecting choices displayed earlier in the list. More recent research has shown that for dichotomous answer choices offered on the telephone, significant recency effects occur about one-fifth of the time but are more likely to occur when choices are longer (Holbrook *et al.*, 2007).

In telephone surveys, the interviewer's voice may provide information about the interviewer (e.g. age, gender and race or ethnicity) and influence people's understandings of questions as well as the kind of answer that is expected. However, research has also shown that not all aural communication is the same. For example, Couper, Singer and Tourangeau (2004) found that IVR respondents exhibited greater disclosure of sensitive information than did respondents interviewed by telephone. At the same time there were no differences in whether the interviewer's voice was male, female or synthesized. They concluded that respondents to IVR do not respond as if they are communicating interactively with a real human being.

In visual surveys, which are more under the control of the respondents, numbers, symbols and graphics (i.e. spacing, size, brightness, etc.) can act as visual paralanguages to provide considerably more guidance in the response process. Not only do these visual features play a major role in how respondents navigate through questions – e.g. the order in which information is attended to – but they also play a significant role in determining the meaning of questions (Jenkins and Dillman, 1997; Redline and Dillman, 2002; Dillman, 2007).

Considerable experimental research has now shown that modifications in numerical, symbolic and graphical features significantly affect answers to both web and paper surveys. For example:

- Labelling scales in a way that includes negative numbers, e.g. – 5 to + 5, produces different answers than when a 0–10 scale without negative numbers is used (Schwarz *et al.*, 1991).
- The size of answer space (a graphical influence) affects whether respondents provide answers consistent with that desired by the surveyors (Couper *et al.*, 2001; Christian *et al.*, 2007).
- Use of a symbol, an arrow, to point from one answer choice to an adjacent blank space increases the likelihood that respondents will see and complete the second item (Christian and Dillman, 2004).
- Adding gratuitous subheadings to categories results in respondents thinking that answers should be selected in each group despite being contrary to the surveyor's intent (Smyth *et al.*, 2006d).
- Linear displays of scales obtain somewhat different answers than when the scalar answers are displayed in multiple columns (Christian and Dillman, 2004; Dillman and Christian, 2005).
- Respondents expect higher numbers to be associated with more positive categories, e.g. greater satisfaction (Tourangeau *et al.*, 2004; Christian *et al.*, 2006).
- Consistent displays of scales from negative to positive obtain the same answers as scales from positive to negative (Christian *et al.*, 2006), but changing the direction within a questionnaire results in response errors (Israel, 2005).
- Changing the perceptual midpoint of scales influences respondent answers (Tourangeau *et al.*, 2004), but changing the spatial distance between adjacent categories, while retaining a spatial midpoint, appears not to affect people's answers (Christian *et al.*, 2006).
- Changing the colour hues of points on scales so that the positive side differs in colour from the negative side (e.g. blue vs. red) produces more positive answers than does use of the same colour for the complete scale, when only the endpoints are labelled and no numbers are used to label points on the scale (Tourangeau *et al.*, 2007).
- The substitution of symbols with embedded numeracy (e.g. MM/YYYY) for words (Month/Year) greatly improves the likelihood that respondents will report answers in an expected two-digit month and four-digit year format (Christian *et al.*, 2008).

While some effects appear quite similar in web and paper surveys, the fundamental technology of constructing each type of survey differs, with the result that some issues and effects are unique to one mode or another. For example, in web surveys respondents can be skipped past some questions based on their answer to a filter question without them knowing that they have skipped the questions that did not apply to them. The same cannot be done on mail surveys and instead respondents must be explicitly instructed whether to answer or skip a question. In this regard, research has shown that using certain graphical manipulations can help respondents successfully navigate the survey. For example, arrows and larger darker print used for branching instructions combined with qualifying instructions in parentheses at the beginning of the next appropriate question (e.g. '(If yes) How often did you. . . .') significantly reduced navigation errors in mail surveys. In an experiment embedded in the 2000 US Decennial Census, branching errors were reduced by the introduction of this format by 20–35 %, depending upon the type of error (Redline and Dillman, 2002; Redline *et al.*, 2003). Branching instructions are a clear instance in which mode-specific construction is important to consider. Refraining from the use of hotlink connections in web surveys would result in substantially greater respondent burden as well as increase the likelihood of branching errors being made because of the more limited visual context (one screen at a time) for web respondents.

Open-ended questions are another area in which mode-specific question formats appear to be advantageous. It has been shown experimentally that the quality of open-ended answers, as measured by number of words and themes, can be improved in mail surveys by increasing the size of the answer spaces (Christian and Dillman, 2004). An expanded replication of this experiment found that the number of words offered increased incrementally with each increase in size for three box sizes (Israel, 2005). Earlier research on Internet surveys has shown a similar increase in words reported (Dennis *et al.*, 2000). However, a web test of the questions used in the mail survey mentioned above did not exhibit an overall effect (Christian and Dillman, 2005). A more detailed analysis of these data showed that late respondents, who seem likely to be less motivated respondents, were clearly affected by the size of the answer box, with more detailed answers being obtained in the larger boxes (Smyth *et al.*, 2006b).

Additional perspective on the use of mode-specific design features emerges from an extensive set of web experiments on finding the best way of getting respondents to insert a two-digit month and four-digit year into answer spaces. The goal of these experiments was to find ways of helping respondents to avoid an error message that indicated an unacceptable answer (e.g. only two digits for the year) had been provided. These experiments showed that a combination of paralanguage changes could improve the reporting of month and year in the desired fashion from 45 % to 95 % of the time. These changes included a month box that was smaller than the year box to signal the appropriate number of digits, substitution of the symbols MM and YYYY with embedded numeracy in place of the words month and year and placing those symbols in natural reading order ahead of the boxes (Christian *et al.*, 2007). Perhaps the most important finding of this research was to reveal the great extent to which respondents to web surveys rely upon how answer categories are presented to determine the exact answer that they report. In contrast, displays of answer categories are not seen by respondents to telephone surveys and thus have no role in determining how questions are answered. Thus, while an initial telephone interview may have simply asked respondents when they began work at a particular company, and used interviewer probes to obtain the month and year, obtaining the same information on web or paper in a follow-up requires more specific question construction that places as much or more emphasis on the visual layout and content of the answer spaces as it does on the question stem itself. This is an example of using mode-specific construction to improve respondent answers.

These research findings raise several issues important to the conduct of longitudinal surveys. First, when one switches from an aural to a visual mode, new sources of question meaning come into play, with numbers, symbols and graphical display conveying additional meaning to the words used to ask the question. Unless conscious attention is accorded to making questions equivalent across modes, it is likely that attempts to measure change accurately will be thwarted. Both unified and mode-specific question design have important roles to play in achieving that equivalence. Further, it is important when the initial data collection is undertaken that careful attention be given to consider the features of questions used at that time and whether steps can be made to ensure the collection of equivalent data when subsequent waves are carried out. The solution may be as simple as offering specific response categories in all data collection modes, which provided a simple solution to the marital status question mentioned earlier, or as complex as thinking through a mode-specific solution when the communication importance of question stems and answer spaces changes significantly, as between aural and visual surveys.

8.3.3 Interviewer Presence

Traditionally, interviewer presence has proved to be a powerful influence for making sure that respondents correctly hear questions and provide appropriate answers. On the one hand interviewers may pose additional queries of respondents to make sure questions are answered, e.g. in response to an answer of 'agree', 'Would that be strongly agree or somewhat agree?' Another positive effect is to act as an intelligent system to translate answers offered in one format into one more amenable to interview form, e.g. changing an answer of 'the first of last month', to a question about when she began working for her current employer, to 'January 2008'. However, interviewer presence also causes problems that may resist easy across-mode solutions.

One such problem stems from differences in how intensely social norms of desirability and agreeableness are evoked by each mode. It has long been recognized that respondents to interviewer-administered surveys are more likely to give answers that they believe are expected from them by the interviewer and that place them in a favourable light (De Leeuw, 1992; Dillman *et al.*, 1996).

Perhaps the most common example of a social desirability effect is demonstrated by the question, 'How do you consider your health to be: excellent, good, fair or poor'. Since the landmark study by Hochstim (1967), it has been found repeatedly that respondents to this question are more likely to answer positively by selecting excellent when the question is administered by an interviewer. In comparison, reports of illegal drug use or other negative behaviours have been found to be significantly lower when interviewers are present (Aquilino, 1994). Recent research has also shown differences in social desirability effects within self-administered surveys. Specifically, findings indicate that social desirability among young respondents is less on computers than in paper and pencil surveys (Chang, 2001; Taylor *et al.*, 2005).

Another social norm that differentially affects responses across survey modes is the expectation that one be agreeable rather than disagreeable (Schuman and Presser, 1981; Javeline, 1999; Dillman, 2007). It has been proposed that interviewed respondents are more likely to agree with interviewers than to disagree, leading to the expectation that larger acquiescence effects will occur in telephone and face-to-face surveys than in self-administered surveys (Schuman and Presser, 1981; Dillman and Tarnai, 1991; De Leeuw, 1992). It has also been suggested that telephone respondents are more likely to acquiesce than are face-to-face respondents (Krosnick, 1999).

A longstanding concern with longitudinal surveys is that respondents may be conditioned by earlier data collection to respond in certain ways to questions in later surveys. Evidence for conditioning is especially evident in web panel surveys where respondents are surveyed frequently (Cartwright and Nancarrow, 2006; Nancarrow and Cartwright, 2007). It also seems plausible that respondents who responded by a particular mode in an earlier data collection may be conditioned to respond similarly to a different mode, but experimental data on such an effect are lacking.

8.3.4 How Answers to Scalar Questions are Affected by Visual vs. Aural Communication

One of the most challenging problems facing sponsors of longitudinal surveys is obtaining equivalent answers to scalar questions, which often rely on vague quantifiers, across visual and aural survey modes. Examples include questions like some of those mentioned at the beginning of the chapter, measuring extent of satisfaction, relative priority and health status.

It has been observed in past studies that respondents answer scalar questions more positively when they are interviewed than on self-administered surveys. For example, Dillman and Mason (1984), Dillman and Tarnai (1991) and Krysan *et al.* (1994) all found that respondents to interview surveys were more likely than mail respondents to report that community and neighbourhood issues were not a problem in their locale when they were presented with the scale points: not a problem, a small problem, a moderate problem, a serious problem. Although it might be argued that there is some social desirability in suggesting the lack of problems, it would seem to be at a low threshold (Dillman, 2007, pp. 229–231).

A more recent study about long-distance carrier satisfaction revealed that respondents to aural survey modes (telephone and IVR) gave more extreme positive answers to all such items than did web or mail respondents. For example, whereas 39 % of both the telephone and IVR respondents expressed complete satisfaction, only 21 % of the mail and 27 % of the web respondents did so (Dillman *et al.*, in press). For the four remaining satisfaction items, telephone and IVR also produced more positive responses than mail and web. In this case a 1–5 scale was used, with respondents being able to use any number from 1 to 5. The appearance of answer spaces with a number beside each on mail and web would seem to be cognitively more prominent than a verbal presentation that adds labels only to the endpoints.

In two mixed-mode surveys simultaneously conducted by telephone and web, a series of experimental comparisons were included in each survey to test whether telephone respondents provide more positive answers than web respondents (Christian, 2005; Christian *et al.*, 2008). The findings from these experiments support the above-reported findings that telephone respondents provide more positive response than web respondents and this finding appears to be robust. This positivity bias on the telephone occurs across a variety of rating scales: five-, seven- and eleven-category scales. It also occurs across fully labelled scales where all categories are verbally labelled, polar point labelled scales where only the endpoints are labelled and branched two-step scales where respondents first report the direction (satisfied or dissatisfied) and then the intensity of their answer (very, somewhat, a little) and across a variety of constructs, including satisfaction, accessibility and confidence. Of the 46 comparisons, 36 were significantly different in the direction of telephone providing the more positive response.

In sum, a highly consistent pattern has emerged. Across a variety of populations (general public to students), a wide variety of question topics (community satisfaction to confidence in being able to graduate on time and perception of advisor's accessibility) and a variety of scalar formats, respondents to the telephone mode are likely to provide more extreme answers on the positive end of the scale than are respondents to mail or web surveys. Although low threshold social desirability remains a possibility for some items, it seems unlikely to exist for all of the items for which this trend has been noted. Whatever the cause of these differences, they bring into question whether it is possible to obtain equivalent answers to scalar questions in aural and visual surveys.

One means of avoiding this problem may be to plan, from the beginning of longitudinal surveys conducted with interviews, for opinion questions to be placed in a self-administered module. Then, if mail or web are used in later waves, the problem of obtaining equivalent measurement will be avoided.

8.4 CONCLUSIONS

The survey environment in which we live is more dynamic than at any time in history. We now have more survey modes, and there is higher motivation to use them in order to improve coverage and response while maintaining lower survey costs. There seems to be no alternative but to use multiple modes for the conduct of longitudinal surveys.

It is also apparent that different survey modes complicate the measuring of change. There is both good news and bad news. In some cases we need to ask questions in the same way using a unified mode procedure for constructing questions. Yet, in other instances better equivalency may be achieved through using a mode-specific approach, whereby accommodations are made because of differences in the ways that aural and visual information is used in the various survey modes.

In addition, it is important to recognize that for scalar questions it appears that ways have not yet been found to avoid differences across aural and visual survey modes. Identifying such means is a high priority for future research.

REFERENCES

Aquilino, W. S. (1994). Interview mode effects in surveys of drug and alcohol use: a field experiment. *Public Opinion Quarterly*, 58, 210–240.

Best, S. J. and Krueger, B. (2004). Internet Data Collection. Thousand Oaks, CA: Sage Publications.

Blumberg, S. J., Luke, J. V. and Cynamon, M. L. (2006). Telephone coverage and health survey estimates: evaluating the need for concern about wireless substitution. *American Journal of Public Health*, 96, 926–931.

Cartwright, T. and Nancarrow, C. (2006). The effect of conditioning when re-interviewing: a pan-European study. In *Panel Research 2006* (pp. 22–28), ESOMAR Publication Series 317. Amsterdam: ESOMAR.

Chang, L. (2001). A comparison of samples and response quality obtained from RDD telephone survey methodology and Internet survey methodology. *Doctoral Dissertation*, Department of Psychology, Ohio State University, Columbus, OH.

Christian, L. M. (2005). Is it possible to obtain the same answers to scalar questions across visual and aural survey modes? *Working Paper*. Athens: University of Georgia.

Christian, L. M. and Dillman, D. A. (2004). The influence of graphical and symbolic language manipulations on responses to self-administered questions. *Public Opinion Quarterly*, 68, 57–80.

Christian, L. M., Dillman, D. A. and Smyth, J. D. (2007). Helping respondents get it right the first time: the relative influence of words, symbols, and graphics in web and telephone surveys. *Public Opinion Quarterly*, 71, 113–125.

Christian, L. M., Dillman, D. A. and Smyth, J. D. (2008). The effects of mode and format on answers to scalar questions in telephone and web surveys. In J. Lepkowski, C. Tucker, M. Brick, E. de Leeuw, L. Japec, P. Lavrakas, M. Link and R. Sangster (Eds), *Advances in Telephone Survey Methodology* (Chapter 12). New York: Wiley-Interscience.

Christian, L. M., Parsons, N. L. and Dillman, D. A. (2006). Understanding the consequences of visual design and layout. *Working paper*. Pullman: Washington State University, Social and Economic Sciences Research Center.

Couper, M. P., Singer, E. and Tourangeau, R. (2004). Does voice matter? an interactive voice response (IVR) experiment. *Journal of Official Statistics*, 20, 551–570.

Couper, M. P., Traugott, M. W. and Lamias, M. J. (2001). Web survey design and administration. *Public Opinion Quarterly*, 65, 230–254.

De Leeuw, E. D. (1992). *Data Quality in Mail, Telephone, and Face to Face Surveys*. Amsterdam: TT Publications.

Dennis, M. J., DeRouvray, C. and Couper, M. (2000). Questionnaire design and respondent training for probability-based web surveys. Paper presented at the *Annual Conference of the American Association for Public Opinion Research*, Portland, OR.

Dillman, D. A. (2007). *Mail and Internet Surveys: The Tailored Design Method 2007 Update*. Hoboken, NJ: John Wiley & Sons, Inc.

Dillman, D. A. and Christian, L. M. (2005). Survey mode as a source of instability in responses across surveys. *Field Methods*, 17, 30–52.

Dillman, D. A. and Mason, R. G. (1984). The influence of survey method on question response. Paper presented at the *Meeting of the American Association for Public Opinion Research*, Delevan, WI.

Dillman, D. A., Phelps, G., Tortora, R., Swift, K., Kohrell, J., Berck, J. and Messer, B. L. (in press). Response rate and measurement differences in mixed mode surveys using mail, telephone, interactive voice response, and the internet. *Social Science Research*. Retrieved June 15, 2008, from http://survey.sesrc.wsu.edu/dillman/papers.htm.

Dillman, D. A., Smyth, J. D. and Christian, L. M. (2009). *Internet, Mail and Mixed-Mode Surveys: The Tailored Design Method*, Third Edition. New York: John Wiley & Sons, Inc.

Dillman, D. A., Sangster, R. L., Tarnai, J. and Rockwood, T. H. (1996). Understanding differences in people's answers to telephone and mail surveys. In M. T. Braverman and J. K. Slater (Eds), *Advances in Survey Research* (pp. 45–62), New Directions for Evaluation Series, No. 70. San Francisco: Jossey-Bass.

Dillman, D. A. and Tarnai, J. (1991). Mode effects of cognitively-designed recall questions: a comparison of answers to telephone and mail surveys. In P. P. Biemer, R. M. Groves, L. E. Lyberg, N. A. Mathiowetz and S. Sudman (Eds), *Measurement Errors in Surveys* (pp. 73–93). New York: John Wiley & Sons, Inc.

Funke, F. and Reips, U. D. (2006). Visual analogue scales in online surveys: non-linear data categorization by transformation with reduced extremes. Presented at the *8th International Conference GOR06*, March 21–22, Bielefeld, Germany. Retrieved June 15, 2008, from http://www.frederikfunke.de/papers/gor2006.htm.

Hauser, R. M. (2003). Stability of methods for collecting, analyzing, and managing panel data. Presented at the *American Academy of Arts and Sciences*, March 26, 2003, Cambridge, MA.

Hochstim, J. R. (1967). A critical comparison of three strategies of collecting data from households. *Journal of the American Statistical Association*, 62, 976–989.

Hoffer, T. B., Girgorian, K. and Fecso, R. (2007). Assessing the effectiveness of using panel respondent mode preferences. Paper presented at *2007 American Statistical Association Annual Conference*, July 27th, Salt Lake City, UT.

Holbrook, A. L., Krosnick, J. A., Moore, D. and Tourangeau, R. (2007). Response order effects in dichotomous categorical questions presented orally: the impact of question and respondent attributes. *Public Opinion Quarterly*, 77, 325–348.

Israel, G. (2005). Visual cues and response format effects in mail surveys. Paper presented at the *Annual meeting of the Rural Sociological Society*, Tampa, FL.

Javeline, D. (1999). Response effects in polite cultures: a test of acquiescence in Kazakhstan. *Public Opinion Quarterly*, 63, 1–28.

Jenkins, C. R. and Dillman, D. A. (1997). Towards a theory of self-administered questionnaire design. In L. Lyberg, P. Biemer, M. Collins, E. de Leeuw, K. Dippo, N. Schwarz and D. Trewin (Eds), *Survey Measurement and Process Quality* (pp. 165–196). New York: John Wiley & Sons, Inc.

Krosnick, J. A. (1999). Survey research. *Annual Review of Psychology*, 50, 537–567.

Krysan, M., Schuman, H., Scott, L. J. and Beatty, P. (1994). Response rates and response content in mail versus face-to-face surveys. *Public Opinion Quarterly*, 58, 381–399.

Miller Steiger, D. and Conroy, B. (2008). IVR: interactive voice response. In E. de Leeuw, J. Hox and D. A. Dillman (Eds), *International Handbook of Survey Methodology* (Chapter 15). Boca Raton, FL: Taylor & Francis.

Nancarrow, C. and Cartwright, T. (2007). Online access panels and tracking research: the conditioning issue. *International Journal of Market Research*, 49, 435–447.

Redline, C. and Dillman, D. A. (2002). The influence of alternative visual designs on respondents' performance with branching instructions in self-administered questionnaires. In R. Groves, D. Dillman, J. Eltinge and R. Little (Eds), *Survey Nonresponse* (Chapter 12). Hoboken, NJ: John Wiley & Sons, Inc.

Redline, C., Dillman, D. A., Dajani, A. N. and Scaggs, M. A. (2003). Improving navigational performance in U.S. Census 2000 by altering the visually administered languages of branching instructions. *Journal of Official Statistics*, 19, 403–20.

Schuman, H. and Presser, S. (1981). *Question and Answers in Attitude Surveys: Experiments on Question Form, Wording and Context.* New York: Academic Press.

Schwarz, N., Knäuper, B., Hippler, H. J., Noelle-Neumann, E. and Clark, L. (1991). Rating scales: numeric values may change the meaning of scale labels. *Public Opinion Quarterly*, 55, 570–582.

Smyth, J. D., Dillman, D. A. and Christian, L. M. (2006a). Does 'Yes or No' on the telephone mean the same as 'Check-All-That-Apply' on the web? Paper presented at the *Second International Conference on Telephone Survey Methodology*, January 12–15, Miami, FL.

Smyth, J. D., Dillman, D. A., Christian, L. M. and McBride, M. (2006b). Open ended questions in telephone and web Surveys. Presented at the *World Association of Public Opinion Research Conference*, May 16–18, Montreal, Canada.

Smyth, J. D., Dillman, D. A., Christian, L. M. and Stern, M. J. (2006c). Comparing check-all and forced-choice question formats in web surveys. *Public Opinion Quarterly*, 70, 66–77.

Smyth, J. D., Dillman, D. A., Christian, L. M. and Stern, M. J. (2006d). Effects of using visual design principles to group response options in web surveys. *International Journal of Internet Science*, 1, 6–16.

Sudman, S. and Bradburn, N. M. (1982). *Asking Questions: A Practical Guide to Questionnaire Design.* San Francisco, CA: Jossey-Bass Publishers.

Taylor, H., Krane, D. and Thomas, R. K. (2005). How does social desirability affect responses?: differences in telephone and online surveys. Paper presented at the *American Association for Public Opinion Research*, May, Miami, FL.

Thomas, R. K. and Klein, J. D. (2006). Merely incidental? effects of response format on self-reported behavior. *Journal of Official Statistics*, 22, 221–244.

Tourangeau, R., Couper, M. P. and Conrad, F. (2004). Spacing, position, and order: interpretive heuristics for visual features of survey questions. *Public Opinion Quarterly*, 68, 368–393.

CHAPTER 9

Using Auxiliary Data for Adjustment in Longitudinal Research

Dirk Sikkel
University of Tilburg, The Netherlands

Joop Hox, Edith de Leeuw
Utrecht University, The Netherlands

9.1 INTRODUCTION

The purpose of this chapter is to present and discuss what can be done to prevent and repair the consequences of nonresponse in longitudinal surveys, by reviewing issues in the collection and use of auxiliary data for statistical adjustment in the context of longitudinal surveys or panels.

The use of longitudinal research designs, such as cohort or panel studies, has become increasingly popular during the last decade. Both cohort and panel designs are based on measuring the same sample of individuals at different points in time, with the objective to study change over time. Cohort studies use subjects who have certain characteristics in common, for example a birth cohort of subjects who were all born between 1980 and 1985. Panel studies are studies that use the same subjects repeatedly. Governmental, semigovernmental and academic agencies use panels for important trend studies. Examples include the PSID (Panel Study of Income Dynamics), CPS (Current Population Survey) and SIPP (Survey of Income and Program Participation) in the USA, the British Household Panel Survey and the Millennium Cohort Study in the UK and Labour Force Surveys in many European countries. While most commercial and market research panels are at present based on the internet, most noncommercial panels are using interview surveys, be it face to face or by telephone, sometimes interspersed with mail questionnaires. For a general discussion of longitudinal surveys and panel designs we refer to Lynn (Chapter 1 of this volume).

Although designed for longitudinal studies, panels are more and more used for cross-sectional research. For instance, in the academic world, panel members may be

Methodology of Longitudinal Surveys P. Lynn

approached for specific interest studies. In the commercial world, in many countries special *access panels* have been established for opinion and market research, which are mainly used for ad hoc cross-sectional studies. An access panel is basically a rich database of willing respondents that is used as a sampling frame for internet studies but may be used for other data collection procedures too. Access panels were established in reaction to the growing nonresponse worldwide (see De Leeuw and De Heer, 2002). The main attraction is the convenience of having a group of available and willing respondents from which subgroups with specific characteristics can be selected for research. The popularity of access panels in market research is illustrated by the database of ESOMAR, the international organization of marketing research agencies. In May 2008, 1307 organizations were listed that carry out panel research on a continuous basis.

In this chapter, we use the term panel to refer to any design in which individual subjects are measured repeatedly. This includes longitudinal research where subjects are followed for many years, but also access panels whose members agree to respond to a certain number of questionnaires. Ideally, a panel consists of subjects who are a probability sample of the population and who remain members during the full lifetime of the panel. However, not everyone agrees to participate in the first place, and all panels suffer from subsequent panel attrition: panel members who cease to cooperate with the study. Furthermore, many internet panels are self-selected volunteer panels, which means that the degree of departure from a probability sample is unknown (see also Couper, 2000; Lee, 2006). If the researcher has auxiliary data on the panel members that can be compared with data from the general population, a number of remedies are possible. In most forms of panel research it is possible to obtain data for a large number of variables during the recruitment process or the first data collection wave. This creates the opportunity to use these auxiliary data in a later stage to adjust for initial nonresponse and panel attrition. During the lifetime of the panel further information can be collected, which can be added to the auxiliary data and used for additional adjustment. Sometimes it is even possible to link administrative data to the panel's database (Calderwood and Lessof, Chapter 4 of this volume).

9.2 MISSING DATA

Statistical sampling theory assumes that data are collected from a probability sample with known and nonzero inclusion probabilities and that there is either full response or 'ignorable' nonresponse. Compared to this ideal, there are in fact several causes of missing data. The most crucial stage in longitudinal research is the *recruitment* stage, where candidates for panel membership may fail to participate for a variety of reasons. An illustrative example is the CentERpanel (see www.centerdata.nl), located at Tilburg University. The recruitment procedure was based on a random sample of phone numbers. Recruitment was carried out on a running basis since 1991. What happened after the first call is described in Table 9.1.

From all selected phone numbers, 98.2 % were usable. From the potential respondents who were contacted, 38.4 % refused the first contact interview and; 61.6 % agreed to participate in follow-up interviews. The data of those who did take part and were prepared to take part in follow-up interviews were stored in a database and retrieved at a later stage when the respondents were asked to take part in the panel. In the end,

Table 9.1 Response rates in different stages of recruitment for the CentERpanel.

	%	Cumulative %
Phone number usable	98.2	98.2
Participation first contact interview	61.6	60.5
Prepared to take part in follow-up interviews	51.2	31.0
Phone number correct in membership interview	98.6	30.5
Participation membership interview	82.4	25.2
Respondent qualifies as a member	91.9	23.1
Prepared to become a member	44.9	10.4

10.4 % of the *initial usable phone* numbers became working panel members. Tortora (Chapter 13 of this volume) describes the procedures for recruitment and retention for the Gallup consumer panel, in which 27 % of those *contacted* by a random digit dialling (RDD) telephone procedure *agree to participate*. Most commercial research institutes, however, do not keep track of the recruitment process in this detail, and it seems plausible that their response rates are at best of the same order of magnitude. Of course, in specialized, noncommercial panels the response rates for the recruitment stage may be higher.

Once respondents have become panel members, new sources of missing data arise. The panel members may fail to answer individual questions (item nonresponse), skip panel waves (wave nonresponse) or drop out of the panel (attrition). A special case of attrition is attrition by design; here the researcher controls how long respondents ideally stay in the panel. A prime example of attrition by design is a rotating panel, in which respondents participate in the panel for a fixed amount of time (Elliot, Lynn and Smith, Chapter 2 of this volume).

When data are *missing by design*, the missingness mechanism is known and accessible to the statistical analyst. However, in most cases data are not missing by design. Instead, missing data is caused by the respondents, who refuse to participate in a panel, do not answer certain questions, fail to complete a questionnaire or drop out of a panel altogether. To successfully treat missing data it is necessary to understand the various types of missingness and to discern the actual patterns. A basic distinction is that data may be missing completely at random (MCAR), missing at random (MAR) or not missing at random (NMAR). Each refers to different processes requiring specific strategies in the data analysis (Little and Rubin, 2002). Auxiliary data are extremely important to help decide which nonresponse mechanism is at work; such data should be made accessible and incorporated in the statistical analysis.

There is a host of techniques for dealing with missing data in longitudinal designs. Three broad strategies can be identified:

- *Prevention by panel management.* Special attention to members of the panel may prevent dropout (De Leeuw, 2005). Such measures may include small presents, but also specific tokens of attention, for instance sending a birthday card at the appropriate time (see also Laurie and Lynn, Chapter 12 of this volume). In addition, keeping in contact and tracking mobile panel members will reduce selective dropout (see also Couper and Ofstedal, Chapter 11 of this volume).

- *Damage repair by imputation*. This can be achieved by finding similar cases in which there are no missing data or by estimating a model $y = f(\boldsymbol{x}) + \varepsilon$, in which y is the target variable for which some respondents have missing values, $\boldsymbol{x}$ is a vector of known auxiliary variables and ε is a random error term. When a model is estimated, there are different strategies for the determination of an imputed value y_i for individual i whose value is missing. The randomness may be ignored by imputing the expected value $f(\boldsymbol{x}_i)$ or accounted for by adding a random term or by multiple imputation. Compared to cross-sectional research, in panel research there are usually many more variables available for the model $f(\boldsymbol{x})$. The best predictor of y_i is often the corresponding value in the previous panel wave.
- *Damage control by calibration*. This involves assigning weights w_i to the individual cases such that known population parameters or unbiased estimators of population parameters are reproduced by the estimation procedure. In this respect, calibration is essentially different from imputation where exact reproduction of population distributions is not required. So for a wave W_τ for auxiliary variable x_k the weights w_i are determined in such a way that

$$\sum_{i \in W_\tau} w_i x_{\tau ik} = t_{\tau k}$$

where x_{ik} is the value of individual i on variable x_k and t_k is the population parameter determined by $x_{k\tau}$ at time τ, such as a mean or a total. In the case when complete records are missing, like wave nonresponse or attrition, calibration may be the most obvious way to handle missing data, as there are no auxiliary variables measured at time τ that can be used to predict the individual value y_i at time τ. The calibration procedure implies the assumption that the missing data are missing at random (see Little and Rubin, 2002), conditional on the variables that determine the weights. In addition, although calibration does not use an explicit model, the choice of the $\boldsymbol{X}$ includes a data-implied model.

In general, longitudinal research offers golden opportunities both for imputation and for calibration because there are so many relevant auxiliary variables available. In this chapter we concentrate mostly on calibration, which is more suited for use in automated routines than imputation and is therefore more likely to be applied routinely on longitudinal surveys. Modern imputation techniques for longitudinal data are described in Little and Zhang (Chapter 18 of this volume).

9.3 CALIBRATION

There are different rationales behind calibration. The simplest is the notion of representativeness. An estimation procedure $\Theta(.)$ is representative with respect to a matrix of a set of auxiliary variables $\boldsymbol{X}$ when $\Theta(\boldsymbol{X})$ reproduces a vector of known population parameters $\boldsymbol{t}$ (Hajek, 1981, p. 40). This definition applies to routine weighting procedures with respect to categorical variables like age (categories), sex and region, but also to totals and averages of numerical variables like savings and income. Basically, weighting is designed to reduce bias, by making the estimation procedure for known population totals unbiased. Under certain conditions, such as in a random sample in which the data satisfy a well-specified model, weighting may also result in variance reduction. The clearest expression of this view is the well-known regression estimator, which is an implicit procedure to weight the data. Let $\boldsymbol{v} = (\boldsymbol{X}'\boldsymbol{X})^{-1}\boldsymbol{t}$, then v is the vector of weight components, i.e. the vector of weights w_i can be written as

$$\boldsymbol{w} = \boldsymbol{X}\boldsymbol{v}, \tag{9.1}$$

or

$$w_i = \sum_k x_{ik} v_k, \tag{9.2}$$

In other words, the weights are linear combinations of the auxiliary variables. Bethlehem and Keller (1987) show that in a simple random sample the variance of a target variable y after weighting is equal to

$$\sigma_{yreg}{}^2 = (1 - \rho^2)\sigma_y{}^2, \tag{9.3}$$

where ρ is the multiple correlation between $\boldsymbol{x}$ and y and $\sigma_y{}^2$ is the unweighted variance of y. As with any weighting scheme based on a set of auxiliary variables, when data are missing at random, given $\boldsymbol{X}$, the weighting procedure yields unbiased estimates of y. Deville and Särndal (1992) show that the regression estimator is a special case of a broad class of weighting procedures that is based on minimizing a distance measure $\Sigma_i D(w_i, \pi_i{}^{-1})$ under the condition $\boldsymbol{w}'\boldsymbol{X} = \boldsymbol{t}'$; this means that the w_i have to stay as close as possible to the reciprocal inclusion probabilities π_i under the condition that the population totals for the auxiliary variables are reproduced. The choice of distance measure $D(.)$ determines the type of weights w_i. For the regression estimator the distance measure D is

$$D(w_i, \pi_i^{-1}) = \pi_i (w_i - \pi_i^{-1})^2 \tag{9.4}$$

Deville and Särndal also show that the well-known procedure of iterative proportional fitting is a special case of this class of weighting procedures, and that the variance (Equation 9.3) is the asymptotic first-order approximation in the total class of weighting procedures. In this chapter we restrict ourselves to the regression estimator, as it is the most flexible and easy to use procedure. It can be applied to both numerical and categorical variables, in the latter case by using dummy variables for each of the categories. The same procedures can also be applied to population averages instead of totals. In that case the weighted auxiliary variables add to $\boldsymbol{t}/N$, and the distance measure D is the function $D[w_i, (n\pi_i)^{-1}]$, where N is the population size and n the sample size.

This is, however, only one part of the story. Equation (9.3) suggests that weighting to calibrate the sample with respect to known population totals always leads to variance reduction. This is true in a simple random sample when the data are missing at random, but what if the sample is not a simple random sample? What if different elements are drawn with different inclusion probabilities? When a target variable is independent from the variable that corresponds to the different inclusion probabilities, all weights will be equal and the best strategy is not to weight at all. For example, consider a sample that consists of 100 randomly drawn men and 300 randomly drawn women. The target variable is 'number of children'. This variable obviously is independent of gender. Now if we weight men by 0.75 and women by 0.25, the weighted sample average of the number of children obviously has a higher variance than the unweighted sample average. Still, both are unbiased estimators of the population average number of children (this obviously is not a simple random sample, so Equation (9.3) does not apply). Similarly, when the target variable is independent of the weights, ρ in Equation (9.3) equals zero and the weights have no impact on the variance. When the target variable y is unrelated to X, the variance of the weighted totals increases compared to the unweighted case by a factor $1 + CV_w^2$, where CV_w is the variation coefficient of the weights w_i (Kish, 1992). This is a rationale for the practice of trimming weights: to change weights that have excessive values into less extreme weights. This can be done after initial weights have been computed (Potter, 1988, 1990, 1993) or by putting restrictions on the weights in the estimation procedure. Such algorithms fall into the class of weighting procedures described by Deville and Särndal (1992). For practical examples see also Kalton and Flores-Cervantes (2003). It should be noted that under some circumstances weight trimming can increase the variance; for a discussion, see Little and Vartivarian (2005). In more complex surveys, such as multistage and stratified samples, Equation (9.3) holds in most cases; for a discussion, including examples, we refer to Lehtonen and Pahkinen (2004).

For prerecruited panels, Equation (9.3) presents an excellent opportunity to calibrate panel waves. What is needed is to collect, within the recruitment process, data with respect to a variety of variables that can be expected to have high correlations with key variables investigated in the panel study. The recruitment process may be a face-to-face survey based on a random sample (e.g. from population registers) or a survey by phone using random digit dialling (Sikkel and Hoogendoorn, 2008). When the response levels in the recruitment process are high, the quality of the estimators can also be high. Of course, some of the key variables of the longitudinal study may be unknown in the recruitment stage, but usually a research organization works in a limited number of fields, so one can anticipate what the key variables will be. For a general data collection organization like CentERdata, which has its main focus on scientific and policy issues, a number of variables collected in the recruitment stage are listed in Table 9.2. The percentages measured in the recruitment stage are compared with those in a panel wave in May 1999. The recruitment stage corresponds to those respondents who, in Table 9.1, were prepared to take part in follow-up interviews. The panel stage corresponds to the bottom row in Table 9.1: those who were prepared to become a member.

The chosen variables provide information about housing, commuting, health, cultural participation and victimization. Table 9.2 shows that members of sport clubs are severely under-represented in the panel, whereas victims of burglary are over-represented. The choice of variables to be measured in the recruitment stage clearly has a large substantive component, but the final criterion is a statistical one: Equation (9.3).

Table 9.2 Distributions of some calibration variables in the recruitment stage and in the CentERpanel, 1999.

	Recruitment (%)	Panel (%)
Number of rooms in the house		
1–3 rooms	13.1	17.7
4 rooms	37.3	31.8
5 rooms	28.3	30.9
6 or more rooms	21.3	19.5
Travelling time to work		
>20 minutes	40.7	49.3
<20 minutes	59.3	50.7
Satisfaction with health (1: low; 10: high)		
1–5	6.0	7.3
6	6.3	10.8
7	22.6	27.5
8	40.0	36.3
9, 10	23.8	18.1
Visited the cinema last year		
Yes	48.6	38.4
No	51.4	61.6
Member of sport club		
Yes	54.7	35.6
No	45.3	64.4
Victim burglary (ever)		
Yes	16.4	22.8
No	83.6	77.2

9.4 CALIBRATING MULTIPLE WAVES

When a research project consists of more than one wave, a straightforward way to proceed is to treat each wave separately and to construct weights for each wave. For very simple forms of analysis this may be a possible approach, but it precludes the simultaneous use of variables from different waves, as each would then have a different set of weights. So it is preferable for respondents that contribute to different waves to have one single weight to be used when combining waves in analysis. For many common applications such a single weight is convenient or even a strict requirement. Simple examples are totals over waves (e.g. consumption of goods) that have to be correlated at the individual level with other individual characteristics, differences between waves (e.g. in attitudes after a commercial campaign), estimation of transition rates (e.g. from unemployed to employed) in a longitudinal survey, etc. For such waves, different auxiliary variables may be available (or different versions of the same auxiliary variable that has changed over time, such as education or health status). This problem arises, of course, only when the relevant waves have been completed, so this weight for multiple waves will be different from weights that may have been assigned to analysis of earlier single waves or combinations of waves.

Only a slight modification of the theory in Section 9.3 is required to accommodate this situation. Let us assume we have waves $W_1, W_2, \ldots W_q$, with $n_1, n_2, \ldots n_q$ respondents. Panel members may have missing values in a subset of the waves. The total number of respondents who participate in at least one of the waves is equal to n. For each wave W_τ a population total $\boldsymbol{t}_\tau$ (or a population mean $\boldsymbol{\mu}_\tau = \boldsymbol{t}_\tau / N_\tau$) is given for a vector of auxiliary variables $\boldsymbol{x}_\tau$, $\tau = 1, 2, \ldots q$. These auxiliary variables may be constant over time (like year of birth) or may vary over time (like income). We formulate the problem in terms of population means. Then the question is: can we construct weights in such a way that the correct expected sampling totals at time τ are reproduced, that is,

$$\sum_{i \in W_\tau} w_i x_{ki\tau} = n_\tau \mu_{k\tau}. \tag{9.5}$$

The solution to this problem is to construct the matrix of observed auxiliary variables. Let $c_{i\tau} = 1$ if respondent i participates in wave τ, and $c_{i\tau} = 0$ if this is not the case. Then let X be the matrix

$$X = [\boldsymbol{c}_1, \boldsymbol{c}_2, \ldots \boldsymbol{c}_q, X_1, X_2, \ldots X_q] \tag{9.6}$$

where $\boldsymbol{c}_\tau$ is the vector of $c_{i\tau}$ and X_τ is the matrix of auxiliary variables at time τ, $\tau = 1, 2, \ldots q$. This matrix implicitly defines the missing data model. For respondents that are missing in wave τ, the auxiliary variables are equal to zero. As a consequence, the expected overall means over all cases are $p_\tau \mu_{k\tau}$ with $p_\tau = n_\tau / n$ (these are the means that have to be specified in a file with population parameters). The weights are determined such that Equation (9.6) is satisfied, with the first q population means (corresponding to the $\boldsymbol{c}_\tau$) equal to n_τ. For every wave W_τ this ensures that

$$\sum_{i \in W_\tau} w_i = n_\tau \tag{9.7}$$

and hence that every weighted total corresponds to the sampling total in wave W_τ.

An example of an application of this theory is the estimation of the total value of the caravans owned by Dutch households. Statistics Netherlands provides data on the percentage of households owning a caravan, which was 14.0, 13.0 and 12.0 in 2001, 2002 and 2003, respectively. In the DNB household survey, which was carried out using the CentERpanel (for more information on the DNB, see www.uvt.nl/centerdata/dhs/),

Table 9.3 Weighting using the percentages of households owning a caravan.

	2001	2002	2003
N unweighted	886	885	808
N weighted	886.0	885.2	808.2
% Households with caravans unweighted	7.4	8.4	9.2
% Households with caravans weighted	14.0	13.0	12.0
Average value of caravans unweighted (€)	873	368	415
Average value of caravans weighted (€)	1639	519	509
Total value of caravans weighted (M€)	11 255	3599	3561

Source: DNB household survey (www.uvt.nl/centerdata/dhs/).

these percentages are considerably lower (see Table 9.3). This has consequences for the average value of a caravan per household (a household that does not own a caravan has value zero). The multiple correlations between the X values (the variables in Table 9.2) and the caravans' values were on average 0.65.

Equation (9.1) shows that calibration using the regression estimator (and many other procedures) involves inversion of the $\boldsymbol{X}'\boldsymbol{X}$ matrix. As a consequence, $\boldsymbol{X}$ may not contain linear dependencies or, more in particular, identical columns. This may be the case when two waves have identical respondents: then the $\boldsymbol{c}_\tau$ vectors are identical. The obvious solution is to delete one of these vectors from the matrix of auxiliary variables. In that case there may exist another problem when one of the auxiliary variables in the dataset is identical (e.g. sex) whereas the population percentages may have changed (e.g. a slight change in the proportion of men in the population). There is no other solution than just to pick one of the values, as a weighted sum can only have one outcome.

9.5 DIFFERENCES BETWEEN WAVES

In commercial and market research panels are often formed and maintained with the specific goal of having easy access to a pool of willing respondents from which special and well-defined subsamples are drawn. These access panels form a rich database of respondents for which demographic and other background characteristics are known. Standing access panels are longitudinal in design, but their main use is for cross-sectional analysis. In contrast, a main *raison d' être* of a scientific panel is measuring differences over time in an efficient way by virtue of the correlation between observations of the same variable in different waves. To illustrate this, we restrict ourselves here to the case of two waves W_1 and W_2. Let $n_{1\cdot}$ be the number of respondents with nonmissing values in $W_1 \backslash W_2$, let $n_{2\cdot}$ be the number of respondents in $W_2 \backslash W_1$ and let n_{12} be the number of respondents in $W_1 \cap W_2$. We are interested in a population parameter $D = Y_2 - Y_1$, the growth between time 1 and time 2. For this parameter we have two estimators: y_{ind}, based on the respondents that are in W_1 or W_2 but not in both, and y_{dep}, based on the respondents in the intersection of W_1 and W_2, for which two correlated observations are available. The standard errors of y_{ind} and y_{dep} are σ_{ind} and σ_{dep}, respectively.

We can combine y_{ind} and y_{dep} to a composite estimator

$$y_c = \alpha y_{dep} + (1 - \alpha) y_{ind} \tag{9.8}$$

such that the variance

$$\text{var}(y_c) = \alpha^2 \sigma_{dep}^2 + (1 - \alpha)^2 \sigma_{ind}^2 \tag{9.9}$$

is minimal. This variance is minimized when

$$\alpha = \frac{\sigma_{ind}^2}{\sigma_{ind}^2 + \sigma_{dep}^2}. \tag{9.10}$$

Assume that y has equal variance ${\sigma_y}^2$ in waves 1 and 2 and let ρ_y be the correlation of y in waves 1 and 2, then we have in the unweighted case

$$\sigma_{ind}^2 = \sigma_y^2 \left(\frac{1}{n_{1.}} + \frac{1}{n_{2.}} \right) \tag{9.11}$$

and

$$\sigma_{dep}^2 = \sigma_y^2 \frac{1 - \rho_y}{n_{12}}, \tag{9.12}$$

which yields

$$\alpha = \frac{n_{1.} + n_{2.}}{n_{1.} + n_{2.} + \dfrac{2 n_{1.} n_{2.} (1 - \rho_y)}{n_{12}}}. \tag{9.13}$$

When the data are unweighted, that is, each case has weight 1, then applying the composite estimator amounts to weighting the cases by $\gamma\alpha$ in $W_1 \cap W_2$ and $\gamma(1 - \alpha)$ in $W_1 \backslash W_2$ and $W_2 \backslash W_1$. When the weights are required to add up to n, γ is the constant $n/\{n_{12}\alpha + (n_{1.} + n_{2.})(1 - \alpha)\}$.

Let us assume that the cases are already weighted as described in Section 9.4. Let us further assume that $\sum_{i \notin W_1 \cup W_2} w_i = n$ and $\sum_{i \notin W_1 \cap W_2} w_i = n_{12}$. It is easily verified that this can be achieved by including a constant column of '1's in X. When the cases in both waves are missing at random the variances can be expected to be proportional to (9.11) and (9.12), respectively. For optimal weighting of y, we may apply the same procedure using an additional weight factor $\gamma\alpha$ and $\gamma(1 - \alpha)$, respectively, with α given by (9.13). The drawback of this procedure, of course, is that for every variable y a new weight has to be calculated because it depends on ρ_y. This may be worthwhile if y is a key variable with a high correlation between waves, such as the change in employment status between two successive months. If there is no such key variable, a value of ρ may be used that is typical for the entire dataset. In practice, α may often be close to 1. In the dataset described in Table 9.3 we found $n_{1.} = 115$, $n_{2.} = 114$, $n_{12} = 771$, $\rho_y = 0.56$, which yielded a value of α of 0.94.

9.6 SINGLE IMPUTATION

A different way to cope with missing data is *imputation*: filling the holes in the dataset with plausible values. Many imputation methods exist, which differ by how they define 'plausible'. Imputation methods can be either model based (parametric) or data based (nonparametric) (Efron, 1994; Little and Rubin, 2002, pp. 20).

Parametric imputation methods include replacing the missing value by the variable's mean, or by the value predicted by a regression analysis on the available complete variables for that case. A model-based technique such as EM estimation (Dempster *et al.*, 1977) can also be used for parametric imputation. The EM method is a regression approach that uses all the information in the available data, both complete and incomplete variables, which makes it a very effective method to impute missing data points.

Nonparametric or data-based imputation replaces the missing values by an appropriate value that is observed elsewhere in the dataset. By using a donor case to provide the imputed value, nonparametric imputation ensures that the imputed value is a value that can actually exist. *Hot deck* imputation sorts the respondents and nonrespondents on a set of auxiliary variables into imputation classes. Missing values are then imputed, using randomly chosen observed values from donors that are in the same imputation class. To create imputation classes, we need auxiliary variables that are related to either the imputed variable or to the missingness mechanism. It should be noted that nonparametric imputation does involve a model; the implied model is that the relationships observed in the observed data also hold for the missing-data units, in other words the choice of the X assumes a data-implied model.

9.7 MULTIPLE IMPUTATION

Parametric imputation methods will often result in imputation of a value that is optimal for the respondent according to some criterion. This is especially the case with model-based methods, such as regression imputation. As a result, models fitted on imputed datasets have a tendency to produce results that are too good; the strength of relationships is exaggerated and models fit too well. A remedy is to add an appropriate random error to the imputed value. This random error can come from a model or from the observed data. For example, with regression imputation it is possible to add a random error term, either from the appropriate error distribution (normal or Student distribution) or a randomly chosen residual from the complete cases. With imputation methods such as hot deck this is not necessary. The hot deck method imputes observed values, and thereby includes the error terms that are present in those observations.

Since adding error terms or using randomly chosen donor cases in a hot deck procedure incorporates a chance mechanism, the resulting imputed data files will vary. When error is added to the imputed values, the logical next step is multiple imputation. Multiple imputation means that the missing data are imputed m times (typically $3 \leq m \leq 5$), with a different randomly chosen error term added or a differently chosen donor case. This leads to m separate datasets. These completed datasets are analysed using standard analysis methods and the results are combined. Resulting parameter estimates are combined by taking their average across the multiple imputed datasets:

$$\overline{Q} = \frac{1}{m}\sum \hat{Q}_i \,. \tag{9.14}$$

The variance of the analysis results across the multiple imputed data sets provides an estimate of the imputation error (cf. Rubin, 1987). The appropriate formula is

$$T = \overline{U} + (1 + m^{-1})B \,, \tag{9.15}$$

where $\overline{U}$ is the mean of the squared standard errors given by $\overline{U} = \frac{1}{m}\sum U_i$, and B is the variance of the parameter estimates between the imputed datasets given by

$$B = \frac{1}{m-1}\sum(\hat{Q}_i - \overline{Q})^2.$$

While combining the multiple estimates is relatively straightforward, constructing the multiply imputed datasets is difficult. Again, there is a parametric and a nonparametric approach. In the parametric approach, the multiply imputed (MI) datasets are simulated draws from a Bayesian predictive distribution of the missing data. This requires a model for the complete data, and properly adding uncertainty about both the missing values and the parameters of the predictive distribution. Schafer (1996) describes such procedures for normally distributed data and for other data models, including categorical and mixed normal–categorical data and panel data. In the nonparametric approach hot deck imputation is used. The approach is to use a bootstrap procedure to generate the random imputations. First, a bootstrapped logistic regression is used to predict the nonresponse. Next, the regression equation is used to compute the propensity to have missing data for each case in the sample. The complete and incomplete cases are then sorted into subgroups based on these propensity scores. Finally, missing data are replaced by values taken from a randomly chosen donor in the same imputation class. This procedure is repeated a number of times to generate multiple imputations. The advantage of the nonparametric approach is that it does not require an explicit data model; the implicit model for the missing data is provided by the observed data. In addition, it will always generate imputed values that actually exist in the observed data.

Multiple imputation may be somewhat laborious, but it is a powerful approach. The power of multiple imputation, especially in longitudinal research, is based on the notion that the imputation model need not be the same as the analysis model. The most important imperative for the imputation model is that it should be at least as complex as the analysis model to be used. Thus, if it is anticipated that nonlinear curves will be fitted to a longitudinal dataset, nonlinear terms must also be used when the incomplete data are imputed. When interactions are tested, interaction terms should be included in the imputation model. In nonparametric imputation, where existing values are imputed from a donor case, this happens implicitly, but there is no rule against the imputation model being more complex than the analysis model. As we stated before, in longitudinal research there is a potential richness of auxiliary variables, both initially measured and time-varying variables. Not only can missing data at time point t (y_t) be imputed using the information from previous measurement occasions (y_{t-1}, y_{t-2}), but also later observations (y_{t+1}) can – in due course – be used in the imputation process. In multiple imputation, the preferred strategy would be to use as many variables as possible. Collins, Schafer and Kam (2001) investigated whether a liberal use of available auxiliary variables improves the results. They conclude that a liberal strategy is clearly superior to a restrictive strategy. Using many auxiliary variables for imputation not only decreases the chance of overlooking an important cause of missingness, but also increases the accuracy of the imputed values. If it is necessary to select a smaller number of auxiliary variables, Collins *et al.* recommend looking for auxiliary variables that predict the incomplete variables well, and/or auxiliary variables that correlate with the missingness mechanism.

Little and Vartivarian (2005) make the same recommendation in the context of weighting. For diagnostic procedures that help to identify such variables, we refer to De Leeuw, Hox and Huisman (2003).

9.8 CONCLUSION AND DISCUSSION

Over the past decade the number of panel surveys has increased dramatically. This is partly due to the growing popularity of online panels and the potential of the Internet, which makes large-scale national and international panels affordable. When assessing the field one sees two different worlds: the world of governmental and academic research and the world of commercial, market and opinion research. In the first world, the (semi)governmental and academic world, mostly single purpose panels are used in which one major research topic is studied over a longer period of time with the emphasis on measurement of change. The data are processed and analysed in detail by specialists, both substantively and methodologically. In the second world, the world of market and opinion research, the majority of the panels are multipurpose panels or access panels run by commercial organizations working for a large variety of clients. For many commercial projects only cross-sectional data from the panels are used, but there is also a substantial class of research problems for which the longitudinal aspect is essential. Obvious examples are tracking studies to evaluate the effect of advertising campaigns and investigating trends in markets such as the IT market or the leisure time market. In this chapter, we attempt to bring these worlds closer together by reviewing how auxiliary data can be used for statistical adjustment in both kinds of panels. The emphasis is on longitudinal surveys and we discussed calibration in general and over multiple waves. We also discussed the use of imputation and multiple imputation for handling missing data.

Although most effort in longitudinal research is focused on the prevention and adjustment of panel attrition, one of the most crucial parts is the recruitment stage, when the majority of nonresponse is encountered. It is during recruitment and panel formation that selective nonresponse and the opportunity for nonresponse bias are greatest. In the described case of the CentERpanel, only 10 % of the initial sample agreed to become a panel member; in the Gallup consumer panel 27 % of those contacted agreed to participate. In our opinion both in the commercial world and the scientific world far more attention should be devoted to the recruitment process; this should be carefully described in terms of inclusion probability and amount and type of nonresponse. But one should go one step further and also collect data in the recruitment process with the explicit goal of using these to adjust for initial nonresponse and differences between those approached, those willing to grant a recruitment interview and the final panel members.

This approach goes much further than the standard adjustment procedures in which demographic variables are used, such as age and sex. It is desirable to have datasets that are representative with respect to standard demographic variables, but this will only reduce nonresponse bias if these demographic variables are strongly related to the missingness pattern and the target variables of the study. In other words, is that one old gentleman in an online panel like all the other old gentlemen not in the panel, or does he differ for instance in computer knowledge or technological interest? In a survey on technological changes and society, weighting to age and gender will probably be only 'cosmetic' and will not remove the nonresponse bias. One needs other variables that have stronger relationships with the target variables. With high initial nonresponse in the recruitment phase, the availability of a choice of variables for model-based weighting or imputation remains the only possible solution.

The choice of what auxiliary data to collect is obviously important, and partly shifts the responsibility of the quality of the estimates from the survey statistician to the substantive expert. In the discussion of missing data in the context of item-nonresponse Schafer and Graham (2000) state that two types of variables are important: firstly variables that help to predict the missing values, and secondly variables that help to predict the missingness patterns. In addition, they present simulations that suggest that including too many variables in the imputations is less likely to lead to bias than including too few. This conclusion would also cover the choice of a set of weighting variables. To the extent that covariates, including auxiliary variables, predict the missingness and are able to mediate possible relationships between the missingness and the (non)observed values, we have a situation described by Rubin (1987) as missing completely at random (MCAR). Given the importance of potential predictors of missingness, Schafer and Graham (2000) recommend including a question like 'How likely is it that you will remain in this study through the next measurement period?' in the longitudinal data collection. To the extent that the recorded value of an outcome variable predicts the unobserved missing value on the same variable, the appropriate assumption is missing at random (MAR; Rubin, 1987). Since repeated measures tend to be highly correlated, this appears a plausible assumption. However, to convert an essentially not missing at random (NMAR) situation to MAR or MCAR, we need not only the right set of auxiliary variables but also a correct model (Fitzmaurice *et al.*, 2004). In this respect, nonparametric models appear to have an important advantage. Because in this approach the model is implicitly provided by the observed data, nonlinear relationships and interaction effects are automatically included in the model.

If in longitudinal data the missingness pattern is monotone, weighting provides a good solution; however, when nonmonotone patterns are encountered it is generally better to use imputation. For commercial panels, the main argument for choosing between weighting and imputation is one of convenience. The easiest way to construct a dataset with correction for missing values is to weight the data. Weighting simultaneously for different waves is no exception, as the same mechanisms apply for which there are routine software solutions. In the resulting dataset missing data are still visible and have to be accommodated in the analysis. When complex multiwave, multivariate models are concerned multiple imputation or explicit use of auxiliary variables in models that include a missingness model may be preferable. Imputation has the advantage that it results in 'clean' datasets that can be used without the user being aware of the missing values, even if they are flagged, as the user is free to ignore the flags. Imputation in wave t based on a series of prior waves $1, 2, \ldots t-1$ may lead to a dilemma when subsequent waves are obtained, as an improved imputed value for wave t may be established. Although the improvement in estimation is real, for research agencies revision of previously published results or data may be a problematic policy.

REFERENCES

Bethlehem J. G. and Keller, W. J. (1987). Linear weighting of sample survey data. *Journal of Official Statistics*, 3, 141–153.

Collins, L. M., Schafer, J. L. and Kam, C. M. (2001). A comparison of inclusive and restrictive strategies in modern missing data procedures. *Psychological Methods*, 6, 330–351.

Couper, M. P. (2000). Web surveys: A review of issues and approaches. *Public Opinion Quarterly*, 64, 464–494.

De Leeuw, E. D. (2005). Dropout in longitudinal data. In B. S. Everitt and D. C. Howell (Eds), *Encyclopedia of Statistics in Behavioral Science* (Vol. 1, pp. 515–518). Chichester: John Wiley & Sons, Inc.

De Leeuw, E. D. and De Heer, W. (2002). Trends in household survey nonresponse: a longitudinal and international comparison. In R. M. Groves, D. A. Dillman, J. L. Eltinge and R. J. A. Little (Eds), *Survey Nonresponse* (pp. 41–54). New York: John Wiley & Sons, Inc.

De Leeuw, E. D., Hox, J. and Huisman, M. (2003). Prevention and treatment of item nonresponse. *Journal of Official Statistics*, 19, 153–176.

Dempster, A. P., Laird, N. M. and Rubin, D. B. (1977). Maximum likelihood from incomplete data via the EM algorithm. *Journal of the Royal Statistical Society, B*, 39, 1–38.

Deville, J. C. and Särndal, C. E. (1992). Calibration estimators in survey sampling. *Journal of the American Statistical Association*, 87, 376–382.

Efron, B. (1994). Missing data, imputation, and the bootstrap. *Journal of the American Statistical Association*, 89, 463–475.

Fitzmaurice, G. M., Laird, N. M. and Ware, J. H. (2004). *Applied Longitudinal Analysis*. New York: John Wiley & Sons, Inc.

Hajek, J. (1981). *Sampling from a Finite Population*. New York: Marcel Dekker.

Kalton, G. and Flores-Cervantes, I. (2003). Weighting methods. *Journal of Official Statistics*, 19, 81–97.

Kish, L. (1992). Weighting for unequal P_i. *Journal of Official Statistics*, 8, 183–200.

Lee, S. (2006). Propensity score adjustment as a weighting scheme for volunteer panel web surveys. *Journal of Official Statistics*, 22, 329–349.

Lehtonen, R. and Pahkinen, E. (2004). *Practical Methods for Design and Analysis of Complex Surveys*. New York: John Wiley & Sons, Inc.

Little, R. D. and Vartivarian, S. (2005). Does weighting for nonresponse increase the variance of survey means? *Survey Methodology*, 31, 161–168.

Little, R. J. A. and Rubin, D. B. (2002). *Statistical Analysis with Missing Data* (2nd edn.). New York: John Wiley & Sons, Inc.

Potter, F. J. (1988). Survey of procedures to control extreme sampling weights. In *Proceedings of the American Statistical Association, Section on Survey Research Methods* (pp. 453–458). Washington, DC: American Statistical Association.

Potter, F. J. (1990). A study of procedures to identify and trim extreme sampling weights. In *Proceedings of the American Statistical Association, Section on Survey Research Methods* (pp. 225–230). Washington, DC: American Statistical Association.

Potter, F. J. (1993). The effect of weight trimming on nonlinear survey estimates. In *Proceedings of the American Statistical Association, Section on Survey Research Methods* (pp. 758–763). Washington, DC: American Statistical Association.

Rubin, D. B. (1987). *Multiple Imputation for Nonresponse in Surveys*. New York: John Wiley & Sons, Inc.

Schafer, J. L. (1996). *Analysis of Incomplete Multivariate Data*. New York: Chapman and Hall.

Schafer, J. L. and Graham, J. W. (2000). Missing data: our view of the state of the art. *Psychological Methods*, 6, 147–177.

Sikkel, D. and Hoogendoorn, A. (2008). Panel surveys. In E. D. de Leeuw, J. J. Hox and D. A. Dillman (Eds), *The International Handbook of Survey Methodology* (pp. 479–499). New York: Taylor & Francis.

CHAPTER 10

Identifying Factors Affecting Longitudinal Survey Response

Nicole Watson and Mark Wooden
University of Melbourne, Australia

10.1 INTRODUCTION

This chapter examines the factors that influence continued participation by sample members in longitudinal surveys. It is structured into two distinct parts. First (Section 10.2), evidence from previous research that has modelled the response process within a multivariate framework is reviewed. Second (Section 10.3), estimates of predictors of response from a national household panel survey – the Household, Income and Labour Dynamics in Australia (HILDA) Survey – are presented. Following other recent treatments in the literature, the estimation model treats survey participation as involving two sequential events, contact and response.

Like all sample surveys, longitudinal (or panel) surveys are affected by unit nonresponse. A distinctive feature of longitudinal surveys, however, is that nonresponse is not a one-off event and tends to accumulate over time as further waves of interviewing are conducted. Longitudinal surveys also face the problems of tracking sample members who relocate and of dealing with the respondent fatigue that is associated with repeated survey participation (Laurie *et al*., 1999).

As a consequence, many longitudinal surveys find, even after just a few waves of interviewing, that relatively large fractions of the responding sample from the initial wave are no longer participating. This has certainly been the case for the world's leading household panel surveys. The University of Michigan's Panel Study of Income Dynamics (PSID), for example, had lost just over one-quarter of its original 1968 sample by 1975 (i.e. wave 8) (see Fitzgerald *et al*., 1998, table 1).[1] More recent household panel

[1] This figure includes deaths. If deaths are excluded, the accumulated attrition rate declines to about 22 %.

Methodology of Longitudinal Surveys P. Lynn

studies, typically employing more complicated designs, report higher rates of sample attrition. After eight years of interviewing, the German Socio-Economic Panel (GSOEP), which commenced in 1984, and the British Household Panel Survey (BHPS), which commenced in 1991, both reported the loss of about 34 % of their original sample (Spieß and Kroh, 2004, figure 9; Taylor *et al.*, 2005, table 20), and in the case of the Dutch Socioeconomic Panel the rate of sample loss at the same stage was almost certainly in excess of 50 %.[2] Relatively high rates of sample loss have also been reported in the European Community Household Panel (ECHP), a multicountry study conducted over the period 1994–2001. Watson (2003), for example, reported five-year retention rates that varied from a high of 82 % in Portugal to a low of 57 % in Ireland (after excluding deaths and other movements out-of-scope). Finally, the Survey of Income and Program Participation (SIPP), run by the US Census Bureau, has reported cumulative rates of sample loss of up to 35 % (of households) over a four-year period (Westat, 2001, table 2.5, pp. 2–19).

Of course rates of attrition might be different in other types of longitudinal surveys employing different designs and covering different populations. Birth cohort studies, for example, often report very high response rates many years after the original sample was drawn (e.g. Wadsworth *et al.*, 2003; Hawkes and Plewis, 2006). Such studies, however, are distinctive in that interviewing is relatively infrequent, and hence respondent burden tends to be far less than in other panel surveys where interview waves are more frequent. Nevertheless, it is also true that frequent (e.g. annual) survey waves do not need to be associated with high rates of sample attrition. The National Longitudinal Study of Youth (NLSY), which follows cohorts of young people in the USA until well into adulthood, for example, obtained a rate of attrition after eight years of interviewing from its 1979 cohort of just 8 %, and even after 21 years the rate of sample loss was still under 20 % (Bureau of Labour Statistics, 2003). Nevertheless, the NLSY experience appears to be the exception and not the norm, with most other youth cohort panels (e.g. the Youth Cohort Study of England and Wales and the various cohort studies that comprise the Longitudinal Surveys of Australian Youth) recording much higher rates of attrition.

There is also mounting evidence indicating that the extent of sample attrition has been increasing over time (Atrostic *et al.*, 2001; De Leeuw and De Heer, 2002; Tourangeau, 2003). The recent experience of the NLSY seems to be in line with this conclusion, with the rates of attrition for the 1997 cohort noticeably higher than the rates of attrition recorded for the 1979 cohort. After the first five waves, the overall sample retention rate, while still a healthy 88 %, was over 8 % lower than the rate reported at the comparable stage of the 1979 cohort.[3] A similar deterioration over time has also been reported for the SIPP. Involving relatively short overlapping panels (ranging from 24 months to 48 months long), rates of cumulative sample loss over eight waves (32 months) averaged around 21 % for the panels commencing between 1984 and 1991 (Westat, 1998, table 5.1, p. 45). For the 1992 and 1993 panels the rate of sample loss rose to around 25 % over

[2] Winkels and Withers (2000) reported that after 11 years of interviewing only 42 % of the original sample remained.

[3] The initial samples for the two cohorts, however, were not identical. Specifically, the 1997 cohort was younger, ranging from 12 to 16 years, whereas the 1979 cohort was aged between 14 and 21 years.

the same time span, and for the 1996 panel the rate of loss was over 31 % (Westat, 2001, table 2.5, pp. 2–19).

In general, a high rate of sample attrition poses a serious problem for longitudinal studies. At a minimum, attrition reduces the precision of survey estimates, and at sufficiently high levels can threaten the viability of continuing a panel, especially if the initial sample size was relatively small. Of greater concern, since attrition tends not to be random it may impart bias to population estimates. Survey administrators thus face pressures both to ensure that they employ design features and fieldwork procedures that will maximize sample retention and, since some nonresponse is unavoidable, deliver as much information as possible about nonrespondents to assist data analysts to make inferences in the presence of missing data. Achieving both of these objectives requires good knowledge of the response process and the factors that give rise to sample attrition, and this is the subject of this chapter. More specifically, this chapter reviews the growing body of empirical evidence on the determinants of response to, and attrition from, longitudinal surveys. We are especially interested in those factors that are most amenable to manipulation by survey designers and administrators. As a result, the emphasis here is more on the role of *survey design features* and the *interview situation*, and less on the demographic characteristics of nonrespondents, information that can be readily distilled from all longitudinal survey datasets and is regularly used in the construction of population weights that adjust for attrition. We then estimate a model predicting response over the course of the first four waves of the Household, Income and Labour Dynamics in Australia (HILDA) Survey, a relatively new household panel survey that has experienced significant attrition.

10.2 FACTORS AFFECTING RESPONSE AND ATTRITION

Following Lepkowski and Couper (2002), the response process can be divided into three conditional stages: the sample member is located; contact with the sample member is established; and the sample member provides an interview.

10.2.1 Locating the Sample Member

Locating a member of a longitudinal sample is a straightforward exercise where the sample member has not moved. Determining the whereabouts of movers, however, is often far from straightforward. Moreover, changing address has generally been found to exert a highly significant negative influence on the likelihood of response at the next wave (Becketti *et al.*, 1988; Lillard and Panis, 1998; Zabel, 1998; Lepkowski and Couper, 2002; Watson, 2003). Sample unit characteristics, of course, have a major bearing on mobility (Buck, 2000) and hence the likelihood of establishing location. Studies of attrition, however, do not explicitly model the decision to move and thus cannot separate the effect of a variable on attrition via mobility from its effect via other stages in the response process. Nevertheless, studies that model the location/contact outcome separately from the cooperation outcome generally find that movement per se does not have any negative effects on cooperation once the sample member has been found and contacted (Gray *et al.*, 1996; Lepkowski and Couper, 2002). Such findings are consistent with findings from cross-sectional surveys, which have found slightly lower refusal rates among recent movers compared with members of households that have not moved (Groves and Couper, 1998, p. 139).

Where movement has taken place, the likelihood of successful location can be influenced by the tracking procedures that are implemented, the length of time between survey waves and the extent and nature of contact between waves. Tracking is especially important (see Couper and Ofstedal, Chapter 11 of this volume) and all well-designed longitudinal surveys will ensure that information is collected during each interview that can be subsequently used to locate the respondent if they move (such as work, home and mobile numbers, e-mail addresses and contact details of relatives or friends). How successful interviewers are at extracting this information is potentially very important for subsequent attrition. For example, Laurie *et al.* (1999) analysed attrition over the first four rounds of the BHPS and found a strong positive association between response and the provision of a contact name at wave 1. Of course, even when armed with details of contact persons, it does not automatically follow that the details provided at the previous interview will be current or that the contact person knows the current location of the sample member. Other tracking mechanisms will thus need to be employed, such as searching telephone directories and electoral rolls, or asking neighbours or other local people who might know where the sample member may have moved to.[4]

10.2.2 Contacting the Sample Member

In many respects, the contact task in longitudinal surveys is little different from that in cross-sectional surveys and, as discussed in Groves and Couper (1998, chapter 4), will be affected by such factors as the accessibility of the dwelling (in the case of face-to-face interviews), the use of answering machines (in the case of telephone surveys) and the availability of the sample member. Survey design features that can influence the rate of successful contact include the number of calls attempted, the timing of those calls (over the day, week or year) and the duration of the fieldwork period. Cross-sectional surveys, however, are often not well placed to assess how important some of these influences are because of both lack of information about nonrespondents, though this problem can be overcome if survey samples can be matched to some external data source such as Census records (e.g. Foster, 1998; Groves and Couper, 1998), and the potential endogeneity of some design features (e.g. number of calls made). In contrast, longitudinal surveys typically contain a lot of information about nonrespondents after the initial wave, and the panel nature of the data permits stronger inferences about causality.

Nevertheless, empirical studies that both employ longitudinal data and distinguish the contact stage from the cooperation stage remain relatively scarce. The major studies that we have identified are: Gray *et al.* (1996), who analysed response to the two waves of a health and lifestyle survey conducted in Britain in 1984–1985 and 1991–1992; Thomas *et al.* (2001), who reported on attrition in the four-year follow-up to the 1993 Indonesian Family Life Survey; Lepkowski and Couper (2002), who examined response in two separate two-wave longitudinal surveys conducted in the USA; Nicoletti and Buck

[4] For a review of the variety of tracking mechanisms employed in panel studies, see Ribisl *et al.* (1996).

(2004), who analysed survey response over three waves of both the BHPS and GSOEP panels; and Nicoletti and Peracchi (2005), who examined response behaviour over the first five waves of the ECHP. Variables that have featured prominently in the models of contact estimated in one or more of these studies include: residential mobility (discussed earlier); socio-demographic characteristics hypothesized to be associated with the likelihood of finding someone at home (such as age, sex and household size and composition); characteristics of the region in which the sample member resides; measures of community attachment (such as home ownership and caring responsibilities); interview workloads; interviewer continuity; the number of calls made; and the length of the fieldwork period. Space precludes a detailed discussion of these findings and so we focus here on data collection characteristics.

It is axiomatic that making additional attempts to contact a sample member who has yet to be found should raise the probability of contacting that sample member. The evidence from longitudinal surveys, however, indicates that the sample members requiring the most effort in terms of the number of calls made at wave $t - 1$ are at greatest risk of attrition at wave t (Branden *et al.*, 1995; Lillard and Panis, 1998; Zabel, 1998; Nicoletti and Buck, 2004; Nicoletti and Peracchi, 2005). The usual explanation for this finding is that the number of calls needed is indicative of both how difficult it is to find the respondent at home and how evasive the respondent might be when setting up interview appointments (though only Nicoletti and Buck (2004) and Nicoletti and Peracchi (2005) specifically identify attrition due to noncontact).

Of course, it is not just the *number* of calls made that matters, but *when* those calls are made (see Groves and Couper, 1998). In longitudinal surveys this is less of an issue, given that interviewers will typically have available to them a lot of information about sample members, including the call pattern and interview times from the previous wave, to determine appropriate call times. What interviewers often have less control over is how long they spend in the field. Longer fieldwork periods should enhance the likelihood of making contact: for example, by increasing the probability of reaching sample members who are away temporarily from their home address. Nicoletti and Peracchi (2005), in their analysis of response behaviour over the first five waves of the ECHP, were able to include a measure of the duration of the fieldwork period, since this varied both over time and across the different countries participating in the ECHP. As expected, this variable was found to be associated with higher contact probabilities.

It is also widely recognized that interviewers can have a major influence on survey response, though only rarely do researchers distinguish between effects at the contact stage and at the cooperation stage. Many researchers, for example, point to the benefits of interviewer continuity in enhancing survey response (more on this below), but only Nicoletti and Buck (2004) explicitly test whether such effects matter more for contact probabilities than for response. The results they report appear to suggest that interviewer continuity matters at both stages, but the relative magnitude is greater at the contact stage.

Finally, Nicoletti and Buck (2004) have also examined whether differences in interviewer workloads might affect contact probabilities. The estimated coefficients are mostly negatively signed when using the BHPS data but positively signed when using data from the GSOEP. The magnitude of the coefficients, however, is always quite small and only has marginal significance at best.

10.2.3 Obtaining the Cooperation of the Sample Member

The factors that affect response once contact has been made are highly varied and often out of the interviewer's control. For example, willingness to participate in further waves of interviews can be expected to vary with individual personality traits and with how busy sample members are. Nevertheless, the way surveys are designed and administered can also have a marked effect on response. For example, cooperation with surveys has been found to be influenced by the use of incentives, interviewer experience and continuity, respondent identification with the study, the survey topic and the interview experience in prior waves.

Incentives

Panel studies often use incentives, both in the form of cash payments and gifts, to improve the initial response rate and reduce subsequent attrition. The experience with incentives is reviewed elsewhere in this volume (see Laurie and Lynn, Chapter 12) and so we do not devote much space to that issue here. Suffice to say that incentives can enhance response rates, though their effectiveness varies widely depending on how they are designed and administered.

Interviewer Effects

It is widely recognized that interviewers play a vital role in gaining cooperation from respondents. In cross-sectional surveys, for example, response rates have been found to improve with interviewer age and experience and are also influenced by interviewer attitudes and confidence (e.g. Couper and Groves, 1992; Groves and Couper, 1998; Japec and Lundqvist, 1999; Martin and Beerten, 1999). In longitudinal surveys the role of the interviewer is arguably even more important given that the task of maintaining respondent interest is likely to become more difficult as respondents start to lose interest in the study or even question its salience. Interestingly, Nicoletti and Buck (2004) reported evidence that high interviewer workloads (households per interviewer) were negatively associated with response probabilities (conditional on contact being made). Such findings suggest that concentrating workloads on the most experienced interviewers may not necessarily bring any response gains if those workloads are too high.

As previously noted, one factor that is regularly reported as being beneficial for response in longitudinal studies, at least those administered on a face-to-face basis, is interviewer continuity (Waterton and Lievesley, 1987; Zabel, 1998; Laurie *et al*., 1999; Hill and Willis, 2001; Nicoletti and Buck, 2004; Behr *et al*., 2005; Nicoletti and Peracchi, 2005; Olsen, 2005). The size of the effect, however, is highly variable across studies, and in some cases contradictory. For example, both Behr *et al*. (2005) and Nicoletti and Peracchi (2005) analysed the ECHP experience, with the latter finding small insignificant effects and the former finding highly significant and large relationships. More importantly, the coefficient on a simple dummy variable identifying interviewer continuity will almost certainly be biased away from zero. This is because attrition also occurs among interviewers, and interviewer attrition can be expected to be higher in areas where survey response rates are lower. In other words, the interviewer continuity variable – for face-to-face surveys – will often capture part of any systematic difference across areas in response. Nevertheless, there is still disagreement on just how large this bias is. Campanelli and O'Muircheartaigh (2002), using data from the first four waves of the

BHPS, the same data used by Laurie *et al.* (1999), distinguished between regions with and without interviewer attrition and reported that nonrandom interviewer attrition explained all of the interviewer continuity effect reported by Laurie *et al.* (1999). In contrast, Nicoletti and Buck (2004), also using BHPS data (as well as data from the GSOEP), reported that interviewer continuity remained an important and valid predictor of response even after controlling for the presence of a random component linked to interviewers.

Very differently, Olsen (2005) dealt with the endogeneity issue by including, in his analysis of response to the NLSY, two variables that measured the ability of the fieldwork company to exploit interviewer continuity in the current round, and not whether interviewer continuity actually held. While he reported evidence of a response advantage to interviewer continuity, the magnitude of the effect was quite small, indicating an improvement in the attrition rate of only 0.7 %, which was only apparent after the respondent had been interviewed twice by that interviewer.

The Interview Experience

One influence about which there is little disagreement is respondents' previous interview experience. Interviewer assessments of how cooperative the respondent was, how much they enjoyed the interview and/or the ease with which the survey questions could be answered have invariably been found to be highly predictive of cooperation at the next survey wave (e.g. Kalton *et al.*, 1990; Branden *et al.*, 1995; Laurie *et al.*, 1999; Hill and Willis, 2001; Lepkowski and Couper, 2002; Olsen, 2005).

It is also generally assumed that missing data, or item nonresponse, is indicative of an unpleasant or negative interview experience (Loosveldt *et al.*, 2002). It thus follows that item nonresponse at one wave should also be predictive of unit nonresponse at the next, and again the weight of evidence suggests that this is so (Burkam and Lee, 1998; Zabel, 1998; Laurie *et al.*, 1999; Loosveldt *et al.*, 2002; Schräpler, 2002; Lee *et al.*, 2004; Hawkes and Plewis, 2006), though there are exceptions. Serfling (2004), for example, in contrast to Schräpler (2002), could find no evidence of a monotonic relationship between item nonresponse on income-related items and unit nonresponse in data from three waves of the GSOEP. He did, however, report weak evidence of an inverse U-shaped relationship. The proposed explanation here is that some respondents will only participate in the survey because they know they do not actually have to answer questions on sensitive or difficult to answer topics.

Another aspect of the interview experience often asserted to be an important influence on sample attrition is the time taken to administer the interview, with the likelihood of respondent cooperation expected to be inversely related to expected interview length. A number of longitudinal surveys have included interview length as a predictor of response in the next wave, but have found little evidence of a significant negative effect. Indeed, the estimated coefficient is often positively signed (Branden *et al.*, 1995; Zabel, 1998; Hill and Willis, 2001). With hindsight such findings are easily explained. Interview length is not exogenous to the interview process, and is instead a product of how willing respondents are to talk to the interviewer. Thus the respondents who are most engaged by the study and find it most enjoyable will typically have longer interviews. Testing the effect of interview length thus requires experimental evidence, and there are few examples of such experiments being conducted on longitudinal data. Nevertheless, Zabel (1998) reported on the experience of the PSID, which made an explicit attempt to reduce the interview length in 1972 and appeared to have had the effect of reducing attrition, though the magnitude of the effect was not large. This finding is consistent with the weight of evidence from experimental research conducted on cross-sectional surveys, which, according to the review by Bogen (1996, p. 1024), leads to the conclusion that the 'relationship between interview length and nonresponse is more often positive than not, but it is surprisingly weak and inconsistent'.

It is also widely accepted that survey delivery mode has a significant impact on response rates, with methods involving greater levels of personal contact between the survey administrator and the respondent usually found to have higher rates of respondent cooperation, but for a greater cost (see Yu and Cooper, 1983). This thus suggests that changing modes of collection during the administration of a panel could affect sample attrition. Unfortunately, while a number of large panel studies (including, for example, the PSID, the UK National Child Development Study, the 1992–1997 British Election Panel Study and the 1984–1991 Australian Longitudinal Survey) have changed survey modes, rarely has the impact of this change on response rates been reported. Zabel (1998), however, reported that the switch to telephone interviewing in the PSID following five successive years of personal interviewing had a positive, but statistically insignificant, impact on attrition.

Very differently, providing respondents with multiple modes for responding, as is the practice in the GSOEP, is also expected to reduce respondent reluctance, a hypothesis that has received support from experiments conducted in conjunction with cross-sectional surveys (e.g. Voogt and Saris, 2005). We, however, are unaware of any experimental evidence gathered from a longitudinal design. Indeed, in many longitudinal studies employing predominantly face-to-face survey methodologies, alternative response modes such as telephones are often used as a 'last resort' strategy to obtain responses. In such situations, a telephone interview in one year is likely to be indicative of a relative lack of interest in the study and thus predictive of nonresponse in the next. Analyses of attrition in the NLSY (Branden *et al.*, 1995), the SIPP (Zabel, 1998) and the ECHP (Nicoletti and Peracchi, 2005) data support this hypothesis.

10.2.4 The Role of Respondent Characteristics

As mentioned in Section 10.1, associations between respondent characteristics and survey response are not the main focus of this chapter. This stands in marked contrast to the many studies of the correlates of sample attrition where personal and household characteristics represent most, if not all, of the explanatory variables included (e.g. Becketti *et al.*, 1988; Gray *et al.*, 1996; Burkam and Lee, 1998; Fitzgerald *et al*, 1998; Kalsbeek *et al.*, 2002; Watson, 2003; Behr *et al.*, 2005; Hawkes and Plewis, 2006). The emphasis placed on respondent characteristics in these studies should not be surprising. First, and most obviously, detailed information about nonrespondents is something that is readily available for all persons who exit longitudinal surveys – it is only nonrespondents at wave 1 about whom relatively little is usually known. Second, differences in attrition propensities may be suggestive of possible attrition bias, and thus can be used to help correct for such bias. The importance of respondent characteristics thus cannot be ignored. In any case, the effect of survey design features may be conditional upon respondent characteristics, so it is appropriate to control for respondent characteristics when attempting to identify the effects of survey design. This is the approach taken in Section 10.3.

Before that, however, we summarize what the literature tells us about the characteristics of nonrespondents to longitudinal surveys and in what ways they differ from respondents. Note that the scope of this review is limited to studies that have analysed sample attrition, though we draw on insights from the much larger body of evidence based on single-wave cross-sectional surveys where appropriate. Further, we restrict our attention to those studies that have employed multivariate methods. Where possible we attempt to distinguish between contact and cooperation but, as observed earlier, we are limited by the small number of studies that have explicitly separated out these two stages in the response process. Finally, we limit our attention to the major demographic and socio-economic variables that are regularly included in attrition models.

Sex

Studies of survey response nearly always find that response rates are higher among women than among men, and this is no less true of analyses of survey attrition. The main reason usually cited for this is that women are at home more frequently. Nevertheless, there is limited evidence that, even conditional on contact, men may be slightly more likely to discontinue survey participation (Lepkowski and Couper, 2002; Nicoletti and Buck, 2004).

Age

A widely reported result in cross-sectional surveys is that response rates are lowest for both the youngest and oldest members of the population. The evidence from longitudinal surveys confirms that survey response rates tend to be relatively low among young people. At the other end of the age spectrum the results are less clear. We would expect the older population to be less mobile and thus easier to reach, and a number of studies have reported evidence consistent with this hypothesis (e.g. Gray *et al.*, 1996; Thomas *et al.*, 2001; Lepkowski and Cooper, 2002). In aggregate, however, the evidence is more mixed. Some studies have found that overall attrition propensities are rising in old age (e.g. Becketti *et al.*, 1988; Fitzgerald *et al.*, 1998), others have reported the reverse (e.g. Hill and Willis, 2001), while still others have reported no clear evidence in either direction (e.g. Nicoletti and Buck, 2004; Behr *et al.*, 2005; Nicoletti and Peracchi, 2005). Perhaps it is not age per se that matters for survey cooperation but the salience of the survey content. Hill and Willis (2001), for example, speculated that the positive relationship found between age and survey response in their study of attrition from wave 3 of the US Health and Retirement Study (HRS) might reflect the greater relevance of the HRS to the eldest members of their sample.

Race/Ethnicity

Studies that have included controls identifying racial and ethnic minority groups have generally found negative relationships with response probabilities (e.g. Burkam and Lee, 1998; Zabel, 1998). Those studies that distinguish between contact and cooperation, however, find that this relationship is mainly due to lower rates of contact and not higher rates of refusals (Gray *et al.*, 1996; Lepkowski and Cooper, 2002). That said, an important mediating factor that appears to have been largely ignored (but see Burkam and Lee, 1998) is language-speaking ability. In English-speaking countries, for example, cross-sectional surveys almost always report higher rates of survey nonresponse among non-English speakers.

Marital Status

Marital status is another variable where there is widespread consensus; single people have a higher propensity to drop out than married people. Whether this higher rate of attrition is due to lower rates of contact or higher rates of refusal is less clear. The results reported by Gray *et al.* (1996) suggest that it is mainly a function of lower contact probabilities, but supportive evidence for this from other studies of survey attrition is lacking.

Household Size and Composition

The evidence from cross-sectional surveys indicates that single-person households will typically have lower contact probabilities than larger households and may even have higher refusal rates, though the evidence here is more mixed. The results of both Gray *et al.* (1996) and Thomas *et al.* (2001) suggest that it is only the impact on contact propensity that matters for attrition. A small number of studies also distinguish between households not just on the basis of the number of family members, but whether those family members are children or adults. More often than not it is the presence of children that seems to be most strongly, and negatively, associated with attrition probabilities (e.g. Kalton *et al.*, 1990; Fitzgerald *et al.*, 1998; Zabel, 1998; Nicoletti and Peracchi, 2005). Again, the most plausible explanation for this finding is that the presence of children increases the likelihood of the respondent being home. Somewhat perplexing are the results reported by Nicoletti and Buck (2004). They found in both their BHPS and GSOEP samples that the number of adults in the household was associated with both lower contact rates and higher refusal rates. Watson (2003) also reported higher attrition rates in larger households, but only after single-person households and the number of children had been controlled for.

Education

Education is usually thought to be positively associated with survey response, mainly because those with higher educational attainment are likely to better appreciate the utility of research and information-gathering activities (Groves and Couper, 1998, p. 128). The evidence from attrition studies is mostly in line with this hypothesis (e.g. Gray *et al.*, 1996; Fitzgerald *et al.*, 1998; Lillard and Panis, 1998; Lepkowski and Couper, 2002; Watson, 2003; Behr *et al.*, 2005), though the magnitude of the relationship is arguably quite small.

Home Ownership

Home ownership is frequently included in models of survey attrition and in nearly all cases attracts a negative coefficient (Kalton *et al.*, 1990; Fitzgerald *et al.*, 1998; Zabel, 1998; Lepkowski and Couper, 2002; Watson, 2003). The rationale for the inclusion of this variable offered by Lepkowski and Couper (2002) is that it is an indicator of community attachment, and so is expected to be positively associated with contact propensity. In fact, their results suggest that cooperation may also be positively affected, but other studies have found that the effect mainly works through enhanced contact probabilities (Gray *et al.*, 1996; Thomas *et al.*, 2001; Nicoletti and Peracchi, 2005).

Income

A widely held view is that response rates, at least for cross-sectional surveys, tend to be lowest in both tails of the income distribution. The evidence from longitudinal surveys, however, is mixed. While a number of studies have reported evidence of a quadratic (inverted U-shaped) relationship between income measures and re-interview rates

(Becketti *et al.*, 1988; Hill and Willis, 2001), others have reported no significant decline at high income levels (e.g. Kalton *et al.*, 1990; Fitzgerald *et al.*, 1998; Watson, 2003). The magnitude of the estimated effects, however, is typically small and, given the number of studies that have found no evidence of any significant relationship (e.g. Gray *et al.* 1996; Zabel, 1998; Lepkowski and Couper, 2002; Nicoletti and Peracchi, 2005), it can probably be concluded that income is relatively unimportant for attrition.

Labour Force Status

It might be expected that, compared with the unemployed and the economically inactive, employed persons will be both harder to contact, since they spend more time away from the home, and more likely to refuse, given the greater opportunity cost of their time. The analyses of pooled data from the ECHP by Nicoletti and Peracchi (2005) and Watson (2003) both report evidence supportive of this hypothesis, with survey re-interview rates highest among the economically inactive. The country-specific analysis of these same data by Behr *et al.* (2005), however, found labour market inactivity significantly and positive associated with survey response in only 4 of the 14 participant countries. Very differently, analysis of PSID data suggests no strong association between employment status and attrition. The analysis of Fitzgerald *et al.* (1998) initially revealed higher attrition rates among nonworkers than among workers, but this effect became insignificant in the presence of other controls. Zabel (1998) also reported an insignificant relationship. In contrast, and counter to expectations, Gray *et al.* (1996) found attrition rates to be lowest among the employed, the result of both lower refusal rates and a greater likelihood of tracing employed respondents. Similarly, Lepkowski and Couper (2002) found employment to be positively and strongly associated with contact propensity and positively but weakly associated with cooperation propensity in one of their samples (in the other, the relationships were insignificant). To further confuse matters, Nicoletti and Buck (2004) reported a significantly higher cooperation propensity among the economically inactive in BHPS data, but significantly lower contact probabilities for both the unemployed and the inactive in GSOEP data.

Location

Finally, it is common to allow for regional differences in survey response, with many studies distinguishing between urban and rural localities. The usual expectation is that residents in large cities will be both less available and harder to reach (e.g. due to the security features of their housing) (Groves and Couper, 1998, p. 85). Social isolation explanations for survey participation also suggest that cooperation rates might be superior in smaller rural communities. For the most part the evidence is consistent with these hypotheses, with numerous studies reporting higher survey attrition rates among sample members living in urban locations (Kalton *et al.*, 1990; Gray *et al.*, 1996; Burkam and Lee, 1998; Fitzgerald *et al.*, 1998; Zabel, 1998). Only Lepkowski and Couper (2002) report contrary evidence.

10.3 PREDICTING RESPONSE IN THE HILDA SURVEY

We now examine the experience with sample attrition in the HILDA Survey in order to test some of the hypotheses and relationships summarized in Section 10.2. There are at least three reasons why the HILDA Survey is well suited to this task. First, it has experienced significant sample attrition over its first four waves (see Section 10.3.1). Second, the data available explicitly distinguish between the contact and cooperation stages. Third, and perhaps most significantly, the range of explanatory variables available exceeds what has been reported in any previous analysis of attrition from a longitudinal survey.

10.3.1 The HILDA Survey Data

The HILDA Survey is a nationwide household panel survey with a focus on employment, income and the family. Modelled on household panel surveys undertaken in other countries, and described in more detail in Watson and Wooden (2004), it began in 2001 with a large national probability sample of Australian households occupying private dwellings. All members of those responding households in wave 1 form the basis of the panel to be pursued in each subsequent wave (though interviews are only conducted with those household members aged 15 years or older), with each wave of interviewing being approximately one year apart. Like many other household panels (including the PSID, the GSOEP and the BHPS), the sample is extended each year to include any new household members resulting from changes in the composition of the original households. With the exception of children of original sample members (OSMs) and persons who have a child with an OSM, new sample members only remain in the sample for as long as they live with an OSM. During waves 1 to 4, households were paid either A$20 or A$50 each year they participated, with the higher amount only paid when interviews were completed with all in-scope household members.

After adjusting for out-of-scope dwellings and households and multiple households within dwellings, the number of households identified as in-scope in wave 1 was 11 693. Interviews were completed with all eligible members at 6872 of these households and with at least one eligible member at a further 810 households. Within the 7682 households at which interviews were conducted, 13 969 persons were successfully interviewed.

Details about the evolution of the responding sample over the first four waves are provided in Table 10.1. This table shows that 10 565, or 76 %, of those persons initially interviewed in wave 1 were re-interviewed in wave 4. If deaths and movements out of scope are excluded the four-wave sample retention rate rises to 78 %. The rates of nonresponse among persons joining the sample at later waves appears much larger, but this is largely because many of these new sample members are only added to the sample on a temporary basis. The total wave-on-wave attrition rates (calculated as the proportion of in-scope previous wave respondents who did not provide an interview) for waves 2, 3 and 4 were 13.2 %, 9.6 % and 8.4 %, respectively.

Table 10.1 Individual response (*N*) by wave, HILDA Survey.

Wave first interviewed	Wave 1	Wave 2	Wave 3	Wave 4
Wave 1	13 969	11 993	11 190	10 565
Wave 2	—	1048	705	594
Wave 3	—	—	833	543
Wave 4	—	—	—	706
Total	13 969	13 041	12 728	12 408

10.3.2 Estimation Approach

Pooled data from the first four waves of the HILDA Survey are now used to estimate a model that employs information about the respondents and their interviews in wave $t-1$ to predict response at wave t. Earlier we argued that the response process involves three stages. Distinguishing between location and contact, however, is empirically difficult in the HILDA Survey data. We thus model survey response as the occurrence of two sequential events: establishing contact with the sample member; and obtaining the cooperation of the sample member once contacted. Our model is thus essentially the same as that used by both Nicoletti and Buck (2004) and Nicoletti and Peracchi (2005).

We estimate three separate model specifications. Model I assumes the two events are independent and thus two separate probit equations are estimated. Following Nicoletti and Peracchi (2005), Model II relaxes this assumption and allows for conditional correlation in the error terms in the two equations. This is achieved by estimating a bivariate probit with sample selection. Model III again imposes the assumption of conditional independence, but following Nicoletti and Buck (2004) an unobserved random effect for the interviewer is introduced into both probit specifications.

The units of analysis are all individuals who were interviewed at wave $t-1$ and deemed as in-scope at wave t (i.e. units are individual-wave combinations). Individuals were deemed as out-of-scope if they are known to have died or moved abroad, or were new (i.e. temporary) sample members who no longer live with an OSM. All wave 1 nonrespondents are excluded since no details of nonrespondents at wave 1 are known. The pooled dataset contains 38 831 observations on 15 313 individuals (2423 are observed once only, 2262 are observed twice and 10 628 are observed three times). Of these 38 831 person-wave observations, 38 118 were contacted in the next wave and 34 751 were interviewed in the next wave. As many individuals from identical or like households are observed across multiple waves, the estimated standard errors in Models I and II assume that the outcome variables are correlated across observations on individuals from the same wave 1 households, but are independent across individuals from different households.[5]

10.3.3 Explanatory Variables

The probability of making contact with an individual is assumed here to be a function of: whether the individual has changed address; the likelihood that the respondent is at home when the interviewer calls; the willingness of the individual to be found (as in Lepkowski and Couper, 2002); and interviewer effort and knowledge.

The first influence is relatively straightforward to measure and represented by a variable identifying whether the individual relocated between survey waves.

[5] This requires linking all individual respondents in waves 2–4 to a wave 1 household. For some households this assumption is unrealistic. The alternative assumption would be to only allow for correlated errors within individuals and not households. The assumption adopted here results in larger standard errors on most estimates.

The likelihood of finding the respondent at home is assumed to be a function of the number of calls made by the interviewer to the household in the previous wave.[6] In addition, it is expected to be correlated with various respondent characteristics, including: age, sex, marital status, the number of people in the household and the age of the co-residents (the presence of young children is expected to be associated with a greater likelihood of finding the respondent at home);[7] employment and labour force status; the type of dwelling; home ownership (i.e. living in rental accommodation compared with living in a home that the respondent owned and was purchasing); and the presence of a serious long-term health condition (defined as any health condition or disability that had lasted or was expected to last at least six months and prevented the respondent from undertaking any form of work). Area characteristics might also be relevant. We thus include a set of geographic dummies that identify whether the respondent lives in one of the major cities and, if not, how remote they are from a major urban centre.[8] Also included is a measure of the relative socio-economic disadvantage of the neighbourhood in which the respondent lives.[9]

Willingness to be found is represented by a range of variables describing the experience of the previous interview. We hypothesize that being from a partially responding household (at least one other member in the household refused to participate in the survey), not returning the self-completion questionnaire[10] and being assessed by the interviewer as relatively uncooperative, suspicious and not having a good understanding of the questions will all be negatively associated with the likelihood of making contact at the next wave. Willingness to be found might also be a function of respondent characteristics, most notably country of birth and English language ability.

We are unable to include any direct measures of interviewer effort but, following Nicoletti and Buck (2004), include a measure of the interviewer's workload in wave t (the number of previous wave respondents allocated to the interviewer at the start of the fieldwork period). Nicoletti and Buck hypothesized that large workloads will reflect overworked interviewers and thus be negatively associated with both the probability of contact and response. In the HILDA Survey, however, better interviewers tend to be allocated larger workloads, which should work to offset this effect. We thus experiment with both linear and quadratic specifications for this variable. We also include variables

[6] Total number of calls consists not only of the number of calls needed to make contact, but additionally the number of calls to complete all interviews. However, the latter component can be largely explained by other variables in the model such as household size and economic activity status.

[7] The presence of older household members might also be expected to increase the likelihood of the respondent being at home. We, however, experimented with a measure of the presence of household members over the age of 64 years and then 74 years and could find no evidence of any significant association with contact probabilities.

[8] The major cities are the mainland State capitals – Sydney, Melbourne, Brisbane, Perth, Adelaide – as well as Canberra, Newcastle, the Central Coast region of New South Wales, Wollongong, the Gold Coast in southern Queensland and Geelong.

[9] Designed by the Australian Bureau of Statistics (2003), this variable has a mean of 1000 and a standard deviation of 100. The variable, however, is designed only to have ordinal meaning and so we have divided cases into quintiles, with the lowest quintile being the most disadvantaged.

[10] All persons completing a personal interview are also given an additional self-completion questionnaire. Interviewers attempt to collect the completed questionnaire at subsequent visits to the household, but where this is not possible respondents are asked to return it by mail. The proportion of interviewed respondents who return it (completed) averaged 92 % over the first four waves.

identifying whether the interviewer conducting the interview at wave t is the same as at previous waves. It is expected that using the same interviewer as in the previous wave should enhance the probability of making contact, given the interviewer's prior knowledge of the household. We interact interviewer continuity with the survey wave in an attempt to identify whether the impact of interviewer continuity changes with the duration of that continuity.

The probability of an individual providing an interview once they have been contacted is a function of both their ability and willingness to respond. The list of variables used to proxy these influences, however, includes most of the variables included in the contact model.[11] In addition, we also include measures of: the length of the personal interview; the length of the household interview; the proportion of missing responses in each section of the questionnaire; whether the interview was conducted by telephone; whether an indigenous Australian; educational attainment; and equivalized household income.

10.3.4 Results

The results of our model estimation are presented in Table 10.2. Before focusing on specific coefficients we make four general observations. First, a Wald test suggests that the assumption of conditional independence should be rejected ($\rho = 0.356$; chi-squared = 7.47; $P = 0.006$). The estimated coefficients, however, are little different when the conditional independence assumption is relaxed. Second, unobserved interviewer effects, while statistically significant, are relatively small in magnitude. In both the contact and response models the estimated proportion of the total variance attributable to an unobserved random interviewer effect is less than 7 %.[12] The comparable percentages reported by Nicoletti and Buck (2004) varied from 13 % to 51 % in the contact models and from 13 % to 31 % in the response models. The implication again is that focusing on Model I results will not be misleading. Third, most of the covariates (but not all) have coefficients that are either in line with expectations or are, *ex post*, intuitively sensible (and, given the large sample size, most are highly significant). They are also extremely robust to model specification. Fourth, despite the array of significant coefficients, the overall explanatory power of these models is relatively poor. This is a desired outcome and presumably reflects the large random component in survey nonresponse.

What then do our estimates reveal? Focusing first on the probability of making contact, we observe the following:

(1) The optimal interviewer workload for maximizing contact rates is 124 previous wave respondents. With larger or smaller workloads, the likelihood of contact is reduced.

(2) Interviewer continuity does not appear to be of any great benefit to contact probabilities. Only by wave 4 is there any evidence that interviewer continuity is

---

[11] To help identify Model II, dwelling type (in addition to the variable identifying individuals that move) was excluded from the response equations in all models. It was not significant when included in the response equation in Model I.

[12] The estimated values of ρ (with standard errors in parentheses) were as follows – contact: $\rho = 0.067$ (0.016); response: $\rho = 0.062$ (0.009).

beneficial for contact probabilities, and even then the positive differential is both small (statistically insignificant) and restricted to cases where the interviewer has been visiting the same respondent for all four survey waves.

(3) Indicators of the interview experience tend to be strong predictors of response at the next wave. The number of calls made, belonging to a part-responding household and not returning the self-completion questionnaire in the previous wave are all negatively associated with making contact with the respondent in the next wave. A cooperative respondent is also much more likely to be contacted at the next wave.

(4) Consistent with other studies, moving house is a strong negative predictor of the propensity to make contact with a respondent in the next wave. The estimates in Model I suggest that the mean predicted probability of making contact with a mover is 95.5 %, compared with 99.1 % for nonmovers.

(5) Renters are typically harder to establish contact with than homeowners. This negative effect is even greater if living in a unit or flat.[13]

(6) Contact probabilities are higher for women, married persons and English speakers, and rise with age. The presence of a severe long-term health problem attracts a large positive coefficient but the estimate is very imprecise and not statistically significant.

(7) Counter to expectations, the number of children is a negative predictor of contact, whereas the number of adults in the household has no significant association with the likelihood of making contact. The negative effect of children is a perplexing result but nevertheless is rigorous to alternative specifications. For example, similar results are obtained when using a simple dummy variable representing the presence of children, or when disaggregating children based on their age.

(8) With the exception of persons working very long hours (55 hours or more per week), it is *easier* establishing contact with employed persons than with the unemployed and other nonworkers. This finding stands in contrast to results often reported in cross-sectional surveys and possibly reflects more extensive social networks among the employed, which, in turn, makes it easier to trace and contact sample members.

(9) Clear straightforward associations with location are a little hard to find, though it is very clear that contact rates are, other things equal, relatively low in the remotest parts of Australia. This almost certainly reflects the high cost of getting interviewers to these areas. We also find that people living in the areas of least socio-economic disadvantage have significantly higher contact rates.

One other influence that is expected to influence contact probabilities, but not included in Table 10.2, is the amount of contact information collected at the preceding wave. In particular, apart from the usual contact details (e.g. mobile telephone numbers, e-mail addresses, business contact details), all respondents were asked to provide contact details for friends or relatives who might know their whereabouts in the event that we cannot easily find them in the future. We thus included the number of names provided in a separate model (not reported here) using data from waves 2 and 3 (this information is not available for wave 1). Respondents who provided two contacts instead of one were

[13] As might be expected, contact rates are lowest for people living in ‘other dwellings’, which consists mainly of highly mobile structures such as caravans, tents and houseboats. These, however, represent a very small fraction of the sample (less than 1 %).

Table 10.2 Coefficients for contact and response probit models, HILDA survey.

Variable	Contact			Response		
	Model I	Model II	Model III	Model I	Model II	Model III
Survey design features						
Interviewer workload (/10^2)	0.624***	0.629***	0.625**	0.203	0.245	0.287
Interviewer workload squared (/10^4)	−0.251**	−0.253**	−0.281**	−0.009	−0.026	−0.100
Interviewer continuity (base = T = 2, ivwr not same as last wave)						
T = 2, ivwr same as last wave	−0.069	−0.062	−0.054	0.013	0.015	−0.043
T = 3, ivwr not same as last wave	0.418***	0.409***	0.406***	−0.002	0.016	0.001
T = 3, ivwr same as last wave only	0.302***	0.303***	0.274***	0.071	0.087	0.046
T = 3, ivwr same as last 2 waves	0.345***	0.351***	0.359***	0.203***	0.219***	0.147***
T = 4, ivwr not same as last wave	0.502***	0.489***	0.502***	0.021	0.041	0.047
T = 4, ivwr same as last wave only	0.412***	0.411***	0.368***	0.140***	0.159***	0.112**
T = 4, ivwr same as last 2 waves only	0.441***	0.449***	0.412***	0.266***	0.283***	0.246***
T = 4, ivwr same as last 3 waves	0.647***	0.639***	0.669***	0.379***	0.400***	0.323***
Previous wave interview situation						
Number of calls made	−0.022***	−0.024***	−0.022***	−0.029***	−0.029***	−0.030***
Part-responding household	−0.215***	−0.218***	−0.214***	−0.530***	−0.535***	−0.538***
Cooperative	0.336***	0.353***	0.339***	0.307***	0.320***	0.328***
Understanding	0.085	0.091	0.081	0.099**	0.101**	0.093**
Suspicious	−0.060	−0.067	−0.058	−0.272***	−0.271***	−0.298***
Didn't return SCQ	−0.372***	−0.373***	−0.385***	−0.382***	−0.397***	−0.399***
Prop. missing in labour force section				−1.115	−1.124	−1.527
Prop. missing in income section				−2.082***	−2.059***	−2.009***
Prop. missing in family/relationship section				−0.202	−0.214	−0.166
Prop. missing in ivwr obs section				−0.823	−0.768	−0.537
Prop. missing in special section				0.108	0.120	0.204

Table 10.2 (Continued).

Variable	Contact			Response		
	Model I	Model II	Model III	Model I	Model II	Model III
Prop. missing in satisfaction/moving section				– 0.427	– 0.390	– 0.277
Prop. missing in SCQ if returned				– 1.029***	– 1.017***	– 1.021***
Prop. missing in childcare/housing				– 0.758**	– 0.754**	– 0.659**
Prop. missing in household roster				1.132	1.119	1.440
PQ length (/10^2)				0.644**	0.625**	0.601*
PQ length squared (/10^4)				– 0.848**	– 0.829**	– 0.783**
PQ length missing				0.170	0.162	0.133
HQ length (/10^2)				0.897*	0.891*	0.893**
HQ length squared (/10^4)				– 0.678	– 0.680	– 0.754
HQ length missing				0.228**	0.226**	0.239***
Telephone interview				0.156*	0.156*	0.159**
Respondent characteristics						
Moved	– 0.754***	– 0.757***	– 0.760***			
Dwelling type (base = Separate/ semi-detached house)						
Unit/flat	– 0.112*	– 0.108*	– 0.145**			
Nonprivate dwelling	– 0.059	– 0.052	– 0.133			
Other dwelling	– 0.397**	– 0.411**	– 0.378***			
Missing dwelling type	0.309	0.314	0.322			
Renter	– 0.230***	– 0.223***	– 0.225***	– 0.003	– 0.027	– 0.011
Female	0.122***	0.120***	0.121***	0.006	0.011	0.008
Age (/10)	0.065***	0.066***	0.069***	0.183***	0.188***	0.186***
Age squared (/10^2)				– 0.018***	– 0.018***	– 0.018***
Country of birth (base = Australia)						
Main English-speaking country	– 0.134**	– 0.133**	– 0.136**	– 0.068*	– 0.072*	– 0.082**
Other o/s; Speaks English well	– 0.262***	– 0.269***	– 0.268***	– 0.160***	– 0.169***	– 0.157***
Other o/s; Not speak English well	– 0.499***	– 0.493***	– 0.499***	– 0.221***	– 0.237***	– 0.212***

Indigeneous Australian				– 0.033	– 0.033	– 0.038
Marital status (base = Married)						
Defacto	– 0.252***	– 0.245***	– 0.243***	0.029	0.018	0.036
Separated	– 0.320***	– 0.316***	– 0.339***	0.139**	0.125*	0.143**
Divorced	– 0.366***	– 0.359***	– 0.390***	0.063	0.050	0.056
Widowed	– 0.035	– 0.031	– 0.043	0.113*	0.111*	0.103*
Single	– 0.346***	– 0.344***	– 0.352***	0.052	0.038	0.044
Number of children aged 0–14	– 0.050**	– 0.048**	– 0.052***	– 0.001	– 0.003	– 0.001
Number of adults	– 0.003	– 0.005	– 0.004	– 0.108***	– 0.107***	– 0.109***
Education (base = Year 11 and below)						
Year 12				0.061**	0.059*	0.057*
Certificate				0.060**	0.059**	0.052*
Diploma				0.171***	0.168***	0.167***
Graduate				0.265***	0.262***	0.258***
Equivalized household income ($/10^5$)				0.045	0.043	0.037
Equivalized household income squared ($/10^{10}$)				– 0.009	– 0.009	– 0.009*
Employment / labour force status (base = Not in labour force)						
Unemployed	– 0.084	– 0.087	– 0.075	– 0.033	– 0.044	– 0.034
Employed part time (1–34 hrs)	0.200***	0.201***	0.213***	– 0.002	0.006	0.002
Employed full time (35–54 hrs)	0.187***	0.191***	0.197***	– 0.135***	– 0.128***	– 0.131***
Employed full time (55 + hrs)	0.068	0.064	0.083	– 0.142***	– 0.138***	– 0.140***
Area of residence (base = Major city: Sydney)						
Major city: Melbourne	0.022	0.024	– 0.037	– 0.037	– 0.034	0.060
Major city: Brisbane	0.042	0.031	– 0.038	0.088	0.088	0.059
Major city: Adelaide	0.189*	0.175*	0.232*	0.024	0.032	– 0.038
Major city: Perth	– 0.169*	– 0.162*	– 0.303***	0.042	0.036	0.031
Major city: other	0.202*	0.190*	0.179	– 0.016	– 0.010	0.056
Inner regional	– 0.061	– 0.063	– 0.094	0.048	0.044	0.067
Outer regional	– 0.114	– 0.117	– 0.138	0.032	0.026	0.026
Remote and very remote	– 0.352***	– 0.360***	– 0.423***	– 0.036	– 0.054	– 0.035

Table 10.2 (Continued).

Variable	Contact			Response		
	Model I	Model II	Model III	Model I	Model II	Model III
Index of disadvantage (base = Lowest quintile)						
2nd lowest quintile	– 0.030	– 0.021	– 0.009	– 0.025	– 0.026	– 0.009
Middle quintile	0.062	0.069	0.069	– 0.056	– 0.053	– 0.066*
2nd highest quintile	0.080	0.086	0.084	– 0.039	– 0.036	– 0.025
Highest quintile	0.147**	0.159**	0.169**	0.036	0.041	0.029
Serious long-term health condition	0.347	0.353	0.356	– 0.216**	– 0.204**	– 0.227**
Constant term	1.567***	1.544***	1.682***	0.656***	0.582***	0.698***
Pseudo log-likelihood	– 2745	– 13054	– 2722	– 10314	– 13054	– 10133
Pseudo R^2	0.228			0.090		
N	38778	38756	38761	38043	38756	38026

Notes: ivwr = interviewer; SCQ = Self-Completion Questionnaire; PQ = Person Questionnaire; HQ = Household Questionnaire; o/s = overseas.
*$0.10 \geq p > 0.05$, **$0.05 \geq p > 0.01$, ***$0.01 \geq p$

found to be more likely to be contacted in the next wave, while respondents who refused to provide any contacts were less likely to be contacted in the next wave. However, contrary to the BHPS experience (Laurie *et al*, 1999), neither of these differences is statistically significant.

Turning now to the likelihood of response conditional on making contact, we make the following observations:

(1) Unlike the GSOEP and BHPS experience, the interviewer workload does not appear to have had a detrimental impact on the likelihood of gaining an interview with a respondent. Indeed, interviewers with the highest workloads have tended to achieve the best response rates, but these differences are statistically insignificant. As hypothesized earlier, we believe that the absence of a negative association reflects the way work is allocated to interviewers on the HILDA Survey.

(2) Interviewer continuity is positively associated with response and the strength of this relationship increases with each wave the interviewer remains the same. Further, this result is only slightly weakened by controlling for a random interviewer effect in Model III. Nevertheless, the magnitude of this effect is relatively small. Based on the parameter estimates in Model III, the mean predicted probability of response in wave 4 when the interviewer is not the same as in wave 3 is 91.3 %. This compares with 92.2 % if the interviewer is the same for the last two waves, 93.8 % if the same for the last three years and 94.6 % if the same for all four waves.

(3) The number of calls made at wave $t - 1$ has the expected negative association with survey response at wave t. Thus a respondent from a household that required just one call at $t - 1$ is calculated to have a mean predicted response of 92.8 %. If five calls are required (the mean number in the sample) the predicted response rate falls to 91.2 %. At households requiring as many as 20 calls (the maximum in the data is 24 calls) the response probability at wave t is just 83 %.

(4) Possibly the most important predictor of response in the next wave is whether the household was partly responding in the last wave. The estimates in Model I indicate the mean predicted probability of response for individuals in partly-responding households is 81.9 %, compared with 92.1 % for individuals in fully responding households.

(5) Another important predictor is whether the individual returned the self-completion questionnaire after their interview in the last wave or not. The mean predicted probability of response when this questionnaire was returned is 91.8 %, compared to 85 % when it was not returned.

(6) As has been found in other studies of attrition, the interviewer's assessment of the respondent in one wave is predictive of response in the next. Respondents who were cooperative, not suspicious and appeared to understand the questions were more likely to provide an interview in the next wave.

(7) Also consistent with most previous research, the proportion of missing data items tends to be negatively associated with the likelihood of responding in the next wave. However, this effect is not uniform across all components of the survey. Specifically, these negative associations are only significant when the item

nonresponse occurs in the income or childcare and housing sections of the interview, or if it occurs in the self-completion questionnaire.

(8) In contrast to the analysis of attrition in the NLSY, SIPP and ECHP, conducting the previous wave interview by telephone was not predictive of attrition in the next wave. Indeed, the coefficient is positive (though only of marginal significance).

(9) The highest level of participation in the next wave is observed with a personal interview of around 38 minutes, which is slightly in excess of the duration targeted (35 minutes). The implication thus is that both very short and very long interviews will result in relatively low rates of cooperation next wave. We believe that short interviews signal either the lesser relevance of the survey content to the respondent or respondent disinterest in the survey. Long interviews, on the other hand, obviously impose a greater time cost on the respondent. Household interview length, on the other hand, is positively associated with the likelihood of participation in the next wave, a result we believe reflects the importance of topic salience.

(10) Response probabilities are lowest among both the young and elderly (people aged around 50 are the most cooperative). Response probabilities are also relatively low for persons born overseas (and especially those who speak English poorly), the least educated and those with a serious long-term health condition, and decline with the number of adults living in the household. Sex, marital status, the number of children, Aboriginality, (equivalized) household income and location are all insignificant predictors of the probability of response.

(11) While employed people are somewhat easier to establish contact with, they are less likely to respond if they work full-time hours (35 or more per week).

10.4 CONCLUSION

There are perhaps four main conclusions that can be distilled from the review and empirical analysis reported here. First, it cannot automatically be assumed that experience of nonresponse in cross-sectional surveys will always be relevant when considering the correlates with attrition from longitudinal surveys. This is most obvious with respect to the role of the interview experience. In single wave surveys the way the interview is conducted can only adversely affect interview duration and not whether an interview is obtained. In contrast, the evidence presented here suggests that the respondent's perception of the interview experience is possibly the single most important influence on cooperation in future survey waves.

Second, the factors that influence contact are quite distinct from those that influence cooperation, and the empirical modelling process should reflect this. That said, it should still be borne in mind that contact rates in most well-resourced surveys are likely to be very high (in the HILDA Survey they average around 98 % each wave). Inevitably it is the cooperation stage where the risk of sample loss is greatest.

Third, while this chapter was partly motivated by interest in identifying survey design features that influence response probabilities, it is actually very difficult to test the influence of design features without controlled experiments. Thus, while a number of potentially important influences were identified in our literature review (e.g. a variety of tracking procedures, the length of the fieldwork period, respondent incentives, multiple modes and interview length), we were unable to say anything directly about their efficacy from the HILDA Survey experience. Our analysis, however, is not without any implications for survey design. Most obviously, we found evidence that quite large interviewer workloads are beneficial for response. This is almost certainly a function of the practice in the HILDA Survey of assigning more work to better performing interviewers. Our results also support the contention that interviewer continuity improves response, though the magnitude of the effect is small and restricted to response conditional on contact. More surprising, we found no evidence that obtaining more contact names at one wave significantly improves either contact or response probabilities at the next, though the direction of effect is in the expected direction.

Fourth, while there is undoubtedly (and thankfully) a large random component to survey nonresponse, it is nevertheless clear that there are strong associations between many observable characteristics of both respondents and the interview process and experience that are predictive of nonresponse. Indeed, arguably the most striking feature of the analysis reported here is just how many different variables are relevant. Such information potentially can be used to assist survey managers in tailoring approaches and targeting special measures at the next wave. It also can provide variables for inclusion in attrition models used in the construction of population weights or as instruments at the analysis stage.

REFERENCES

Atrostic, B. K., Baytes, N. and Silberstein, A. (2001). Nonresponse in U.S. Government household surveys: consistent measures, recent trends, and new insights. *Journal of Official Statistics*, 17, 209–226.

Australian Bureau of Statistics (2003). *Information Paper: Census of Population and Housing – Socio-Economic Indexes for Areas, Australia* (ABS cat. no. 2039.0). Canberra: Australian Bureau of Statistics.

Becketti, S., Gould, W., Lillard, L. and Welch, F. (1988). The Panel Study of Income Dynamics after fourteen years: an evaluation. *Journal of Labor Economics*, 6, 472–492.

Behr, A., Bellgardt, E. and Rendtel, U. (2005). Extent and determinants of panel attrition in the European Community Household Panel. *European Sociological Review*, 21, 489–512.

Bogen, K. (1996). The effect of questionnaire length on response rates – a review of the literature. In *Proceedings of the American Statistical Association, Survey Research Methods Section* (pp. 1020–1025). Washington, DC: American Statistical Association.

Branden, L., Gritz, R. M. and Pergamit, M. (1995). The effect of interview length on attrition in the National Longitudinal Study of Youth, *National Longitudinal Surveys Discussion Paper,* No. 28. Washington, DC: Bureau of Labor Statistics, US Department of Labor. http://www.bls.gov/osmr/nls_catalog.htm.

Buck, N. (2000). Using panel surveys to study migration and residential mobility. In D. Rose (Ed.), *Researching Social and Economic Change: The Uses of Household Panel Studies.* London: Routledge.

Bureau of Labour Statistics (2003). *NLS Handbook, 2003.* Washington, DC: US Department of Labour, Bureau of Labour Statistics.

Burkam, D. T. and Lee, V. E. (1998). Effects of monotone and nonmonotone attrition on parameter estimates in regression models with educational data. *Journal of Human Resources*, 33, 555–574.

Campanelli, P. A. and O'Muircheartaigh, C. (2002). The importance of experimental control in testing the impact of interviewer continuity on panel survey nonresponse. *Quality and Quantity*, 36, 129–144.

Couper, M. P. and Groves, R. M. (1992). The role of the interviewer in survey participation. *Survey Methodology*, 18, 263–277.

De Leeuw, E., and De Heer, W. (2002). Trends in household survey nonresponse: a longitudinal and international comparison. In R. M. Groves, D. A. Dillman, J. L. Eltinge and R. J. A. Little (Eds), *Survey Nonresponse* (pp. 41–54). New York: John Wiley & Sons, Inc.

Fitzgerald, J., Gottschalk, P. and Moffitt, R. (1998). An analysis of sample attrition in panel data: the Michigan Panel Study of Income Dynamics. *Journal of Human Resources*, 33, 251–299.

Foster, K. (1998). Evaluating nonresponse on household surveys, *GSS Methodology Series* No. 8. London: Government Statistical Service.

Gray, R., Campanelli, P., Deepchand, K. and Prescott-Clarke, P. (1996). Exploring survey non-response: the effect of attrition on a follow-up of the 1984–85 health and life style survey. *The Statistician*, 45, 163–183.

Groves, R. M. and Cooper, M. P. (1998). *Nonresponse in Household Interview Surveys*. New York: John Wiley & Sons, Inc.

Hawkes, D. and Plewis, I. (2006). Modelling non-response in the National Child Development Study. *Journal of the Royal Statistical Society (Series A)*, 169, 479–492.

Hill, D. H. and Willis, R. J. (2001). Reducing panel attrition: a search for effective policy instruments. *Journal of Human Resources*, 36, 416–438.

Japec, L. and Lundqvist, P. (1999). Interviewer strategies and attitudes. Paper presented at the *International Conference on Survey Non-Response*, October 28–31, Portland, Oregon.

Kalsbeek, W. D., Yang, J. and Agans, R. P. (2002). Predictors of nonresponse in a longitudinal survey of adolescents. In *Proceedings of the American Statistical Association, Survey Research Methods Section* (pp. 1740–1745). Washington, DC: American Statistical Association.

Kalton, G., Lepkowski, J., Montanari, G. E. and Maligalig, D. (1990). Characteristics of second wave nonrespondents in a panel survey. In *Proceedings of the American Statistical Association, Survey Research Methods Section* (pp. 462–467). Washington, DC: American Statistical Association.

Laurie, H., Smith, R. and Scott, L. (1999). Strategies for reducing nonresponse in a longitudinal panel survey. *Journal of Official Statistics*, 15, 269–282.

Lee, E., Hu, M. Y. and Toh, R. S. (2004). Respondent non-cooperation in surveys and diaries: an analysis of item non-response and panel attrition. *International Journal of Market Research*, 46, 311–326.

Lepkowski, J. M. and Couper, M. P. (2002). Nonresponse in the second wave of longitudinal household surveys. In R. M. Groves, D. A. Dillman, J. L. Eltinge and R. J. A. Little (Eds), *Survey Nonresponse* (pp. 259–272). New York: John Wiley & Sons, Inc.

Lillard, L. A. and Panis, C. W. A. (1998). Panel attrition from the Panel Study of Income Dynamics: household income, marital status, and mortality. *Journal of Human Resources*, 33, 437–457.

Loosveldt, G., Pickery, J. and Billiet, J. (2002). Item nonresponse as a predictor of unit nonresponse in a panel survey. *Journal of Official Statistics*, 18, 545–557.

Martin, J. and Beerten, R. (1999). The effect of interviewer characteristics on survey response rates. Paper presented at the *International Conference on Survey Non-Response*, October 28–31, Portland, Oregon.

Nicoletti, C. and Buck, N. (2004). Explaining interviewee contact and co-operation in the British and German household panels. In M. Ehling and U. Rendtel (Eds), *Harmonisation of Panel Surveys and Data Quality*. (pp. 143–166). Wiesbaden: Statistisches Bundesamt.

Nicoletti, C. and Peracchi, F. (2005). Survey response and survey characteristics: microlevel evidence from the European Community Household Panel. *Journal of the Royal Statistical Society, Series A*, 168, 763–781.

Olsen, R. J. (2005). The problem of respondent attrition: survey methodology is key. *Monthly Labour Review*, 128, 63–70.

Ribisl, K. M., Walton, M. A., Mowbray, C. T., Luke, D. A., Davidson II, W. S. and Bootsmiller, B. J. (1996). Minimizing participant attrition in panel studies through the use of effective retention and tracking strategies: review and recommendations. *Evaluation and Program Planning*, 19, 1–25.

Schräpler, J. P. (2002). Respondent behavior in panel studies – a case study for income-nonresponse by means of the German Socio-Economic Panel (GSOEP). Paper presented at the *5th International Conference of German Socio Economic Panel Users*, July 3– 4, Berlin.

Serfling, O. (2004). The interaction between unit and item nonresponse in view of the reverse cooperation continuum: evidence from the German Socioeconomic Panel (GSOEP). Paper presented at the *6th International Conference of German Socio Economic Panel Users*, June 24–26, Berlin.

Spieß, M. and Kroh, M (2004). Documentation of sample sizes and panel attrition in the German Socio Economic Panel (GSOEP), *Research Note*, No. 28a. Berlin: German Institute for Economic Research (DIW).

Taylor, M. F., Brice, J., Buck, N. and Prentice-Lane, E. (2005). *British Household Panel Survey User Manual Volume A: Introduction, Technical Report and Appendices*. Colchester: University of Essex.

Thomas, D., Frankenberg, E. and Smith, J. P. (2001). Lost but not forgotten: attrition and follow-up in the Indonesia Family Life Survey. *Journal of Human Resources*, 36, 556–592.

Tourangeau, R. (2003). The challenge of rising nonresponse rates. In *Recurring Surveys: Issues and Opportunities*, Report to the National Science Foundation based on a Workshop held on March 28–29.

Voogt, R. J. J. and Saris, W. E. (2005). Mixed mode designs: finding the balance between nonresponse bias and mode effects. *Journal of Official Statistics*, 21, 367–387.

Wadsworth, M. E. J., Butterworth, S. L., Hardy, R. J., Kuh, D. J., Richards, M., Langenberg, C., *et al.* (2003). The life course prospective design: an example of benefits and problems associated with study longevity. *Social Science and Medicine*, 57, 2193–2205.

Waterton, J. and Lievesley, D. (1987). Attrition in a panel study of attitudes. *Journal of Official Statistics*, 3, 267–282.

Watson, D. (2003). Sample attrition between waves 1 and 5 in the European Community Household Panel. *European Sociological Review*, 19, 361–378.

Watson, N. and Wooden, M. (2004). The HILDA Survey four years on. *Australian Economic Review*, 37, 343–349.

Westat (1998). *SIPP Quality Profile 1998* (SIPP Working Paper No. 30). Washington, DC: US Department of Commerce, US Census Bureau.

Westat, in association with Mathematica Policy Research (2001). *Survey of Income and Program Participation Users' Guide* (3rd edn.). Washington, DC: US Department of Commerce, US Census Bureau.

Winkels, J. W. and Withers, S. D. (2000). Panel attrition. In D. Rose (Ed.), *Researching Social and Economic Change: The Uses of Household Panel Studies*, (pp. 79–95). London: Routledge.

Yu, J. and Cooper, H. (1983). A quantitative review of research design effects on response rates to questionnaires. *Journal of Marketing Research*, 20, 36–44.

Zabel, J. E. (1998). An analysis of attrition in the Panel Study of Income Dynamics and the Survey of Income and Program Participation with an application to a model of labour market behavior. *Journal of Human Resources*, 33, 479–506.

CHAPTER 11

Keeping in Contact with Mobile Sample Members

Mick P. Couper and Mary Beth Ofstedal

Survey Research Center, University of Michigan, USA

11.1 INTRODUCTION

Panel surveys are an increasingly important tool for studying a wide variety of social phenomena. They are particularly useful for studying dynamic processes. For this reason, panel attrition is of particular concern, because of the risk of those most likely to experience change being lost to follow-up, potentially producing biased estimates of change measures. While panel attrition was long viewed as a unitary phenomenon, recent research has focused on different sources of panel attrition and their implications for data quality. Lepkowski and Couper (2002) offered a general model of attrition in panel surveys, including the location, contact and cooperation process. This chapter narrows the focus to the first of these, namely the process of locating or tracking[1] sample persons in subsequent waves of a panel survey. A tracking problem is generally triggered by a change of address, telephone number, e-mail address or other contact information. This change in turn is usually triggered by a move – whether of an individual or an entire household – and such moves are often associated with the very dynamics one is interested in measuring in panel surveys. By their very nature, moves are associated with change, and those who move are likely to be different from stayers (nonmovers) on several dimensions. Panel member mobility may not only affect data quality, but the efforts

[1] The terms 'locating,' 'tracking' and 'tracing' are variously used in the literature. We use all three interchangeably.

Methodology of Longitudinal Surveys P. Lynn

expended to find such mobile members may also increase costs. However, panel surveys can – and often do – collect a variety of information that can help predict the likelihood of moving, and reduce the likelihood of location failure given a move, thus mitigating the effect on key survey estimates and costs. To date, however, the problem has received scant attention in the research literature, often being relegated to an operational issue. Our view is that the tracking problem is worthy of research attention, and the purpose of this chapter is to identify gaps in the literature and encourage research on the issue of tracking mobile sample members.

How big a problem is it to locate those who move between waves in a panel survey? Based on their review of the literature, Groves and Hansen (1996, p. 5) concluded that 'with adequate planning, multiple methods, and enough information, time, money, skilled staff, perseverance, and so on, 90 % to 100 % location rates are possible'. Similarly, Call, Otto and Spenner (1982) noted that in most studies it should be possible to locate addresses for all but 2–4 % of the panel members. For example, The German Socio-Economic Panel (SOEP) needed to track 14 % of its sample from 2003–2005, and successfully located 96 % of these. Similarly, the British Household Panel Survey (BHPS) attempted tracking for 15.1 % of the sample from 2003 to 2004, and located 93.7 % of these. Of course, time and resources are not limitless, and even if the proportion of nonlocated units can be minimized it may be costly to do so. For this reason, advancing our understanding of the tracking problem may inform the optimal allocation of resources for these and many other competing activities in panel surveys.

In this chapter, we first offer an overview of the location problem, as distinct from other sources of panel attrition or nonresponse. We then offer a framework for understanding the location propensity to guide fieldwork strategies and identify likely sources of bias. Then we briefly describe two cases studies, examining the extent of the location problem and correlates of (non)location. We end with a brief discussion of the changing role of technology in facilitating or impeding the ability to locate sample persons who have moved or changed contact information, and offer some thoughts for research and practice.

11.2 THE LOCATION PROBLEM IN PANEL SURVEYS

A feature of panel surveys is that they (attempt to) interview the same individuals at multiple time points, regardless of where those persons currently reside. If a sample person has not moved since the previous interview, locating them is generally not a problem and the effort turns to making contact with the sample person at the given location. Thus, locating sample persons is different from making contact with them. The former is a necessary but not sufficient condition for the latter. Once a sample unit has been located and contacted, attention turns to gaining cooperation, a process addressed in detail elsewhere.

The location problem thus involves two related propensities,[2] the first being the propensity to move (or to change contact information) and the second, conditional on

[2] We use the terms 'propensity' and 'likelihood' interchangeably to refer to the conditional probability of an event (e.g. moving, being located) given a set of covariates.

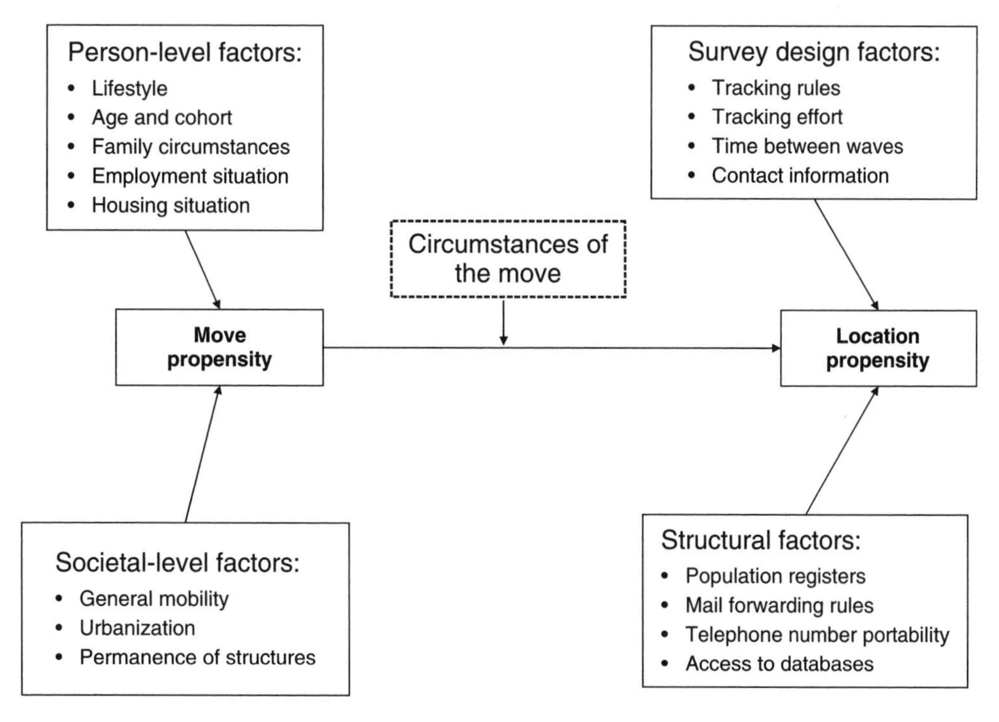

Figure 11.1 Factors affecting tracking and location propensity.

the first, being the propensity to locate a person who has moved. Figure 11.1 offers a framework for understanding the elements of the location problem, in the spirit of Groves and Couper (1998) and Lepkowski and Couper (2002). The figure is not meant to be exhaustive, but illustrates some of the factors at work in the propensity to move and, given a move, the propensity to be located.

11.2.1 The Likelihood of Moving

The extent of geographical mobility varies both within and between countries. For example, 13.7 % of the US population moved in 2004 (US Census Bureau, 2005). The majority of these moves were within the same county (57.8 %), and an additional 20.1 % were within the same state. However, 18.8 % of movers moved to a different state, and 3.3 % moved to a different country. The reasons for moving in 2004 were dominated by housing-related reasons (57.8 % of movers), followed by family-related reasons (24.3 %) and employment-related reasons (17.0 %).

The level of geographic mobility is somewhat lower in most Western European countries. In 1996–1997, Finland and Denmark had annual mobility rates ranging from 11 to 13 %, France, The Netherlands and Belgium had more moderate rates (7–9 %) and Ireland, Italy, Greece, Spain, Portugal and Austria all had rates of 4.5 % or less (Gregg *et al.*, 2004). In England, 10.7 % of the population moved per year on average between 1996 and 1998 (Donovan *et al.*, 2002). About 63 % of the moves were short-distance moves (i.e. within a local authority district), 21 % were medium distance (between local-authority districts in the same region) and the remainder were long distance (between regions or beyond) (Buck, 2000). Reasons for moving were similar to those in the USA, but varied according to the distance of the move. Short-distance moves were generally related to family and housing concerns, medium-distance moves to job changes and long-distance moves to job changes and/or educational pursuits.

The causes and correlates of geographical mobility are an area of research in its own right, and we do not intend an exhaustive review here. Rather, we highlight a few points pertinent to the tracking issue. Whether a person or family will move depends on many circumstances. There are more stable societal-level influences such as residential and job mobility and urbanization. These affect the overall level of moves in a given society. In general, societies with less mobility and greater permanence of structures (housing units, addresses) will exhibit fewer moves. Similarly, macro-level factors such as the differential availability of jobs in different regions, changes in housing costs and other economic changes may increase the number of moves (Gregg *et al.*, 2004). Quality of schools, crime rates and housing density are also important, particularly for families with young children (Frey and Kobrin, 1982).

While the socio-economic climate may predispose people to move, the decision to move is generally made at the level of the individual person or family. The likelihood that a given person will move depends in large part on their current family circumstances, their current life stage, job security, and so on (Rossi, 1955; Speare *et al.*, 1975; De Jong and Gardner, 1981; Landale and Guest, 1985). For example, younger people are more mobile, as are those with higher education levels (Long, 1988). On the other hand, those with more tenuous employment or low-skilled workers may move more frequently to find work. The number of years at a residence, household tenure (i.e. whether the dwelling is owned or rented) and community attachments through family and friends may also influence the likelihood of moving. In later life, retirement, widowhood and major health events are important triggers of residential mobility (Walters, 2000).

One can distinguish between the overall propensity to make a move and the likelihood of a specific move. A person or family may be predisposed towards moving, but the move itself is usually precipitated by a particular set of circumstances. Specific events such as births or deaths, marriage or divorce, job changes and crime victimization affect the likelihood of moving. These proximate factors affect not only the likelihood of moving, but also the circumstances of the move itself. Some moves are planned many months in advance, and the timing of the move and intended destination may be known at the time of the current wave interview. Other moves are more a response to an unanticipated change in family circumstances, employment, health, etc. The latter types of moves are likely to present more challenges for tracking and location than the former.

Moves also vary in the number and type of family members they involve. The move of a child out of the parental home to attend college or to enter the labour market is a quite different event from a relocation involving the whole family. Moves due to divorce or separation involve the creation of two new households, while marriage or cohabitation involves the creation of a single new household from parts of others. Moves due to external factors (e.g. job loss, changes in the housing market, natural disasters) may be less well-anticipated, and may leave fewer traces as a consequence.

The distance of a move will also have consequences for the individuals involved and may be an important factor in locating sample members. Moving to a different house in the same city is likely to leave many social connections intact, while moving across country or abroad may diminish the strength of those ties. On the other hand, some long-distance moves are undertaken for the purpose of living nearer to family members, and may be precipitated by a specific event such as a decline in health or death of a spouse.

11.2.2 The Likelihood of Being Located, Given a Move

Different types of movers are likely to exhibit different propensities for being located following a move, and both the reason for the move and the circumstances of the move are likely to affect the success with which movers are located. For example, a name change in marriage or following divorce may make it more difficult to track and locate someone who has moved since the last panel wave. A move following a bankruptcy may leave fewer traces because of the desire to avoid creditors. Unlisted telephone numbers and other means to ensure privacy may be associated with such life changes.

The degree of family or community attachment may not only affect the likelihood of moving but also provide more data on units that do move. Individuals with large extended families and strong family ties have many potential sources through which their current location can be ascertained. Those with strong community ties are likely to leave many traces to their new address: they are likely to be politically, socially and economically engaged in their new community. Their lives are accessible through public databases such as telephone directories, credit records, voter registration, library registration, membership in churches or religious organizations or children in schools. Socially isolated individuals will be more difficult to track following a move, as will those who withdraw from community attachments. In other words, both the characteristics of the movers and the life events that precipitated the move may have consequences for the ease with which movers are found.

In addition, two other sets of factors are likely to be at work in the location propensity. The first relates to structural factors facilitating location and the second to survey design elements.

Structural or Societal-level Factors

At a structural level, societies differ in features that may affect the resources available for tracking. For example, countries with population registers usually require individuals to update their address within a specified time after a move. To the extent that such registers are available to survey organizations, they provide an accurate and up-to-date source of tracking information. On the other hand in countries such as the USA and the UK filing a change of address is voluntary. The National Change of Address Register (NCOA) in the USA and equivalent registers elsewhere are only as good as the extent to which people comply with providing updates and the alacrity with which they do so. Privacy laws prevailing in a country may limit the amount of publicly available information on the whereabouts of its members. There are a number of commercial vendors in the USA that consolidate information from a variety of sources (e.g. credit card bills, tax rolls, voting records, etc.) and make these available for a fee. Similarly, legislation such as the Telephone Number Portability Act in the USA and the Communications Act in the UK, which permit subscribers to transfer telephone numbers across devices (landline and mobile) as well as across locations, may facilitate the tracking of individuals as the use of such options becomes more widespread. The public climate related to privacy may, in addition, affect the rate with which telephone numbers are listed, e-mail addresses are publicly accessible, and so on. Of course, how these various resources are used is part of survey design, discussed below.

Survey Design Factors

Panel surveys vary on many dimensions, including the tracking or following rules employed, the temporal distance between waves, the time and effort expended in tracking movers, the amount and type of contact information collected in the prior wave, and so on. All of these are likely to affect location propensity.

Other things being equal, the longer the time between data collection waves, the greater the likelihood that sample persons may have moved and the greater the difficulty in tracing those who did (see Duncan and Kalton, 1987). For example, Schejbal and Lavrakas (1995) report that response rates to a second-wave telephone interview were lower for those cases assigned to a 17-month lag (56.1 %) than those assigned to a 14-month lag (65.2 %) when controlling for demographic differences between the groups. Specifically, the percentage of nonworking numbers was higher in the 17-month group (12.1 % versus 5.0 %), as were the cases coded as 'no new number available' (4.9 % versus 2.1 %) and 'another household has number' (8.8 % versus 7.5 %), which are all examples of a change of telephone number status. Hampson *et al.* (2001) report on an effort to track elementary school children originally assessed in the early 1960s in Hawaii. Using an extensive tracking process, they located 85 % of the original sample. Clarridge, Sheehy and Hauser (1978) report on a 17-year follow-up of high school seniors, with a 97.4 % tracing success rate. At the other extreme, panel surveys with monthly or quarterly data collection are likely to have fewer tracking problems.

The mode of data collection likely also has an impact on tracking success. In face-to-face surveys, an empty dwelling or new occupants can lead the interviewer to ask the current residents or neighbours of the whereabouts of the sample members. A nonworking telephone number is much less informative. A mailed survey that is returned without a forwarding address (or not returned), or an e-mail message that is undelivered, similarly indicates a contact problem but is uninformative about the current location of the sample person. For example, Hermann and Grigorian (2005) report a mixed-mode (predominantly mail and web) survey of doctoral recipients, with over 30 % of the sample requiring locating and 11 % of the sample being classified as unlocatable.

The population under study is similarly important to success in locating sample persons. For example, school-leavers, immigrants and other highly mobile or unique populations present special challenges for the tracking process. Indeed, many of the papers on tracking in panel surveys focus on such special populations (e.g. Wright *et al.*, 1995; Coen *et al.*, 1996; McKenzie *et al.*, 1999; Hampson *et al.*, 2001; Menendez *et al.*, 2001; Scott, 2004).

The tracking rules directly affect the need for tracking and the success of such efforts. Some surveys do not track individuals, but rather addresses or housing units. The Current Population Survey in the USA and the Labour Force Survey in the UK are of this type. People who move between waves of data collection are not tracked, but the new residents of the sampled addresses are interviewed. Other surveys limit the range or distance of tracking, for example not following up on those who had moved abroad or out of state. With the increased use of multiple modes of interview for panel surveys, such exclusions may become rarer. However, other things being equal, the more inclusive the tracking rules, the greater the resources expended to locate and interview those who have moved.

Other features of the survey design are more directly related to the process of tracking itself. These include the amount and type of contact information collected in the prior wave, and any efforts to maintain contact with sample members between waves of data collection. Finally, of course, the resources, strategies, time and effort put into the actual tracking operation itself are likely to have important consequences for the success of the tracking. In fact, much of the published literature on tracking focuses on operational aspects of the process. However, this literature is dominated by anecdotal reports of what works and what does not, with very few experimental studies of alternative methods or empirical evaluations of the cost-effectiveness of alternative procedures. Examples of papers describing various strategies for tracking include Coen *et al.* (1996), Ribisl *et al.* (1996) and Scott (2004).

A distinction can be made between prospective (or what Burgess (1989) calls forward tracing) and retrospective tracking methods (see Laurie *et al.*, 1999). The former attempts to update address or contact information prior to the current wave. The latter begins at the point at which the interviewer discovers the panel member is no longer at the designated address or the contact information is no longer valid. We can also distinguish between tracking done in the field by interviewers and tracking done by a centralized team of specialists. While interviewers may have direct access to neighbours, family members, etc., and can observe possible indicators of the move, such tracking is typically done on a case-by-case basis and is thus likely to be expensive. Centralized tracking by a specialized team may make most efficient use of expensive resources such as commercial databases. In addition, databases can be queried for a large number of cases simultaneously, making for a more efficient process.

The particular methods and procedures used vary from country to country, depending on the availability of information. The sequence of steps followed in tracking a case also varies from one organization to the next, and appears to be based on experience rather than empirical research. A good example of the British Household Panel Survey (BHPS) tracking procedures can be found in Laurie *et al.* (1999). Table 11.1 summarizes various tracking strategies used in panel surveys. This list is not exhaustive, but reflects common strategies used in ongoing panel surveys, such as the BHPS, the English Longitudinal Study of Ageing (ELSA), the German Socio-Economic Panel Survey (SOEP), the European Community Household Panel (ECHP) and the Panel Study of Income Dynamics (PSID) and Health and Retirement Study (HRS) in the USA. Most surveys employ a combination of proactive, interviewer or field and centralized tracking procedures, in that order. Confidentiality restrictions differ across countries and surveys, and tracking methods must be in compliance with those restrictions.

It is not our intent to review all of these procedures here, nor are we able to comment on their relative effectiveness. Our point is simply to note that that there is a large (and growing) array of tracking resources and methods available but (to our knowledge) these have not been subject to systematic research about which works best for which types of tracking problems. Understanding more about the predisposing and precipitating circumstances behind a move may be the first step in learning how to use these tools more effectively and efficiently. We believe that prospective methods are likely to be more cost-effective than retrospective ones, given interviewer involvement in the latter. This is especially true if better-targeted efforts are possible. Given this, understanding which panel members are most at risk of a move between waves may help us direct our tracking resources to where they may do the most good.

Table 11.1 Tracking strategies used in panel surveys.

Proactive techniques

- Ask respondents to provide address updates; requests can be made at prior interview, by postcard mailing or e-mail; provide multiple options for respondents to update address (via telephone at toll-free number, prestamped postcard, e-mail, participant website); small incentive may increase compliance
- Obtain address/phone information for one or two contact persons at time of interview
- Record pertinent information from prior interview to help identify respondents that may require tracking (e.g. plans for moving, contact information for alternative residences, successful tracking techniques for prior wave)
- Maintain contact with respondents between waves (via newsletters, greeting cards for special occasions, postcard mailings, mail surveys); returned mail will signal need for tracking

Field/interviewer tracking (face-to-face contacts)

- Obtain new address from present occupants or neighbours at respondent's former address
- Leave letter for present occupants or neighbours to forward to respondent if they are reluctant to provide new address to interviewer
- Send letter to respondent requesting new address and phone number

Field/interviewer tracking (telephone contacts)

- Try all telephone numbers available for respondent (cell phone, phones at primary and any other residences)
- Check current listing for any changes in area/city/region codes
- Check with operator or directory assistance for any changes in phone numbers
- Call people designated by respondent as potential contact person(s); conduct same phone number updates if unable to reach
- Send letter to respondent requesting new address and phone number

Centralized tracking

- Search available databases for respondent, spouse or other known household members or relatives (e.g. Insight is a primary database used in the USA)
- Internet search for respondent name and city, especially for an unusual name
- Check birth/marriage/divorce records if information is available online
- Check property ownership of last known address(es)
- Search records at administrative level (state/region), e.g. voter registration, motor vehicle registration, court records, etc.
- Use available administrative resources to help locate respondents (e.g. health authority, governmental department)

With this in mind, we examine the location problem in the context of two example surveys, both of which are biennial data collection efforts that invest considerable time and effort in locating sample members and maximizing sample retention over time.

11.3 CASE STUDY 1: PANEL STUDY OF INCOME DYNAMICS

The Panel Study of Income Dynamics (PSID), begun in 1968, is a longitudinal study of a representative sample of US individuals (men, women and children) and the family units in which they reside (see http://psidonline.isr.umich.edu/ for details and data; see also Hill, 1991).

From 1968 to 1996, the PSID interviewed members of family units from the original (1968) core sample every year, whether or not they were living in the same dwelling or with the same people. Adults have been followed as they have grown older, and children have been observed as they advance through childhood and into adulthood, forming family units of their own. In 1997, the study changed to biennial data collection. Starting in 1972, an increasing number of interviews were conducted by telephone, with 97 % of interviews done by phone in 2003. Computer-assisted interviewing was introduced in 1993. Unlike its counterparts such as the BHPS and the SOEP, the PSID follows families rather than individuals. The central focus of the PSID is economic and demographic, with substantial detail on income sources and amounts, employment, family composition changes and residential location. We use the 2003 PSID family data to explore the correlates of tracking and location in the 2005 data collection. All analyses use unweighted estimates.

Without the 2005 wave interview data, we do not know for sure whether or not sample members have moved between 2003 and 2005. Thus as an indirect indicator of move status we use any cases that were flagged for tracking on one or more call records during 2005 data collection.[3] Overall, about 18 % of families interviewed in 2003 required tracking in 2005. Of these 1441 families, only 48 were not located after all tracking efforts had been exhausted. The final dispositions of cases by whether tracking was needed are presented in Table 11.2.[4] Families for whom tracking was needed were more likely than those not in tracking to result in a noninterview, due to refusal, inability to locate or other reasons. Of particular note, however, is the high proportion of families for whom tracking was needed who ended up completing an interview in 2005 (89.5 %). This suggests that tracking pays off. However, as will be shown later, tracking can also have high costs.

First, we examine the propensity to require tracking, as an indirect indicator of the propensity to move. We examined a variety of predictors of the need for tracking, and retained those that remained significant in a multivariate logistic regression model. Table 11.3 (column 1) presents the unadjusted percentage of families that needed tracking in 2005, by various characteristics of the family and/or household head as measured in 2003.[5] All of these bivariate associations are statistically significant ($P < 0.001$). An additional set of variables were significantly related to the need for tracking in bivariate analyses, but not when controlling for other variables in a multivariate model. These include tenure (whether the home is owned or rented), ownership of

[3] A case is flagged for tracking if the interviewer learns that the family has moved and no forwarding information is provided, or if the first five call attempts yield no contact. We acknowledge that tracking status is an imperfect indicator of moves, as not all moves require tracking and tracking may occur for sample members or households who have not moved. In the former instance, a sample member may notify the survey organization of a move prior to data collection or a call made to an address or former telephone number may yield information about the sample member's current whereabouts, thus preventing the need for tracking. In the latter instance, temporary absences due to travel, hospitalization, etc. and/or a change in telephone number may give rise to the need for tracking. Nevertheless, there is likely to be a strong correlation between residential moves and tracking status and, in the absence of complete information on moves, tracking status is the best indicator available for analysis.

[4] A total of 164 cases found to no longer be eligible in 2003 are excluded from the table and further analysis.

[5] A complication of the PSID design is that split-offs from families are identified during data collection; for example, a family that shares a single ID and data record in 2003 becomes two families in 2005. A total of 413 such splits occurred, but we restrict our analyses to the 8104 families interviewed in 2003 and eligible in 2005. We do so because the determination of a split-off is conditional upon successful tracking and contact.

Table 11.2 PSID: 2005 final dispositions for 2003 interviewed families, by need for tracking.

Disposition	Tracking needed		No tracking	
	No.	%	No.	%
Interview	1289	89.5	6528	97.8
Refusal	98	6.8	124	1.9
Tracking exhausted	48	3.3	0	0.0
Other non interview	6	0.4	11	0.2
Total	1441	100.0	6663	100.0

other real-estate (second home), family income (in quartiles), employment status of the household head, receipt of food stamps, self-rated health, whether the head has any chronic health conditions and Hispanic origin. The variables in Table 11.3 remained statistically significant ($P < 0.01$) in the multivariate model, and the odds ratios and their 95 % confidence intervals are also presented in Table 11.3. The Nagelkerke (1991) max-rescaled R^2 value for this model is 0.17.

The results suggest that the propensity to require tracking is higher among racial minorities and those with lower education, as well as among families who are at an earlier life stage (younger age of head, presence of young children) or have less stable family circumstances (never married, divorced, separated, families who have split or experienced some other change in composition). While female-headed households exhibit a higher need for tracking in the bivariate analysis, they have significantly lower odds of needing tracking when controlling for covariates in the multivariate model. Surprisingly, renters do not appear to be at higher risk of tracking than home owners, after controlling for other variables, nor do those in the lowest income quartile or in receipt of food stamps (both indicators of greater risk of tracking in bivariate analyses). Furthermore, both a recent move and the reported likelihood of moving in the next two years are significant predictors of the need for tracking. These variables in particular, as well as others in the model, could serve as leading indicators of the propensity to move, and hence the propensity to require tracking.

Next we are interested in the propensity to be located, given that tracking was needed for a case. For the PSID sample, this indicator is highly skewed, with only 3.2 % of those needing tracking remaining unlocated after tracking is exhausted. As a result, analysis of this outcome is not informative. As an alternative, we examine the total number of interviewer call attempts as an indicator of differential effort required for those cases needing tracking.[6] After truncating positive outliers (at three standard deviations above

[6] We differentiate between interviewer call attempts and tracking calls. Interviewer call attempts include all calls made by the interviewer prior to and/or subsequent to formal tracking of a case. Tracking calls are calls made while the case is flagged for tracking. As noted previously, a case is flagged for tracking if the interviewer learns that the respondent has moved and no forwarding information is provided, or if the first five calls made to the respondent result in no contact. Once a respondent has been located, the tracking flag is removed. Interviewer call attempts are almost always completed by telephone, though in special circumstances they may be done in person. Tracking calls typically involve calls to alternative phone numbers, contact persons or directory assistance. Initial tracking steps are conducted by the interviewers, but if these fail the case is transferred to the National Tracking Team for more intensive tracking.

Table 11.3 PSID: Percentage of families needing tracking in 2005, and odds ratio estimates of need for tracking in 2005 from multivariate logistic regression model, by 2003 variables.

Characteristic	Unadjusted % needing tracking	Predicted odds ratios (OR)	95 % Confidence interval for OR
Age of head			
<30	31.1	—	—
30–39	18.9	0.68	(0.56, 0.83)
40–49	17.9	0.71	(0.57, 0.88)
50–59	12.7	0.67	(0.52, 0.86)
60+	9.5	0.45	(0.33, 0.62)
Sex of head			
Male	16.1	—	—
Female	22.7	0.77	(0.65, 0.93)
Race of head			
White	11.9	—	—
Black	27.9	2.02	(1.75, 2.33)
Other	22.3	1.80	(1.45, 2.22)
Marital status of head			
Married	12.6	—	—
Never married	28.1	1.17	(0.95, 1.44
Widowed	12.5	1.12	(0.79, 1.60)
Divorced	22.2	1.56	(1.26, 1.93)
Separated	28.8	1.41	(1.04, 1.91)
Age of youngest child			
No children	15.0	—	—
<7	24.0	1.33	(1.13, 1.58)
7–12	19.0	1.33	(1.09, 1.62)
12–17	19.7	1.55	(1.24, 1.93)
Split-off indicator[a]			
Reinterview family	16.4	—	—
Split-off from reinterview family	34.6	0.90	(0.66, 1.24)
Recontact family	34.3	1.82	(1.34, 2.48)
Family composition change			
No change	15.1	—	—
Change, not in head or wife	18.5	1.08	(0.92, 1.76)
Same head, different wife	24.9	1.34	(0.98, 1.84)
Wife is now head	26.7	1.51	(1.07, 2.12)
Other change	33.1	1.73	(1.30, 2.32)
Education of head			
Less than high school	24.2	1.31	(1.12, 1.53)
High school	18.3	—	—
Some college	18.4	0.99	(0.83, 1.17)
College graduate	9.6	0.63	(0.49, 0.80)
Postgraduate	10.4	0.76	(0.57, 1.03)
Number of vehicles			
None	30.8	—	—
One	22.0	0.83	(0.69, 0.99)

Table 11.3 (Continued).

Characteristic	Unadjusted % needing tracking	Predicted odds ratios (OR)	95 % Confidence interval for OR
Two	14.4	0.71	(0.57, 0.87)
Three or more	10.4	0.57	(0.45, 0.74)
Type of dwelling unit			
One-family house	13.6	—	—
Two-family house, duplex	26.5	1.44	(1.11, 1.88)
Apartment	30.1	1.39	(1.18, 1.64)
Mobile home, trailer	23.3	1.52	(1.21, 1.90)
Row house, townhouse	22.3	1.41	(0.96, 2.05)
Other	20.6	1.31	(0.75, 2.31)
Head hospitalized in 2002			
Yes	22.3	1.33	(1.09, 1.61)
No	17.6	—	—
Moved in last 3 years			
No move	13.9	—	—
Moved in 2001	21.3	1.14	(0.92, 1.41)
Moved in 2002	25.9	1.34	(1.09, 1.65)
Moved in 2003	30.9	1.32	(1.11, 1.56)
Move in next 2 years			
Definitely will move	32.1	1.72	(1.46, 2.03)
Probably will move	21.5	1.25	(1.03, 1.52)
Uncertain	21.0	1.33	(1.08, 1.65)
Will not move	13.6	—	—

[a] Reinterview families are continuing families that were interviewed in 2001 and again in 2003. Split-offs are one or more family members interviewed in 2001 who moved out of the main (reinterview) family to form a new family in 2003. Recontacts are nonrespondents in 2001 who were interviewed in 2003.

the mean) we find that those requiring tracking needed an average of 13.6 call attempts, compared to 9.1 for those not requiring tracking (the respective medians are 8 and 5). Given the positive skew of this variable, we model the log of the number of calls, controlling for all the variables in Table 11.3 and the results of the prospective tracking effort (see Table 11.4). Controlling for these variables, the predicted mean number of call attempts is 8.95 for cases needing tracking and 6.92 for those not needing tracking. Tracking thus significantly increases the number of call attempts made by interviewers. However, tracking efforts are recorded in a separate process. In addition to the interviewer call attempts, the cases needing tracking required an additional 10.3 tracking calls on average (a median of 7 additional calls), bringing the total number of call attempts for cases needing tracking to a mean of 24 (and a median of 17). This does not include additional tracking efforts such as Internet searches and other lookup operations. Again, this suggests that tracking adds significant costs to data collection, as reflected in the number of additional call attempts made to resolve cases.

The PSID employs prospective tracking methods between waves. These include sending addresses to the NCOA for change of address or address verification, followed by the mailing of a postcard to all PSID families with the promise of a $10 incentive for

Table 11.4 PSID: Effect of prospective tracking efforts on 2005 need for tracking and predicted number of total calls in 2005.

Result of prospective tracking	No.	% Needing tracking	Predicted mean no. of total calls[a]	
			Tracking needed	No tracking
No update	1398	30.2	20.92	8.12
NCOA update	2002	27.9	21.17	8.62
Postcard returned: updated	1171	12.1	18.81	5.60
Postcard returned: verified correct	3257	5.5	19.38	5.76
Other (PMRs, etc.)[b]	276	51.1	20.15	8.21
(*N*)	(8104)	(8104)	(1441)	(6663)

[a] Estimated from the antilog of the adjusted (least-squares) means from a multivariate model predicting the log of the total number of call attempts (field calls + tracking calls), controlling for the variables in Table 11.3.
[b] PMR = Post Master Returns.

a returned postcard verifying or updating their address. The postcard is sent to the new address if the NCOA reports a change, otherwise it is sent to the last known address. How effective are these methods at reducing the need for tracking? The outcome of these efforts is presented in Table 11.4, along with the proportion needing tracking for each group and the total number of calls by need for tracking. The NCOA tracking alone does not appear to produce many gains, but is a relatively cheap and easy batch operation.

More than half of the respondents (4428 out of 8104) returned the postcard for the $10 incentive. These prospective efforts had a significant effect on tracking during 2005 data collection. Less than 6 % of respondents who verified their address as correct needed tracking, compared to 30 % of those for whom no address updates were obtained, 28 % of those for whom the NCOA update was the only source of information and 51 % of those where the postcard was returned undelivered (Post Master Returns or PMRs) or some other problem with the address was present.

In addition, those who returned the postcard needed significantly fewer calls to reach final disposition in 2005. The remaining two columns in Table 11.4 are based on the total number of calls (field and tracking calls) from a multivariate model of the log-transformed number of calls, controlling for all the variables in Table 11.3 (overall model, $F = 65.7$, d.f. = 48 and 8055, $R^2 = 0.28$). The interaction term between the result of the prospective tracking efforts and need for tracking is statistically significant ($P < 0.01$) and the antilog of each least-squares mean from the interaction is presented in Table 11.4. Not shown in Table 11.4 is that these effects remain significant when only examining field call attempts, although the magnitude of the differences between the estimated calls by tracking status is smaller.

We see that those who returned the postcard require fewer call attempts, both for those who needed subsequent tracking and for those who did not, in both cases of the order of two or three calls per case. Whether the cost of the postcard plus incentive is offset by the reduction in the number of calls depends on the cost of each additional call. The fact that the return of the postcard is associated both with less need for tracking and with fewer call attempts suggests that this may be a cost-effective method of keeping in contact with mobile sample members. For example, if we simply assume that the 4428 who did return the postcard were the same (i.e. the same proportion needing tracking; the same expected number of total calls) as the 1398 for whom no update was obtained, we estimate over 23 000 additional call attempts, or about 5.3 for each of the 4428 families who returned postcards. Restricting this just to field calls, we estimate over 11 000 additional field call attempts, or 2.6 calls per postcard return family. To reduce the costs of the postcard mailing further, targeting the mailing to those shown to be at higher risk of moving or greater tracking efforts could be considered. In this case, models like those in Table 11.3 could be used to guide the decision. We caution, however, that the above results are not from an experimental design and the anticipated gains may not be realized in practice.

11.4 CASE STUDY 2: HEALTH AND RETIREMENT STUDY

The Health and Retirement Study (HRS) is an ongoing panel survey of a nationally representative sample of over 20 000 men and women over the age of 50 in the USA. The HRS began in 1992 as a longitudinal study of a preretirement cohort of individuals born in 1931–1941 and their spouses of any age. It was joined in 1993 by a companion study, the Study of Asset and Health Dynamics of the Oldest Old (AHEAD), comprised of a cohort of persons born before 1924 and their spouses of any age. In 1998, the study design was modified to convert the HRS sample from a set of specific cohorts into a steady-state sample that would represent the entire US population over age 50. This was achieved by adding new cohorts in 1998 to fill in the age range (over 50) and by adding a new six-year cohort of persons entering their 50s every six years thereafter.

The HRS conducts core interviews every two years using a mixed-mode design of telephone and face-to-face interviews (supplemental studies are conducted in the off-years using mail and Internet modes). The primary mode used for baseline interviews and for sample members age 80 + is face to face. Up through 2002, the primary mode for follow-up interviews with sample members under age 80 was telephone. Starting in 2004, the proportion of face-to-face interviews increased substantially so that 65–75 % of all interviews are now conducted in person.

Although the study began with a sample of community-dwelling individuals residing in the USA, follow-up interviews are attempted with all sample members regardless of where they live, including those in institutions, as well as those who have moved out of the country. In the event of divorce or separation of an original sample couple, both members remain in the sample. In addition, interviews are obtained (or attempted) with new spouses in the case of a new marriage or remarriage.

The primary focus of the HRS is on the intersection between health, retirement and economic status in later life. The survey provides detailed information on each of these topics, as well as on employment history and availability of pensions, work disability and related benefits, family composition and resource transfers and health insurance and utilization of health services. The HRS focuses on middle-aged and older individuals and thus provides a nice complement to the PSID, which includes families and individuals of all ages.

As with the PSID analyses, we use measures from the most proximate prior wave (2002) to examine correlates of tracking and effort in the latest wave of data collection (2004). All analyses presented here are unweighted.

A total of 12 502 households (representing 18 167 individual respondents) who were interviewed in 2002 were in the sample for 2004. A total of 1294 households (10.7 %) required tracking during the 2004 field period. Of these, all but 17 (1.3 %) were located. Table 11.5 presents the final 2004 dispositions for primary respondents in HRS households by tracking status. Results are similar to those for the PSID, although interview rates for sample members who required tracking is slightly higher for the HRS (91 % vs. 86 % for the PSID).

We conducted a parallel set of analyses as presented for the PSID to examine correlates of need for tracking and effort (represented as number of calls made), by tracking status. There is substantial, though not complete, overlap in the correlates examined in the two case studies. The HRS sample is restricted to households for which an interview was completed in 2002 and is comprised of the primary (i.e. originally sampled) respondent from each household. Unless otherwise specified, the measures represent respondents' status as of the 2002 interview. Several measures were included to capture change in marital, employment and health status, as well as residential moves. These measures reflect change during the interval prior to the 2002 interview. For most respondents, this is a two-year interval; however for those not interviewed in 2000, the interval is four or more years.

Table 11.6 presents results of bivariate and multivariate analyses of need for tracking. We used the same strategy as with the PSID of first estimating a full model including a variety of demographic, socio-economic and survey design factors, and then estimating a reduced model in which only the significant variables in the full model were retained. Variables in the reduced model are shown in Table 11.6. All of the bivariate results are statistically significant ($P < 0.01$) with the exception of having a second residence.

For the most part, the findings echo those from the PSID. Focusing first on the bivariate results in the first column, female respondents and those who are not married and/or have experienced a recent change in marital status are more likely to require tracking than their respective counterparts. This is also the case for respondents with lower socio-economic status (as measured by education and household assets), those who have less stable housing conditions and those in poorer health. As with the PSID, a recent move (either local or nonlocal) between 2000 and 2002 is a significant predictor of the need for tracking in 2004.

Table 11.5 HRS: 2004 final dispositions for 2002 interviewed households, by need for tracking.

Disposition	Tracking needed		No tracking	
	No.	%	No.	%
Interview	1182	91.4	10 319	95.9
Refusal	79	6.1	421	3.9
Tracking exhausted	17	1.3	0	0.0
Other noninterview	16	1.2	18	0.2
Total	1294	100.0	10 758	100.0

Table 11.6 HRS: Percentage of households needing tracking in 2004, and odds ratio estimates of need for tracking in 2004 from multivariate logistic regression model, by 2002 variables.

Characteristic	% Needing tracking	Predicted odds ratios (OR)	95 % Confidence interval for OR
Age			
< 50	20.5	2.01	(0.89, 4.54)
50–59	11.6	1.39	(1.11, 1.74)
60–69	10.0	1.12	(0.94, 1.33)
70–79	9.8	—	—
80–89	11.6	0.85	(0.69, 1.04)
90 +	16.9	1.11	(0.82, 1.50)
Sex			
Male	9.8	—	—
Female	11.4	0.83	(0.73, 0.96)
Marital status			
Married	6.3	—	—
Partnered	15.8	1.76	(1.18, 2.63)
Separated	19.0	1.76	(1.12, 2.76)
Divorced	16.8	2.00	(1.63, 2.46)
Widowed	13.5	1.96	(1.65, 2.34)
Never married	14.6	1.80	(1.33, 2.42)
Other	12.5	1.65	(0.36, 7.58)
Divorced in past 2 years			
Yes	32.8	—	—
No	10.5	0.46	(0.30, 0.70)
Education			
Less than high school	14.0	0.84	(0.68, 1.04)
High School	9.3	0.70	(0.58, 0.85)
Some college	11.3	—	—
College graduate	8.7	0.88	(0.67, 1.16)
Postgraduate	8.7	0.91	(0.68, 1.23)
Household assets			
Lowest quartile	17.3	1.87	(1.496, 2.34)
2nd quartile	10.0	1.38	(1.111, 1.71)
3rd quartile	7.6	1.12	(0.900, 1.40)
Highest quartile	6.4	—	—
Type of dwelling unit			
One-family house	8.3	—	—
Two-family house, duplex	9.2	0.83	(0.57, 1.22)
Apartment	19.4	1.44	(1.17, 1.76)
Mobile home, trailer	11.9	1.09	(0.82, 1.47)
Other	16.8	1.48	(1.17, 1.87)
Housing tenure			
Own	7.7	—	—
Rent	20.8	1.50	(1.22, 1.85)
Other	14.6	1.16	(0.93, 1.44)
Has second home			
Yes	11.1	—	—
No	10.7	0.62	(0.48, 0.81)

Moved in last 2 years			
No move	9.7	0.76	(0.60, 0.95)
Local move	16.7	—	—
Non-local move	16.2	1.26	(0.96, 1.66)
Self-assessed health			
Excellent or very good	8.0	—	—
Good	10.3	1.21	(1.03, 1.42)
Fair or poor	14.3	1.37	(1.16, 1.60)
Participation in 2003 supplements			
Not contacted for supplement	12.2	—	—
Contacted, did not participate	17.4	1.28	(1.05, 1.56)
Participated in 1 + supplements	7.4	0.57	(0.50, 0.65)
Mode in 2002			
Face to face	13.5	1.34	(1.14, 1.57)
Telephone	10.0	—	—

One finding not examined in the PSID (due to insufficient sample size) relates to the U-shaped pattern observed for age in the HRS. Those at the youngest (< 50) and oldest (80 +) ages are significantly more likely than those in the middle ages (50–79) to require tracking, with the lowest propensity occurring for those aged 70–79. In addition, the HRS conducted several supplemental studies on different subsamples of respondents between the 2002 and 2004 core interviews. These included two mail surveys and an Internet survey. Outcomes from the supplemental studies are highly predictive of the need for tracking in 2004. Respondents who participated in any supplemental study were significantly less likely to require tracking than were those who were either contacted but did not participate or were not contacted for a supplemental study. These supplemental studies can be viewed as serving a similar function for the HRS with regard to tracking as the postcard reminder served for the PSID.

A number of other variables were statistically significant in the bivariate analysis but not the multivariate model. These include race, Hispanic ethnicity, foreign born, recent widowhood and marriage, household income, employment status and recent job changes, residence on a farm and in a nursing home, limitation in Activities of Daily Living (ADL) or Instrumental ADLs, perceived decline in health in the last two years and language in which the 2002 interview was conducted (English vs. Spanish).

Results from the reduced multivariate model are presented in columns 2 and 3 of Table 11.6. The Nagelkerke max-rescaled R^2 for this model is 0.11. For most variables, results are consistent with the bivariate results. One exception is age, for which the association in the multivariate model is diminished and the only statistically significant difference is found for respondents aged 50–59 compared to those aged 70–79. Another exception is sex, for which the direction of the association flips after controlling for demographic, socio-economic and survey design factors. Thus, other things being equal, males are more likely to require tracking than females. In addition, the effect of having a second home is suppressed in the bivariate analysis, and once other factors are controlled this emerges as a significant predictor of need for tracking.

Table 11.7 presents descriptive results for effort, as measured by number of calls. Number of calls are divided into field calls, tracking calls (i.e. calls made by the

Table 11.7 HRS: Means for interviewer, tracking and total calls by tracking status and participation in 2003 supplements.

Characteristic	Tracking status and call type			
	No tracking	Tracking needed		
	Field	Field	Tracking	Total
Participation in 2003 supplements				
Not contacted for supplement	5.44	8.99	7.39	16.38
Contacted, did not participate	7.42	14.24	8.49	22.74
Participated in 1 + supplements	4.56	7.89	6.76	14.65
Mean no. of calls	5.23	9.40	7.36	16.76

interviewer or the centralized tracking team while the case is in 'tracking' status) and the sum of these two (total calls). Positive outliers were truncated at three standard deviations above the mean. The HRS has a large disparity in effort according to whether or not tracking was required, similar to that found in the PSID. Households not requiring tracking in 2004 needed an average of 5.23 field calls, compared to 9.40 for households needing tracking. In addition, households needing tracking required an additional 7.36 tracking calls, yielding an overall differential of 11.53 total calls (16.76 for tracking households vs. 5.23 for nontracking households).[7] This differential is reduced only slightly when examined in a multivariate context, and tracking status remains a significant ($P < 0.001$) predictor of overall effort. Thus, as with the PSID, tracking in the HRS adds significant costs to data collection.

Also presented in Table 11.7 are means for number of calls by participation in the 2003 supplements. In addition to being a strong predictor of need for tracking, this indicator is also highly predictive of effort, controlling for tracking status. Such factors can be effectively used to identify households or respondents who are likely to be difficult to track. Those cases can then be targeted for centralized tracking prior to or at the start of the field period, reducing effort on the part of interviewers during the field period.

11.5 DISCUSSION

The process of tracking or locating sample members in panel surveys is a potentially costly and time-consuming operation. Furthermore, such cases may have a disproportionate effect on measures of change, which is a staple of panel survey design. Despite their potential impact on survey estimates, the process has been largely viewed as an operational matter. For example, Eckland (1968, p. 55) argued that persistence in using a

[7] The variables for number of calls of each type are positively skewed even after truncation. Means are sensitive to extreme values and this can lead to an exaggeration of the levels and subgroup differences in number of calls. Thus, the differences reported here should be interpreted in terms of rough orders of magnitudes rather than exact differences.

particular tracking strategy is more important than the type of strategy, and this sentiment was echoed by Wright, Allen and Devine (1995).

Our view is that many of the sample member moves that occur in panel surveys can be predicted, as can the likelihood of finding such persons following such a move. We argue, consistent with the 'responsive design' perspective (see Groves and Couper, 1998; Groves and Heeringa, 2006), that correlates of the propensity to move and the propensity to be located be measured as part of the data collection activities of a panel survey. Such information could be used not only to reduce the costs and effort associated with tracking, but also for statistical adjustment models to account for the effects of such loss to follow-up on estimates of change. We also believe that pre-emptive efforts to obtain address updates for sample members between waves, such as used in the PSID, may have benefits not only in terms of location propensity but also in terms of cooperation in later waves. A similar effect may be seen with the HRS self-administered supplements.

What impact, if any, has technology had on tracking in panel surveys? We believe these effects to be mixed. On the one hand, we have seen tremendous growth in the number and accessibility of databases to facilitate tracking, many of them online. This means that they are generally cheaper, more efficient and more frequently updated than tracking resources available in the past (see Koo and Rohan, 2000). At the same time, individuals have become more concerned with privacy and, to the extent that they have a choice, are increasingly opting for unlisted addresses and phone numbers. The widespread adoption of cell phones and the increase in cell-phone-only households (e.g. Blumberg *et al.*, 2004) present new challenges for tracking panel members using telephone numbers. However, the introduction of telephone number portability may mean that once a telephone number is obtained from a panel member it is more likely to remain a valid means of contact. E-mail addresses suffer from even more 'churn' or turnover than telephone numbers, and even though they are an inexpensive way to maintain contact with panel members they may require frequent updating.

Despite several dozen papers describing tracking procedures published in the past few decades, this is an understudied area. There are few, if any, experimental tests of alternative approaches to tracking, and few attempts to model the location and tracking process in multivariate analyses. One of the advantages of new technology is the increased collection of paradata (Couper and Lyberg, 2005), including rich call record data. If these data can be expanded to include all of the tracking steps and their outcomes, researchers will be able to better model the process. This will enable us to better understand the relative effectiveness (how successful is each in locating different types of sample persons) and efficiency (how much time and money does each method take, relative to the potential payoff) of different tracking methods. Our hope is that this chapter will encourage others to consider focusing research attention on the tracking process.

ACKNOWLEDGEMENTS

We received assistance from a number of individuals who work on various large-scale panel surveys. In particular, we wish to thank Kate McGonagle, Eva Leissou, Tecla Loup and Mohammad Mushtaq from PSID and Heidi Guyer and Jennifer Arrieta from HRS for their help in preparing and interpreting the data from those surveys; and Hayley

Cheshire and Carli Lessof from the English Longitudinal Study of Ageing, Heather Laurie from the British Household Panel Survey and Gert Wagner and Jürgen Schupp from the German Socio-Economic Panel Survey for providing information on tracking methods and outcomes for those surveys. We are also grateful to Peter Lynn and the reviewers for their helpful suggestions on earlier drafts. All errors remain our own.

REFERENCES

Blumberg, S. J., Luke, J. V. and Cynamon, M. L. (2004). Has cord-cutting cut into random-digit-dialed health surveys? The prevalence and impact of wireless substitution. In S. B. Cohenand J. M. Lepkowski (Eds), *Eighth Conference on Health Survey Research Methods* (pp. 137–142). Hyattsville, MD: National Center for Health Statistics.

Buck, N. (2000). Housing location and residential mobility. In R. Berthoud and J. Gershuny (Eds), *Seven Years in the Lives of British Families: Evidence of the Dynamics of Social Change from the British Household Panel Survey* (p. 149). Bristol: Policy Press.

Burgess, R. D. (1989). Major issues and implications of tracing survey respondents. In D. Kasprzyk, G. Duncan, G. Kalton and M. P. Singh (Eds), *Panel Surveys* (pp. 52–73). New York: John Wiley & Sons, Inc.

Call, V. R. A., Otto, L. B. and Spenner, K. I. (1982). *Tracking Respondents: A Multi-Method Approach.* Toronto: Lexington Books.

Clarridge, B. R., Sheehy, L. L. and Hauser, T. S. (1978). Tracing members of a panel: a 17-year follow-up. *Sociological Methodology*, 9, 185–203.

Coen, A. S., Patrick, D. C. and Shern, D. L. (1996). Minimizing attrition in longitudinal studies of special populations: an integrated management approach. *Evaluation and Program Planning*, 19, 309–319.

Couper, M. P. and Lyberg, L. E. (2005). The use of paradata in survey research. Paper presented at the *International Statistical Institute*, April, Sydney, Australia.

De Jong, G. F. and Gardner, R. W. (1981). *Migration Decision Making: Multidisciplinary Approaches to Microlevel Studies in Developed and Developing Countries.* New York: Pergamon.

Donovan, N., Pilch, T. and Rubenstein, T. (2002). Geographic mobility. *Analytical Paper,* 6, Performance and Innovation Unit, Economic and Domestic Secretariat of the Cabinet Office, London, July 2002.

Duncan, G. J. and Kalton, G. (1987). Issues of design and analysis of surveys across time. *International Statistical Review*, 55, 97–117.

Eckland, B. K. (1968). Retrieving mobile cases in longitudinal surveys. *Public Opinion Quarterly*, 32, 51–64.

Frey, W. H. and Kobrin, F. E. (1982). Changing families and changing mobility: their impact on the central city. *Demography*, 19, 261–77.

Gregg, P., Machin, S. and Manning, A. (2004). Mobility and joblessness. In D. Card, R. Blundell and R. B. Freeman (Eds), *Seeking a Premier Economy: The Economic Effects of British Economic Reforms, 1980–2000* (p. 381). Chicago: University of Chicago Press.

Groves, R. M. and Couper, M. P. (1998). *Nonresponse in Household Interview Surveys.* New York: John Wiley & Sons, Inc.

Groves, R. M. and Hansen, S. E. (1996). *Survey Design Features to Maximize Respondent Retention in Longitudinal Surveys.* Unpublished report to the National Center for Health Statistics, University of Michigan, Ann Arbor, MI.

Groves, R. M. and Heeringa, S. G. (2006). Responsive design for household surveys: tools for actively controlling survey nonresponse and costs. *Journal of the Royal Statistical Society: Series A*, 169, 439–457.

Hampson, S. E., Dubanoski, J. P., Hamada, W., Marsella, A. J., Matsukawa, J., Suarez, E. and Goldberg, L. R. (2001). Where are they now? Locating former elementary-school students after nearly 40 years for a longitudinal study of personality and health. *Journal of Research in Personality*, 35, 375–387.

Hermann, A. and Grigorian, K. (2005). Looking for Mr. Right: effective locating techniques for panel surveys. *Paper presented at the Annual meeting of the American Association for Public Opinion Research,* May, Miami Beach, FL.

Hill, M. S. (1991). *The Panel Study of Income Dynamics; A User's Guide*. Thousand Oaks, CA: Sage.

Koo, M. M. and Rohan, T. E. (2000). Use of world wide web-based directories for tracing subjects in epidemiological studies. *American Journal of Epidemiology*, 152, 889–894.

Landale, N. S. and Guest, A. M. (1985). Constraints, satisfaction and residential mobility: Speare's model reconsidered. *Demography*, 22, 199–222.

Laurie, H., Smith, R. and Scott, L. (1999). Strategies for reducing nonresponse in a longitudinal panel survey. *Journal of Official Statistics*, 15, 269–282.

Lepkowski, J. M. and Couper, M. P. (2002). Nonresponse in longitudinal household surveys. In R. M. Groves, D. Dillman, J. Eltinge and R. J. A. Little (Eds), *Survey Nonresponse* (pp. 259–272). New York: John Wiley & Sons, Inc.

Long, L. (1988). *Migration and Residential Mobility in the United States*. New York: Russell Sage Foundation.

McKenzie, M., Tulsky, J. P., Long, H. L., Chesney, M. and Moss, A. (1999). Tracking and follow-up of marginalized populations: a review. *Journal of Health Care for the Poor and Underserved*, 10, 409–429.

Menendez, E., White, M. C. and Tulsky, J. P. (2001). Locating study subjects: predictors and successful search strategies with inmates released from a U.S. county jail. *Controlled Clinical Trials*, 22, 238–147.

Nagelkerke, N. J. D. (1991). A note on a general definition of the coefficient of determination. *Biometrika*, 78, 691–692.

Ribisl, K. M., Walton, M. A., Mowbray, C. T., Luke, D. A., Davidson II, W. S. and Bootsmiller, B. J. (1996). Minimizing participant attrition in panel studies through the use of effective retention and tracking strategies: review and recommendations. *Evaluation and Program Planning*, 19, 1–25.

Rossi, P. H. (1955). *Why Families Move*. Beverly Hills: Sage.

Schejbal, J. A. and Lavrakas, P. J. (1995). Panel attrition in a dual-frame local area telephone survey. In *Proceedings of the Survey Research Methods Section* (pp. 1035–1039). Alexandria, VA: American Statistical Association.

Scott, C. K. (2004). A replicated model for achieving over 90 % follow-up rates in longitudinal studies of substance abusers. *Drug and Alcohol Dependence*, 74, 21–36.

Speare Jr., A., Goldstein, S. and Frey, W. H. (1975). *Residential Mobility, Migration, and Metropolitan Change*. Cambridge, MA: Ballinger.

US Census Bureau (2005). *Current Population Survey, 2004 Annual Social and Economic Supplement*. Internet release date: June 23, 2005. Retrieved July 21, 2008, from www.bls.census.gov/cps/asec/adsmain.htm

Walters, W. H. (2000). Types and patterns of later life migration. *Geografiska Annaler*, 82B, 129–147.

Wright, J. D., Allen, T. and Devine, J. A. (1995). Tracking non-traditional populations in longitudinal studies. *Evaluation and Program Planning*, 18, 267–277.

CHAPTER 12

The Use of Respondent Incentives on Longitudinal Surveys

Heather Laurie and Peter Lynn
University of Essex, UK

12.1 INTRODUCTION

Incentives in the form of a gift or money are given to survey respondents in the hope that this will increase response rates and possibly also reduce nonresponse bias. They can also act as a means of thanking respondents for taking part and showing appreciation for the time the respondent has given to the survey. Survey organisations are keen to make interviewers feel confident as they approach sample members for an interview and having something that interviewers can give to respondents, even if a small token of some kind, helps through a process of reciprocity and social interaction between the interviewer and respondent. The aim in using the incentive is to encourage respondents to see their participation as being important and, as a result, increase response propensities and enhance the quality of the data collected.

There is a considerable literature devoted to the effects of respondent incentives, though most studies are based on cross-sectional surveys (see Section 12.2 below). These studies show that both the form of the incentive, gift or money, and the way in which the incentive is delivered to the respondent have a measurable impact on response rates. A monetary incentive sent to the respondent in advance of the interview has the greatest effect on increasing response, regardless of the amount of money involved. This type of unconditional incentive is thought to operate through a process of social reciprocity where the respondent perceives that they have received something unconditionally on trust so reciprocate in kind by taking part in the research. Published studies present a mixed picture regarding the extent to which the increase in response rate may or may not be associated with a reduction in nonresponse bias. Additionally, some of the literature suggests an improvement in data quality from respondents who are given an incentive, though again some studies conclude the opposite. It is generally felt that

Methodology of Longitudinal Surveys P. Lynn

incentives are more appropriate the greater the burden to respondents of taking part. Longitudinal surveys certainly constitute high burden surveys, but there is little guidance on how and when incentives should be employed on longitudinal surveys.

This chapter reviews the use made of incentives on longitudinal surveys, describing common practices and the rationale for these practices. We attempt to identify the features of longitudinal surveys that are unique and the features that they share with cross-sectional surveys in terms of motivations and opportunities for the use of incentives and possible effects of incentives. In Section 12.2 we review briefly what is known about the effect of incentives on cross-sectional surveys. Section 12.3 then sets out issues in the use of incentives that are specific to longitudinal surveys. Section 12.4 summarises current practice on longitudinal surveys in different countries and with differing designs and Section 12.5 reports experimental evidence on the effect of changing the way in which incentives are used midway in a longitudinal survey. This evidence includes the findings from three experimental studies in the UK carried out on the British Election Panel Survey (BEPS), the British Household Panel Survey (BHPS) and the England and Wales Youth Cohort Study (EWYCS). Each experiment addressed a different type of change in incentive administration.

The BEPS experiment involved introducing an incentive for the first time at wave 6. Three experimental groups were used at both waves 6 and 7, consisting of a zero incentive and two different values of unconditional incentive. The BHPS experiment was carried out at wave 14. BHPS respondents had always received a gift token as an incentive and since wave 6 this had been offered unconditionally in advance of the interview to the majority of respondents. The wave 14 experiment was designed to assess the effect on response of increasing the level of the incentive offered from £7 to £10 for established panel members, many of whom have cooperated with the survey for 13 years. The EWYCS experiment concerned the introduction of incentives at wave 2 of cohort 10 in the context of a mixed-mode design, with the nature of the incentive changing for some groups at wave 3 but repeated unchanged for other groups.

12.2 RESPONDENT INCENTIVES ON CROSS-SECTIONAL SURVEYS

A sizeable body of research into the effects of providing respondent incentives tells us a number of things. Findings regarding effects on response rate are generally consistent, while studies addressing effects on nonresponse bias or on data quality are less numerous and provide somewhat more mixed messages. Overall, the evidence shows that incentives are effective in increasing response rates (Singer and Kulka, 2000) but the effect of incentives varies depending on the mode of data collection, the type of incentive used and the delivery method used (Singer, 2002). Some surveys use a monetary incentive, others give respondents a small gift and some offer entry into a lottery or prize draw. The way in which the incentive is delivered to the respondent also varies from being paid or given unconditionally in advance or promised contingent on response.

12.2.1 Effects of Incentives on Response Rates on Mail Surveys

Church (1993) found that for mail surveys monetary payments are more effective than gifts, prepaid unconditional monetary incentives have a greater effect on increasing response rates than those contingent on response and response rates increase as the monetary amount paid increases. Church found an average increase in response rates to mail surveys of 19.1 % where a prepaid monetary incentive was used compared to 4.5 % where the incentive was conditional on response. In addition, the average increase where a gift was sent unconditionally in advance was 7.9 %. James and Bolstein (1992) found similar results on a mail survey where response rates increased as the incentive amount increased. Sending the payment in advance was also more effective, as amounts of $1 or $5 in cash or a $5, $10, $20 or $40 cheque sent unconditionally in advance had a greater effect on response than an offer of $50 once the questionnaire was returned. Couper *et al.* (2006) also found that a cash incentive in a mail survey yielded a higher response than a gift in kind. The evidence on the use of lotteries for mail surveys is mixed, with some studies reporting a positive effect on response rates (McCool, 1991; Balakrishnan *et al.*, 1992; Kim *et al.*, 2001; Halpern and Asch, 2003) and others finding no effect (Warriner *et al.*, 1996).

12.2.2 Effects of Incentives on Response Rates on Interviewer-Administered Surveys

Singer *et al.* (1999a) analyses the results of 39 experiments with incentives on face-to-face and telephone surveys, concluding qualitatively similar effects on response rates to those found by Church for mail surveys. However, Singer found that the percentage increase in response rates on interviewer-conducted surveys was somewhat smaller than those found on mail surveys. On average, each dollar of incentive paid produced about one third of a percentage point difference in response between incentive and nonincentive groups. Money was more effective than a gift and a prepaid incentive resulted in a significantly higher response rate than a conditional incentive. Singer concluded that for surveys conducted by interviewers there is no evidence that the effect of incentives differs between telephone and face-to-face interviews. A method using (unconditional) promissory notes given by the interviewer to the respondent at the point of interview has also been found to increase response rates. Lynn *et al.* (1998) found that the promissory note increased response by 7.3 %, from 56.0 % for the nonincentive group to 63.3 % for the promissory note group.

Some studies report an interaction between the burden of the interview and response rates for incentive and nonincentive groups (Singer *et al.*, 1999a). As a survey becomes more burdensome, the difference in response rates increases between those paid an incentive and those with no incentive payment. Similar effects were found on an experiment where respondents were offered a monetary incentive for completing a time use diary, typically a high burden type of survey (Lynn and Sturgis, 1997). The leverage-saliency theory proposed by Groves *et al.* (2000) might be expected to predict that incentives will have greater leverage where the salience of the research for respondents is low. This appears to be supported by the evidence. For a survey with low response rates and where the saliency of the research to the respondent is low, the effect of the incentive is likely to be greater than for a survey with high response rates and high saliency (Baumgartner and Rathbun, 1997; Shettle and Mooney, 1999; Groves *et al.*, 2000). There is also some evidence that incentives work primarily by reducing refusals and have little effect on noncontact rates (Shettle and Mooney, 1999; Singer *et al.*, 2000).

12.2.3 Effects of Incentives on Sample Composition and Bias

Couper *et al.* (2006) examined the effect of incentives on sample composition and response distributions. The concern here was whether or not incentives are effective in encouraging groups who are typically under-represented in surveys to respond (such as those on low incomes, ethnic minority groups and those with low levels of education), or whether the additional respondents are similar to the ones who would respond anyway, in which case incentives have no beneficial effect on nonresponse bias. They found that a cash incentive was more likely to increase response than a gift in kind among those with low education levels, single people and those who were not in paid employment. Singer *et al.* (2000) found that on a random digit dialling (RDD) survey a $5 incentive was more likely to increase response among those with low education levels. An earlier review by Singer *et al.* (1999a) had found that in experiments on interviewer-administered surveys three studies had indicated that incentives may be successful in boosting response, particularly amongst low-response groups, while five studies indicated no significant differences between demographic groups in the effects of incentives and one study showed mixed results. A recent study on the German Family Panel pilot (Brüderl *et al.*, 2008) suggests that incentives may reduce nonresponse bias over three waves.

12.2.4 Effects of Incentives on Data Quality

Couper *et al.* (2006) found that the higher response rates produced by the cash and noncash incentive did not translate into lower data quality and there was no evidence of differential measurement errors in the responses for each group. Concerns about reducing data quality through the use of incentives have also been addressed by Singer *et al.* (1999a), who concluded that incentives do not appear to adversely affect data quality as measured by the levels of item nonresponse or the effort expended in the interview measured by the number of words given to verbatim items. On a survey of social attitudes, Tzamourani and Lynn (2000) found that incentives made no difference to the level of item nonresponse. Willimack *et al.* (1995) too found no significant association between the use of a prepaid nonmonetary incentive and data quality. Other studies have found that item nonresponse is reduced when incentives of any kind are used (Singer *et al.*, 2000). James and Bolstein (1990) found that respondent effort increased with the value of the incentive in a mail survey.

12.2.5 Summary: Effects of Incentives

The cross-sectional evidence shows that incentives are effective in increasing response rates, though the extent varies by survey mode and the type of incentive strategy used. It also seems that incentives can improve sample composition, by disproportionately boosting response amongst groups with a relatively low baseline response propensity. There is little evidence of incentives having a significant impact on data quality.

12.3 RESPONDENT INCENTIVES ON LONGITUDINAL SURVEYS

Incentives are particularly likely to be used on longitudinal surveys, due to the inherently burdensome nature of the survey design and the particular value to the researcher of encouraging repeated cooperation. The burden is primarily related to the request for regular participation over a period of time, but is accentuated when the survey involves certain other features that are common in longitudinal surveys. These include long or complex questionnaires, sensitive subject matter, interviews with more than one member of a household and additional measurements such as anthropometric measures or psychological or cognitive tests. Incentives are therefore used to show respondents that the survey organisation recognises the level of burden that is being imposed and wants to thank respondents for their participation. An issue is that payment of any kind may raise the expectations of respondents who will, in future surveys, expect some payment or incentive (Singer *et al.*, 1998, 2000). Certainly, once the decision has been taken to use incentives in a longitudinal survey it may be difficult to withdraw them without having an adverse effect on response rates, something that is discussed further in the sections that follow. The initial decision about the use of incentives for a particular longitudinal survey has long-term consequences for the survey both financially in terms of cost and the expectations of respondents.

Longitudinal surveys differ from cross-sectional surveys in terms of the use of incentives in two ways. First, the *administration* of incentives (type, value, method, etc.) to each sample member can be different at different waves. This leads to a multitude of potential incentive regimes, each consisting of a combination of treatments over waves. Second, the *effects* of incentives may be more complex and may need to be evaluated differently. For example, effects on response rates or on nonresponse bias may be temporary, constant, delayed or cumulative.

A longitudinal survey may choose to change the *value* of incentives offered between waves. The value may be increased, for example to reflect increases in the cost of living or to recognise the increasing value to the researcher of continuing participation, or it may be decreased, perhaps because the survey budget is reduced. Similarly, a survey might choose to vary the *form* the incentive takes, switching from monetary to a gift or lottery, or vice versa. There is little or no evidence on the relative effectiveness of possible combinations over waves.

Another area where there is limited knowledge is the effect of *introducing* an incentive for the first time on a longitudinal survey that has previously not used them. While we might expect that the highest levels of attrition will occur in the early waves of a survey, the introduction of an incentive after one or more waves have been carried out may have a positive effect on cementing the loyalty of sample members for later waves of the survey (Laurie, 2007). Conversely, introducing an incentive may have little effect as the attrition already suffered on the survey may have left a sample that is essentially fairly cooperative and so responds in a limited way to an incentive. The effects of *ceasing* to provide an incentive for the first time where respondents have previously received them, so have an expectation of continued receipt, are largely unknown even though some studies suggest that these may not be significant (Lengacher *et al.*, 1995; Singer *et al.*, 1999b). It is likely that the expectations of respondents would play a key role, but there is limited evidence of how these expectations play out in terms of respondent behaviour in a longitudinal survey.

The longitudinal design, where detailed information is known about respondents' previous response history and characteristics, should be an ideal vehicle for the *tailoring* of incentive strategies. Indicators that may be associated with later attrition, such as the level of item nonresponse at a previous wave (Burton *et al.*, 1999; Schrapler, 2003; Lynn *et al.*, 2005) and interviewer assessments of respondent cooperation, may provide an opportunity to target resources at respondents who have a higher risk of dropping out of the survey. In order to target incentives efficiently, information on the most effective alternatives is required. For example, differing incentive strategies may produce differing responses from subgroups within the population and changes in response behaviour as a result of using an incentive may vary for different subgroups over several waves of data collection.

There may also be a role for preference questions that allow the respondent to choose the form their incentive takes, choosing for example between a cash payment, a gift or making a donation to charity. Lengacher *et al.* (1995) found that charitable giving tends to increase subsequent wave response rates amongst those who are already cooperative respondents but not amongst those who were initially reluctant to take part at the previous wave and were persuaded to do so after a first refusal. They concluded that charitable giving can be viewed as a proxy for altruistic activities more generally, including participation in social surveys. In contrast, Tzamourani (2000) found that offering to make a payment to charity had no beneficial effect on response rates. Providing respondents with a choice of type of incentive may allow the respondent to tailor their incentive to fit their own preferences in terms of altruism or individual benefit but the likely effects on response rates and bias are unknown.

In general, the use of differential incentives where some respondents receive more than others is avoided by most longitudinal surveys, or at least restricted to situations in which different sample members have different response tasks. The reasons for this are several. The first issue is an ethical one of parity and fairness. Surveys requiring ethical approval will be discouraged from using differential incentives, and fieldwork agencies and interviewers may not be willing to implement these (see Lessof, chapter 3 of this volume). In addition, little is known about the longer term effects of tailoring strategies. In particular, a design that offers higher incentives to noncooperative respondents is effectively rewarding them for failing to respond. As such, the payments may be seen as inequitable by cooperative respondents, who, if they become aware of these payments, will see them as being unfair and may refuse to cooperate in future surveys (Singer *et al.*, 1999b).

A potential problem in implementing differential incentives arises on surveys, such as household panel surveys, in which multiple household members take part. It would be difficult, for example, to offer one household member a higher amount than another. Indeed, the design of most household panels leads to the sample including relatives in different households, so even with a consistent incentive treatment within households, sample members receiving different incentives may discuss what they have received. The final problem is raising respondent expectations for the future. If a respondent receives a higher amount at one year of the survey it is reasonable that they will expect this amount the following year and if they do not receive it they may drop out of the survey. There is some evidence that paying higher incentives as part of a refusal conversion programme does not deliver higher participation rates at the following year of the survey (Lengacher *et al.*, 1995). These strategies may therefore have a beneficial effect that is only temporary and may create a longer term problem for the survey organisation, which must choose either to continue to pay at the higher level or risk losing the respondent.

Despite this reticence to using differential incentives, many surveys do have some mechanisms in place for varying what respondents receive. This is generally done by providing additional payments for specific circumstances, such as particular types of outcome or respondent behaviour, or through the provision of small gifts or fees in kind where the interviewer has some discretion in how best to handle particular cases (Martin *et al.*, 2001). One-off payments or ‘end game’ payment strategies to increase response from the least cooperative sample members have also been used (Juster and Suzman, 1995). However, little is known about how successful these are in delivering long-term commitment to the study.

Additional payments for what the survey organisation sees as extra burden are also fairly common. Where additional modules are included for specific types of respondents or the interview length is longer than usual, survey organisations may pay more in recognition of this – even though there is no clear evidence of a relationship between interview length or perceptions of burden on subsequent wave response (Hill and Willis, 2001; Martin *et al.*, 2001). Survey length and complexity have been identified by some studies as being complaints made by reluctant respondents, while a very short questionnaire may reduce response rates on a panel survey (Lynn *et al.*, 2005). Nonetheless, longitudinal survey practitioners are typically concerned that a longer or more burdensome interview may have a greater effect on response at the following rather than current year of the survey. In this sense, higher incentive payments for the current interview are used as something of an insurance policy against higher attrition at the following contact, even though the effects on subsequent response rates are not clear.

12.4 CURRENT PRACTICE ON LONGITUDINAL SURVEYS

Longitudinal surveys vary in their use of incentives for respondents but the use of incentives is generally part of a wider package of measures to encourage participation. The measures typically involve the use of letters and short reports of findings designed to inform and motivate respondents, procedures for tracing respondents who move address and methods for respondents to let the survey organisation know their new address, providing flexibility and adapting to the respondents' constraints regarding when and where interviews take place, mixed-mode data collection approaches and individual personal contacts in response to a bereavement, the birth of a child or a birthday. In the absence of experimental evidence, it is difficult to disentangle the effect of incentives from these other procedures, some of which may have significant impacts on response rates. Nonetheless, it is instructive to consider how longitudinal surveys are currently using incentives and how this process has changed over time for many long-running surveys.

Some major longitudinal surveys such as the Canadian Survey of Labour and Income Dynamics (SLID), the Swiss Household Panel (SHP) and the British Birth Cohort Studies (NCDS, BCS70, Millenium Cohort) have never used a respondent incentive. Surveys that do not use incentives instead rely on appeals to respondents' sense of altruism. The SHP additionally uses collective motivational measures (media communication). In the 2007 wave the SHP gave a small gift to each person who had participated in each of the previous four waves of the survey – the first time that any incentive or reward had been offered on the survey (Zimmerman *et al.*, 2003).

The form that incentives take varies across longitudinal surveys, as do the amounts paid. Surveys may make a financial payment in cash, cheque or money order or by issuing a debit card with a PIN number to respondents. Alternatively, they may use a cash equivalent such as a store or gift voucher, a lottery ticket or provide a small gift of some kind. Table 12.1 summarises current practice in the use of incentives on various longitudinal surveys.

Many surveys change the incentive treatment over time in an attempt to maximise longitudinal response rates. For long-running panel surveys that opt to use financial incentives from the start, it is almost inevitable that the incentive amount will have to increase over time in order to remain meaningful to respondents as the cost of

Table 12.1 Summary of incentive use on various longitudinal surveys (2005/2006).

Survey	Form of incentive	Amount or gift value	Conditionality and timing	Individual level	Household level bonus	Incentive varies for different groups/ households
Panel Study of Income Dynamics	Monetary (cheque or money order)	$60	Mailed 1 week after interview	Yes	No	No
Survey of Income Program Participation	Monetary (cheque or cash via debit card with PIN)	$40	Currently experimenting with unconditional in advance vs. at discretion of interviewer/ targeted at previous wave refusals only	Yes	No	Yes, depends on previous non-response
National Longitudinal Survey of Youth	Monetary (cash, cheque or money order) plus gift in-kind, e.g. meal/pizza. Additional $20 gift card in metropolitan areas	$40 base rate; $60 or $80 if respondent calls in to do interview Gift in-kind up to $20	Face-to-face interviews: cash at point of interview Telephone interviews: sent cheque after interview	Yes	No	Yes, depends on mode of data collection, whether respondent calls in or not, amount given in previous rounds, previous response history, and whether living in a metropolitan area
US Health and Retirement Survey	Monetary (cheque)	$40 $10 extra in 2006 for longer interview	Unconditional in advance	Yes	No	No

The National Longitudinal Survey of Children and Youth	Gift (age specific), e.g. 0–3 giraffe growth chart/4–5 colouring book or stickers/6 + book lights or key chains	$2	Given to child at interview	Yes	No	Yes, given to child only, no incentive for parents
German Socio-Economic Panel	Lottery ticket plus gifts	€1.50 (lottery ticket) €5–7 (gift)	Lottery ticket mailed after interview Gift given at interview.	Yes	No	Yes (mothers receive an extra lottery ticket for child questionnaire/ new 16 year olds receive additional gifts)
Household Income and Labour Dynamics in Australia	Monetary (cheque)	$25 per individual $25 per fully cooperating household	After interview (can take six weeks to process)	Yes	Yes	Yes, household bonus depends on full response
English Longitudinal Study of Ageing	Monetary (gift voucher)	£10	Mailed after interview	Yes	No	No
British Household Panel Survey	Monetary (gift voucher) plus gift (diary/pens)	£10 (voucher) £1 (gift)	Unconditional in advance. Gift given at interview	Yes	No	No

living increases. Increases in the incentive to long-serving sample members may also demonstrate the respondents' continued importance to the survey organisation and even relatively small increases may have some symbolic value in signifying this to respondents.

Panel Study of Income Dynamics (PSID)

The Panel Study of Income Dynamics (PSID) began in 1968 and initially paid respondents \$5 per interview, increasing this amount in fairly small incremental steps over the following years. By 1973 it had been raised to \$7.50, in 1981 it increased to \$10, in 1987 to \$12.50, in 1990 to \$15 and in 1995 to \$20. From 1997 the PSID has been a biennial survey rather than annual and when the interview increased substantially in length in 1999 (partly due to the introduction of an extensive child development supplement) the incentive was doubled to \$40 per interview and in 2005 had reached \$60 per interview. Despite the fact that the incentive is not a payment for time but a token of appreciation, these increases are recognition that in order for the incentive to be effective it must have some value to respondents relative to the current cost of living and also reflect the level of burden being imposed in terms of interview length. Current PSID practice is to pay an incentive of around \$1 per interview minute. A cheque is sent to respondents within a week of the interview being completed. In some cases a money order is sent instead if the respondent is unable to cash a cheque. Additionally, small 'finders fees' are paid to family members who provide a new address for sample members who have moved. Also, respondents are paid \$10 for returning an address confirmation card that is mailed to them each year with details of the next wave of the survey.

National Longitudinal Surveys of Youth (NLSY)

A similar progression of the incentive amount is observed on the National Longitudinal Survey of Youth (NLSY) in the USA, which has used financial incentives for both the 1979 and more recent 1997 samples. On the 1979 sample the incentive payment increased from \$10 to \$20 by 1996. In 1998, as a reaction to response rate concerns, households were offered a bonus of \$100–150 depending on the number of household members, which could be up to seven people. The survey team found that response rate continued to fall but judged that there would have been greater losses without this significant bonus. However, the subsequent round of interviewing suffered when the same incentive was not offered again, demonstrating the difficulties of raising respondent expectations in a longitudinal survey.

From 2002, the NLSY79 employed an alternative strategy where the base rate for an interview rose to \$40 but they attempted to get the more cooperative 'easy' cases at low cost by asking the respondent to call in to do the interview by telephone. Respondents are paid up to \$80 if they call in themselves to do the interview, resulting in reduced costs for these cases. Respondents who are called by the survey organisation receive a maximum of \$40. Telephone respondents receive their incentive as either cash or a cheque following the interview, unless they specifically request a money order (which can be cashed at most grocery stores, the post office or a local bank). Where the interview is conducted face to face the interviewer gives cash to the respondent at the end of the interview. Gifts in-kind and gift cards or food up to \$20 maximum for a household are also offered in some circumstances. For example the interviewer may take a pizza and soft drinks to a family where they are expecting to interview several people at the same visit.

The NLSY97 approach is similar to that of the earlier cohort. In rounds 5 and 6, the rate for all respondents was raised to $20. In round 6, as on the NLSY79 sample, the option to complete the interview by telephone was also offered and resulted in an increase in the overall response rates, an increase that the survey team attribute to mode flexibility rather than the incentive level. In previous rounds of the NLSY97, telephone interviews comprised 5–7 % of interviews completed compared to 15 % at round 6 of the survey. In rounds 7, 8 and 9 (fielded in 2005) the respondent fee was maintained at $20 per interview, but past round refusals or noninterviews received an extra $5 per missed round. The justification for this was greater burden, as the questionnaire collects event data back to the time of the previous interview. Initial analyses suggest that this is most effective with respondents who have been out of the survey for the shortest amount of time. In round 9, an additional $20 gift card was offered to respondents in major metropolitan areas where the cost of living is higher. Offering an alternative site, rather than the home, for the interview combined with the offer of a free meal is also a strategy that the NLSY have found to be effective. This is done by purchasing a light meal and meeting at a coffee house, chain restaurant or fast food service and doing the interview over a lunch hour. While the 'fee' is in-kind, respondents have been found to be quite receptive to this approach.

British Household Panel Survey (BHPS)

The British Household Panel Survey (BHPS) conducts an annual face-to-face interview and provides an incentive in the form of a store gift voucher rather than cash or cheque. From 1991 (wave 1) to 1995 this was £5 per interview and was raised to £7 per interview from 1996 (wave 6). In 1994, children aged 11–15 were interviewed for the first time and received £3 for this interview. In 1996 this was raised to £4 per youth interview. In 2004 (wave 14) the BHPS conducted a split-sample experiment on increasing the incentive to £10 (£5 for the youth interview). The results of this experiment are reported in Section 12.5.2. From 2005 all respondents have received £10 per interview (£5 for the youth interview). The BHPS differs from other studies in that it sends the incentive in advance to respondents who were interviewed at the previous wave. Interviewers have spare vouchers for any new household members or respondents who were not interviewed at the previous wave and hand these to them on conclusion of the interview. The BHPS also offers small gifts to respondents in addition to the gift voucher incentive. In the past these have included pens and diaries embossed with the survey logo and the diary has now become an annual feature that respondents have come to expect to receive. The cost of the diary is around £1 (GBP) per respondent and it is given to respondents by the interviewer when they visit. As with the NLSY, interviewers have some leeway to offer small additional gifts such as a bunch of flowers if they know there has been a bereavement, a box of chocolates for a birthday or a small toy for a new child, and so on. Interviewers also have the option to arrange to meet the respondent in a location other than the home if required.

The Health and Retirement Survey (HRS) and the English Longitudinal Study of Ageing (ELSA)

The US Health and Retirement Survey (HRS), which conducts a bi-annual survey with a sample of the over 50s, used an 'end game' strategy to boost response at their first wave in 1992 (Juster and Suzman, 1995). Reluctant respondents were offered a large financial bonus of $100 for participation and asked for an immediate yes or no decision on whether to take part, a strategy that increased the initial HRS response rate by around 4 %. From 1992 to 2002 the standard amount that respondents received on the HRS was $20, sent unconditionally in advance. In 2004 this increased to $40. Each sample member in a household receives this by cheque in the advance mailing prior to the interview. In 2006, a subsample was asked to do an expanded face-to-face interview and for this each respondent was given $50, reflecting what is seen as the additional burden of the interview length.

Another major survey in the UK, the English Longitudinal Study of Ageing (ELSA), began in 2002 and conducts a bi-annual interview. It has a similar design and aims to the HRS and gives a store gift voucher of £10 per individual interviewed. Unlike the HRS and the BHPS, these are given to the respondent by the interviewer at the point of interview rather than being sent out in advance of the interview.

The Survey of Income and Program Participation (SIPP)

The Survey of Income and Program Participation (SIPP), run by the US Census Bureau since 1984, is an example where the relationship between level of burden and paying an incentive does not apply. The SIPP is relatively burdensome for respondents as it conducts interviews at 4-month intervals, with each panel recruited for up to 12 interviews over a 32-month period. In response to concerns about attrition, the 2001 panel was reduced to nine waves over three years (Weinberg, 2002). The SIPP has never used financial incentives for all sample members, even though it is a survey where experimentation with gifts and financial incentives has been carried out through its history. The results of these experiments are summarised in Section 12.5. The most recent experiment was conducted on the 2001 panel and, in contrast to other surveys where the norm is to give all sample members the same or similar levels of incentive, the SIPP tested a discretionary payment against an unconditional incentive sent in advance to sample members who had been nonrespondents at the previous wave. Interviewers could offer an incentive of $40 if they thought it would be effective, to up to one-tenth of their sample in each one-year cycle.

The German Socio-Economic Panel Survey (SOEP)

The German Socio-Economic Panel Survey (SOEP), which has involved an annual interview with sample members since 1984, provides an incentive in the form of a ticket for the German national lottery rather than cash. Since the early 1990s various small gifts such as pens, bags, an umbrella, etc. have been given to respondents in addition to the lottery ticket. The cost of the lottery ticket for SOEP is €1.50 and the gifts cost an additional €5–7 per respondent. As the main carer of young children (usually the mother) is asked an additional questionnaire about their children, they are given an additional lottery ticket in recognition of the extra burden. The ticket is mailed to each individual respondent after the interview, a mailing that is combined with a follow-up to collect details of any new addresses. The gift is given to respondents by the interviewer when they call. While no experimental data are available on the effect of these incentives, interviewers on the survey report that having something to give on the doorstep makes it harder for respondents to refuse and increases participation through reciprocity. The SOEP has also used additional incentives to encourage response amongst their youth cohorts entering the main panel and being interviewed for the first time, as recruiting these young panel members is critical for the longer term health of the survey.

The Household Income and Labour Dynamics in Australia (HILDA) Survey

The Household Income and Labour Dynamics in Australia (HILDA) survey, which began in 2001, has taken a slightly different approach in that they gave a financial incentive to the household rather than to individual respondents for the first four waves of the survey (2001–2004). Where all eligible household members were interviewed a cheque for AUS $50 was sent out following the interview and for a partial household this was AUS $20. If in the follow-up fieldwork a partially cooperating household was converted into a fully cooperating household, a further AUS $30 was sent to the household. With this method of delivering the incentive it was assumed that the person named on the cheque would be responsible for giving each person their share, but the extent to which this happened is unknown. From 2005, HILDA respondents have received AUS $25 per individual with a bonus of AUS $25 to the household reference person (identified by the interviewer as the main person for communication about the survey) for a fully responding household. The rationale for this change was to encourage response at the individual level as well as gaining complete household cooperation. The previous incentive structure was also felt to be somewhat unfair to larger households who received the same amount as households with fewer members. In addition, there was some anecdotal information that the cheque sent to one person in the household was not always shared between household members but kept by one person. The new incentive structure aimed to ensure that no household was worse off than in previous years while removing some of the problems with the former system. On average it takes about six weeks following the interview for respondents to receive their cheque. As with other surveys, HILDA are not allowed to offer differential levels of incentives even though in practice the whole household completion bonus does this in return for full cooperation.

The National Longitudinal Survey of Children and Youth (NLSCY)

The National Longitudinal Survey of Children and Youth (NLSCY), a long-term child development study of Canadian children conducted by Statistics Canada and Social Development Canada (SDC), began in 1994. The sample consists of several longitudinal age cohorts selected between birth and 11 years of age, some of whom will be followed until they are 25 years old. The NLSCY is conducted face to face and gives small gifts depending on the age of the child(ren) in the sample. The 0–3-year group are given a giraffe growth chart, 4–5-year-olds get stickers or colouring books and older children are given small items such as key chains or book lights. The cost of the items is approximately two dollars per child and anecdotal comments from interviewers suggest that the children and parents generally appreciate the gifts.

Summary: Current Practice on Longitudinal Surveys

This review of current practice shows that longitudinal survey research teams are continually revising and rethinking their incentive structure in terms of the type of incentive offered, the value of the incentive and the delivery of the incentive. In many cases, these decisions are made on the basis of their own experience in the field, comments from interviewers and on the advice of other survey practitioners rather than being based on experimental evidence. While some surveys have conducted formal experiments to test the effects of changing incentive structures, there is surprisingly little evidence about the longitudinal effects of these changes. This may partly be explained by longitudinal survey organisations being reluctant to carry out large-scale experiments on their samples that may risk harming future response in some way. However, given the cost implications of increasing incentives or changing incentive structures and the potential benefits of tailoring, it is somewhat surprising that more information is not available to guide survey practitioners in this area. In the following section we review the results of some experiments that have been carried out on longitudinal surveys and report results from three recent UK studies where incentives have been changed during the survey.

12.5 EXPERIMENTAL EVIDENCE ON LONGITUDINAL SURVEYS

The evidence for the use of incentives on longitudinal surveys suggests that incentives can be effective in reducing attrition over multiple waves of a survey, and that making changes by introducing an incentive, offering higher amounts and targeting of various kinds does affect response, though these effects depend on the survey context. In this section we first summarise the findings of some experiments on longitudinal surveys reported in the literature and then look in more detail at three recent experiments in the UK.

Effects of Incentives on Cumulative Response

A question of interest is how incentives affect response not just at the wave of administration, but cumulatively over multiple waves. An experiment carried out on the SIPP in 1996 found that the effect of a $20 prepaid incentive in lowering nonresponse rates in the initial interview compared with both $10 prepayment and no incentive cases (James, 1997) was upheld over the first three waves of the survey. However, the difference in response between the $10 and no incentive cases was not significant, suggesting that the amount paid does have an independent effect on response over and above whether or not the incentive is unconditional. Mack *et al.* (1998) extended James' analysis to look at the effect of the incentive on response over six waves of the SIPP and also looked at the effect by the poverty status of the household, race and education level. They found that the $20 incentive reduced household, individual and item nonresponse (gross wages) in the initial wave 1 interview and that household response rates remained higher across all six waves for the $20 incentive group and the higher incentive was particularly effective for poor and Black households. The $20 incentive also significantly increased response rates in low education households at all of waves 2–6 and in high education households in waves 2–5. The $10 incentive did not reduce cumulative nonresponse over the six waves.

The positive effect on subsequent wave response does not seem to be limited to monetary incentives paid at an earlier wave. The Swiss Household Panel Survey carried out an experiment where entry into a lottery was offered in conjunction with completion of a biographical questionnaire sent by mail at wave 2 of the survey (Scherpenzeel *et al.*, 2002). Response rates increased amongst the group offered the lottery incentive at wave 2 and this positive effect persisted over the following three waves of the survey. This suggests that the incentive effect was enduring rather than just delaying response for one or two waves.

Effects of Differential Incentives on Response at Subsequent Waves

Practice on some surveys is to offer reluctant sample members increasingly higher incentives to secure cooperation. This is sometimes referred to as the 'end game'. A particular concern on longitudinal surveys is the impact that such a strategy might have on response at subsequent waves.

The effect of the end game strategy during refusal conversion on the first wave of the Health and Retirement Study (1992) on later wave response has been examined by Lengacher *et al.* (1995). Part of the concern at wave 2 of the survey was that respondents who had been paid up to $100 as part of the end game strategy at wave 1 would have significantly lower response at wave 2 when offered the standard $20 incentive due to an expectation effect. Lengacher *et al.* looked at three groups in the sample:

(1) those interviewed at the previous round with no persuasion;
(2) those who were reassigned to a different interviewer or sent a persuasion letter after an initial refusal at the previous round;
(3) those who were part of the nonresponse study and went through refusal conversion last time and were eventually interviewed.

They found that those in groups (2) and (3) were significantly less likely to be interviewed again at wave 2 compared to the compliant group (1) respondents. However, there were no significant differences between the response rates of those in group (2) compared to those in group (3). This suggests that the large payment of $100 at the first wave had no effect on increasing or decreasing later response relative to others who initially refused and were persuaded to take part by other means, nor did the large incentive at wave 1 induce an expectation that large incentives would be offered in later waves of the panel. Despite this, they did find an interaction effect between the level of enjoyment of the wave 1 interview and whether paid a large incentive. While for the sample as a whole the enjoyment of the first interview was associated with increased response propensity at wave 2, amongst those who were paid the large incentive those who enjoyed the interview were less likely to take part at wave 2. The large incentive appeared to cancel out the enjoyment effect, with the memory of the incentive being dominant in respondents' minds when asked to take part at wave 2.

The issue of whether unequal payments for reluctant respondents affects the cooperation of those who were not persuaded by offers of a larger incentive was examined in an experiment on the Detroit Area Study by Singer *et al.* (1999b). They tested the effect of disclosure of unequal incentives on later response. While respondents in the disclosure group perceived these unequal payments as being unfair, there was no significant difference in response rates to the survey one year later between the group who were told about the unequal payments and those who were not told. Singer *et al.* concluded that the factors that motivated participation in the survey were not associated with whether or not the unequal payments had been disclosed to them. Nonetheless, they conclude that this area deserves further enquiry (given that maintaining the goodwill of survey respondents is paramount) as there may be unintended consequences of perceptions of inequity.

Targeting Incentives Based on Response at Previous Waves

Longitudinal surveys offer considerable opportunities for targeting incentives based on the observed response behaviour of sample members at previous waves. Martin *et al.*

(2001) reported the results of an experiment on the SIPP that targeted prepaid incentives at nonresponding households from a previous wave. They found that both a $20 and a $40 incentive significantly improved conversion rates of people who had refused at an earlier wave compared to those who were offered no incentive.

The NLSY79 conducted an experiment in 2002 targeted on relatively cooperative 'easy' cases that were identified using contact data from previous waves. The aim was to establish whether it was possible to reduce data collection costs for these 'easy' respondents by asking them to call in to do the interview by telephone. The standard incentive was $40 but sample members were offered a higher amount if they chose to phone in to do the interview. This amount was randomly varied between $60 and $80 to test whether the amount had an effect on response. Response rate was higher when the opportunity to phone in was offered and the cost per interview was lower. However, the $20 difference in the incentive offered for calling in did not have any significant effect on response. Additionally, the most reluctant respondents were offered $80 and the opportunity to call in, but this had no effect on response rate (Kymn Kochanek, personal communication).

A study by Rodgers (2002), based on an experiment carried out in 2000 on the HRS, addressed the issue of targeting incentives based both on the previous response behaviour of sample members and on previous wave interview data that was felt likely to be predictive of future response propensity. Rodgers concluded that the greatest cost–benefit ratio would have been achieved by offering a higher incentive to households in which there was nonresponse at the previous wave. A similar conclusion was reached by a study based on the SIPP (Martin *et al.*, 2001). The study reported by Rodgers used incentives of $20, $30 and $50 across four strata defined by:

(1) having poor health at the previous round;
(2) proxied respondents at the previous round;
(3) noninterviewed respondents at the previous round;
(4) all other eligible households.

Rodgers found response rates to be consistently and significantly higher for those paid $50 for all groups apart from (2), with the response rates for the $30 payment being intermediate between the $20 and $50 response rates. This suggests that there may be a positive association between the amount of the incentive paid and response rates. Though the incentive was particularly effective for (3), Rodgers concludes that the HRS protocol of not automatically dropping nonrespondents from the survey at the following wave is a more significant factor in maintaining response rates than the incentive level.

Effects of Incentives on Fieldwork Costs

James (1997) found that either a $10 or $20 incentive reduced the number of calls that interviewers needed to make on the SIPP. Similarly, Rodgers (2002) found evidence that the number of calls interviewers had to make to achieve an interview was reduced with a $50 incentive, relative to $20 or $30, leading to some reduction in overall fieldwork costs. Similar effects have also been found on cross-sectional surveys (Lynn *et al.*, 1998). Finally, as reported above, the NLSY79 2002 study found that providing incentives to respondents to phone in at a time convenient to them produced an overall reduction in field costs.

12.5.1 Previous Experiments on UK Longitudinal Surveys

In the UK, two sets of experiments with incentives on longitudinal surveys have been carried out prior to our own experiment, which is reported in Section 12.5.2 below.

An experiment was carried out at waves 7 and 8 of the 1992–1997 British Election Panel Survey (BEPS: Lynn *et al.*, 1997). No incentives had been used on the previous six waves (four of which had been face to face, one telephone and one postal), but at wave 7 a random subset received an unconditional £5 incentive. Of those who received the incentive at wave 7, a random half received the same incentive again at wave 8 while the other half received no incentive. Despite the relative maturity of the panel at the stage when the experiment was carried out, the incentive had a positive effect on the wave 7 response rate. The proportion of sample members who responded at *both* waves 7 and 8 was slightly higher amongst those given an incentive at wave 7 only than those given no incentives, but was considerably higher again amongst those given an incentive at both waves. The proportion of wave 7 responders who responded at wave 8 did not differ between those given no incentives at either wave and those given an incentive only at wave 7. These findings suggest that incentive effects on response rate may be largely independent between waves, with little or no carry-over effect. Interestingly, the BEPS incentives experiment was interleaved with a separate experiment at wave 7 in which some sample members were told explicitly, for the first time, that they were part of a panel and could expect at least two more contacts. Incentive effects were slightly stronger amongst sample members who were *not* told they were in a panel.

The England and Wales Youth Cohort Study is a series of panel surveys of young persons aged 16–23. Each survey in the series samples an age-cohort of 16-year-olds who are then sent questionnaires on between three and five occasions over the following few years. On cohort 10, which used a combination of postal and telephone methods, an experiment with incentives was carried out. No incentives were provided at wave 1 (Spring 2000). Wave 2 (late 2000), which involved both telephone and postal samples, incorporated an experiment whereby within each sample a random subset were sent a £5 voucher while the remainder received no incentive. Furthermore, in the postal sample the incentive treatment group was subdivided into two: incentives were provided either unconditionally (the incentive was sent with the initial mailing) or conditionally (the voucher was promised in the original mailing, but only sent on receipt of a completed questionnaire). At wave 3 (Spring 2002), all incentives were paid unconditionally and all 'lower achievers' (identified from responses to questions about qualifications from earlier waves) were approached in postal mode. At wave 4 (Spring 2003), all respondents were sent postal questionnaires. Although the mode of treatment and the use of conditional or unconditional incentives changed across the waves, the allocation of individuals to either an incentive or control treatment was fixed across the waves.

Analysis of the data from this experiment (Jäckle and Lynn, 2008) showed that the positive effects of incentives on response propensity remained constant across waves: there was no evidence that incentives became less effective at increasing response across waves, for example, because the respondent sample became less sensitive to incentives as potentially less committed sample members dropped out. The difference in cumulative response rates between the incentive and no-incentive groups increased over waves. The positive effects of incentives on response propensity had little effect on sample composition, in terms of a range of characteristics. Incentives reduced nonresponse bias only in term of variables that could in any case be corrected by weighting (Jäckle and Lynn, 2004). The effect of incentives on response rate was stronger in postal mode than telephone, but in both modes there was little evidence of impact on attrition bias. Neither changes in the incentive offered nor changes in the mode of survey administration, conditional on the incentive offered, appeared to influence response propensity. That is, mode/incentive treatment at wave $t - 1$ had no effect on response propensity at wave t, conditional on mode/incentive treatment at wave t.

12.5.2 British Household Panel Survey Incentive Experiment

Respondents on the BHPS have always received an incentive in the form of a store gift voucher that is sent to previously cooperating respondents in advance of each wave of fieldwork. From 1996 (wave 6) to 2003 (wave 13) the value of the voucher was £7 for sample members eligible for the full adult interview (aged 16 or older) and £4 for those eligible for the shorter youth interview (aged 11–15). In 2004 (wave 14) a split-sample experiment was implemented to test the effect of increasing the value of the incentive to £10 for adults and £5 for youths – a relatively small increase in value. All persons in a random half of the sample households received the increased amounts, while those in the other half received the standard amounts. The experimental sample is a national general population sample of Great Britain consisting of just over 5000 households.[1] The design is described in more detail in Laurie (2007).

Laurie (2007) reports initial findings. The wave 14 individual response rate conditional upon full response at wave 13 was 96 % amongst sample members receiving £10 compared to 93 % amongst those receiving £7 ($P < 0.01$), perhaps a surprisingly large effect considering that the sample is one of established cooperative sample members and that the difference in value is small. We speculate that increasing an amount to which sample members have become accustomed may have a beneficial psychological effect independent of the value of the increase.

Effects on Reluctant Respondents

The increased incentive appears to have had a greater effect on response for those who were eligible but not interviewed at wave 13 than for those who were successfully interviewed at wave 13. This suggests that an increased incentive may be an effective strategy for respondents with a lower baseline response propensity. Amongst adults who had refused an interview at wave 13 in households where at least one person was interviewed (known as 'within-household refusals' – many of whom had persistently refused for a number of waves), the percentage providing a full interview at wave 14 was 13 % with the £10 incentive and 6 % with the £7 incentive.

The BHPS carries out telephone interviews with some sample members who cannot be persuaded to provide the full face-to-face interview. These telephone interviews take place as part of the refusal conversion process. Also, proxy interviews are accepted in certain circumstances where it is not possible to interview the sample member personally.

[1] The sample for the experiment included respondents in the original 1991 BHPS sample only, not the extension samples added in Scotland and Wales in 1999 and in Northern Ireland in 2001.

However, the survey instrument for telephone and proxy interviews is a reduced version of the full instrument, so providing a response in one of these two forms is suboptimal and can be considered a form of partial response. It is therefore desirable to find ways not only of increasing the proportion of the sample who respond at all, but also of increasing the proportion who provide a full face-to-face interview rather than a telephone or proxy interview. Amongst sample members who had provided a telephone interview at wave 13, the within-household refusal rate at wave 14 was just 3 % for the £10 group compared to 10 % for the £7 group. The percentage of wave 13 telephone respondents who were converted to a full interview at wave 14 was 19 % for the £10 group compared to 13 % for the £7 group.

Effects on Sample Entrants

At each wave young people turning 16 become eligible for a full adult interview and are effectively recruited into the main panel, making them an important group for the long-term health of the panel survey. Of new 16-year-olds who had completed a youth interview at wave 13 and were eligible for a full adult interview at wave 14 for the first time, the higher incentive increased the response rate significantly, from 91 % to 95 %. It should be noted that for 16-year-olds the £10 incentive represented an increase of £6 over the previous year's incentive, compared to an increase of £3 for those receiving the £7 incentive. This may suggest that for long-running panels where children of original sample members are recruited into the sample at a given age, some form of 'golden handshake' to welcome and encourage them into the main panel at that point may be an effective strategy to ensure as many as possible are recruited into the sample over the longer term. Whether the effect will hold over time for this group can only be assessed as future waves of data are collected.

Effects on Household Response

The rate of household response (meaning that at least the household interview was completed) for eligible households at wave 14 was higher amongst the £10 households (94.4 %) compared to the £7 households (92.4 %, $P < 0.01$). The majority of this difference is accounted for by an increase in the proportion of fully cooperating households (where all eligible household members provided a full interview) from 74.4 % with the £7 incentive to 77.7 % with the £10 incentive. The proportion of whole-household refusals reduced from 2.2 % with the £7 incentive to 1.0 % with the £10 incentive. There was also evidence that the higher incentive increased the chances of tracing and interviewing households that had moved address since the previous interview. Amongst nonmover households, the percentage of households cooperating fully increased from 78.2 % with the £7 incentive to 79.7 % with the £10 incentive, but amongst households where the whole household had moved the response rate increased by 10 %, from 58.7 % to 68.7 %. The increased incentive appeared to improve both location and cooperation rates amongst movers: the household refusal rate was 4 % lower amongst the £10 group of mover households than the £7 group and the proportion of untraced addresses was also reduced by half. As losing sample members through geographical mobility is a significant source of avoidable attrition over time, an incentive strategy that encourages mover households to remain in the survey could have a positive effect on longitudinal response rates, reduce the levels of differential attrition and lessen the potential for bias in the data.

Effects on Individual Response

At an individual level, previous wave response and response history across the life of the survey are significant predictors of current wave response. The increased incentive improved the response rate both for regular responders and for intermittent responders, but in different ways. Amongst sample members who had provided a full interview at all thirteen previous waves, 97.3 % of those receiving £10 gave a full interview, compared to 95.6 % of those receiving £7. This was achieved mainly through a reduction in the proportion providing telephone interviews, from 2.5 % with £7 to 1.2 % with £10. Amongst sample members who had been a nonrespondent to at least one previous wave, the proportion giving a full interview did not differ between the two treatment groups (59.0 % with £7; 59.3 % with £10) but the proportions giving either telephone or proxy interviews were higher with the £10 incentive (proxy: 3.1 % with £7, 4.0 % with £10; telephone: 8.3 % with £7, 14.1 % with £10). This corresponded to a reduction with the increased incentive in both the household refusal rate (from 12.4 % to 9.3 %) and the household noncontact rate (from 5.6 % to 2.6 %).

Amongst all persons known to be eligible for a full interview at wave 14, the effect of the higher incentive on survey outcomes is summarised in Table 12.2, both for the sample as a whole and for a number of important demographic subgroups. Overall, the increased incentive improved the wave 14 response rate from 77.6 % to 81.3 % ($P < 0.001$). The effect was significant, and similar in size, for both men and women. The effect was largest amongst the age group with the lowest response rate, namely 16–24-year-olds, for whom the response rate increased from 69.5 % with £7 to 77.6 % with £10. The result of this was that with the £10 incentive the response rate amongst this group was not much lower than that amongst other age groups. It may be that this group is less likely to be motivated to participate for altruistic reasons. However, the only other age group for which the £10 incentive increased the response rate significantly was the group with the highest response rate, namely 55–64-year-olds. The overall effect of the higher incentive on disproportionate response propensities by age is therefore unclear.

For marital status, the most marked increases in response rate with the higher incentive were for the separated and the never married, two groups who can be difficult to contact and interview – though the difference only reaches statistical significance for the latter due to the small sample size of the former. The response rate for the never married was 70.8 % with the £7 incentive and 79.3 % with the £10 incentive ($P < 0.001$). The effect of the incentive also varied by employment status, appearing most marked for the unemployed, another group who typically have lower response rates than others. The observed response rates amongst unemployed sample members were 70.7 % in the £7 group and 80.3 % in the £10 group, though this difference was not significant due to the modest sample sizes. Sample members whose main activity was looking after the home or family also demonstrated a particularly large effect on response of the higher incentive, from 80.9 % to 87.7 %. Differences in the effect of the incentive by level of education were limited, but the only significant effect was observed amongst the group with the lowest response rates, those with a lower level 'other' qualification. Amongst this group, response rate increased from 83.6 % with £7 to 91.3 % with £10.

Table 12.3 presents predicted coefficients from three logistic regression models where the dependent variable is an indicator of individual wave 14 response (including proxy and telephone interviews). Model 1 includes only the main effects of age and gender plus incentive treatment. The estimates indicate that the main effect of the higher value

Table 12.2 BHPS wave 14 response outcomes by demographic characteristics and incentive treatment.

	Unchanged incentive (£7 voucher)				Increased incentive (£10 voucher)			
	Full interview (%)	Proxy/ phone (%)	No interview (%)	*N*	Full interview (%)	Proxy/ phone (%)	No interview (%)	*N*
All ***	77.6	5.2	17.2	5053	81.3	5.2	13.6	4888
Gender								
Male **	73.4	6.0	20.6	2443	77.6	6.0	16.4	2345
Female **	81.5	4.6	14.0	2610	84.7	4.4	11.0	2543
Age								
16–24 ***	69.5	5.4	25.2	766	77.6	4.9	17.6	722
25–34	73.7	6.2	20.1	889	78.2	4.3	17.5	887
35–44	80.5	4.9	14.6	1022	82.6	5.9	11.5	950
45–54	80.5	5.0	14.5	780	81.7	5.6	12.7	803
55–64 *	81.4	5.1	13.6	708	85.9	5.6	8.5	647
65 and over	79.5	5.0	15.5	888	82.0	4.8	13.2	879
Marital status								
Married **	85.0	6.2	8.9	2482	87.6	5.9	6.5	2355
Cohabiting	81.5	4.8	13.7	664	83.9	4.7	11.5	620
Widowed	83.3	4.2	12.5	287	84.7	3.8	11.6	320
Divorced	84.6	4.0	11.5	253	87.2	5.6	7.3	234
Separated	80.5	2.6	16.9	77	91.1	3.6	5.4	56
Never married***	70.8	5.2	24.0	945	79.3	4.1	16.5	944
Employment status								
Employee***	81.5	5.8	12.8	2487	85.4	5.2	9.4	2480
Self-employed*	80.7	5.7	13.6	352	84.9	7.5	7.5	332

Table 12.2 (Continued).

	Unchanged incentive (£7 voucher)				Increased incentive (£10 voucher)			
	Full interview (%)	Proxy/ phone (%)	No interview (%)	*N*	Full interview (%)	Proxy/ phone (%)	No interview (%)	*N*
Unemployed	70.7	4.9	24.4	164	80.3	2.3	17.4	132
Retired	87.2	4.2	8.6	897	87.4	4.5	8.1	866
Family care*	80.9	5.9	13.3	324	87.7	5.1	7.2	277
Full-time student	76.3	5.4	18.4	261	78.7	5.1	16.2	253
Long-term sick/disabled	78.4	7.4	14.2	176	80.4	7.0	12.7	158
Other*	69.6	5.4	25.0	56	87.0	6.5	6.5	46
Highest qualification								
Degree /higher degree	89.9	3.8	6.4	581	93.3	1.9	4.8	579
Teach/nurse/other higher	90.6	2.8	6.7	1228	92.0	2.3	5.7	1253
A level or equivalent	85.5	2.6	11.9	498	86.8	3.8	9.4	498
GCSE/O level	85.5	3.8	10.6	781	89.1	2.1	8.8	718
Other**	83.6	1.9	14.4	360	91.3	1.9	6.9	321
None	84.1	3.6	12.3	835	86.8	3.9	9.4	779

*** $P < 0.001$; ** $0.001 < P < 0.01$; * $0.01 < P < 0.05$.

Notes: The experiment is restricted to the original wave 1 BHPS sample and their descendants; the Scottish and Welsh boost samples and the Northern Ireland sample were excluded. The base is all sample members believed to be eligible for an interview at wave 14 and issued to field, regardless of response history at previous waves. 'No interview' includes persons who were refusals or noncontacts within an otherwise cooperating household as well as those in nonresponding households. The indicators of marital status, employment status and highest qualification are taken from the most recent interview data available, within the previous six years: for 84.6 % of cases these indicators are from wave 13, 4.2 % from wave 12, 1.8 % from wave 11, 1.3 % from wave 10, 0.8 % from wave 9 and 0.5 % from wave 8; 6.8 % of cases had not completed an interview in the previous six years and are excluded from the analysis by these three variables.

Table 12.3 Logistic regression predicting response at wave 14.

	Model 1		Model 2		Model 3	
	Odds ratio	SE	Odds ratio	SE	Odds ratio	SE
Higher value incentive	1.099***	0.021	1.098*	0.053	1.061	0.054
(vs. lower value)						
Female	1.586***	0.090	1.589	0.511	1.624	0.554
(vs. male)						
16–24 years	0.551***	0.051	0.336*	0.176	0.304*	0.189
25–34 years	0.649***	0.059	0.859	0.443	1.030	0.554
45–54 years	0.962	0.096	1.425	0.808	1.276	0.731
55–64 years	1.203	0.132	0.605	0.377	0.390	0.249
65 years and over	0.872	0.083	1.141	0.620	0.270†	0.208
(vs. 35–44 years)						
Interaction HiVal × Female			1.000	0.038	0.993	0.040
Interaction HiVal × 16–24			1.062	0.066	1.080	0.080
Interaction HiVal × 25–34			0.967	0.059	0.947	0.060
Interaction HiVal × 45–54			0.954	0.064	0.968	0.065
Interaction HiVal × 55–64			1.087	0.081	1.118	0.086
Interaction HiVal × 65+			0.968	0.062	0.993	0.090
Self-employed					0.568	0.432
Unemployed					0.971	0.836
Retired					8.498**	7.025
Family care					0.362	0.301
Full-time student					5.635*	4.472
Long-term sick/disabled					2.308	2.233
Other					0.039†	0.070
(vs. employed)						
Cohabiting					1.425	0.797

Table 12.3 (Continued).

	Model 1		Model 2		Model 3	
	Odds ratio	SE	Odds ratio	SE	Odds ratio	SE
Widowed					0.911	0.796
Divorced					0.591	0.548
Separated					0.067	0.118
Never married (vs. married)					0.430†	0.213
Interaction HiVal × Self-emp					1.157	0.107
Interaction HiVal × Unemployed					0.994	0.103
Interaction HiVal × Retired					0.975	0.095
Interaction HiVal × Family care					1.205†	0.123
Interaction HiVal × Student					0.854†	0.080
Interaction HiVal × LT Sick					0.941	0.107
Interaction HiVal × Other					1.537†	0.352
Interaction HiVal × Cohabiting					1.006	0.067
Interaction HiVal × Widowed					0.983	0.099
Interaction HiVal × Divorced					1.117	0.125
Interaction HiVal × Separated					1.436	0.325
Interaction HiVal × Never married					1.098	0.064

*** $P < 0.001$; ** $0.001 < P < 0.01$; * $0.01 < P < 0.05$; † $0.05 < P < 0.10$.

Notes: The base is all sample members believed to be eligible for an interview at wave 14 and issued to field, regardless of response history at previous waves. The dependent variable takes the value 1 if a full interview, telephone interview or proxy interview was achieved for the sample member at wave 14, 0 otherwise. Demographic variables are defined as in Table 12.2, so the 675 cases with missing values are excluded from this analysis, leaving an analysis base of 9265 cases.

incentive is positive and significant, increasing the odds of responding by a factor of 1.10. The significant associations of gender and age with response are also apparent.

In model 2, interaction terms are introduced in order to test whether the effect of the incentive varies across age groups or between the sexes. However, the results provide no evidence of any such differential effects between the groups. The predicted odds ratios (increased incentive vs. unchanged incentive) are highest for 16–24-year-olds (1.17) and for 55–64-year-olds (1.19), but neither of these estimates are significantly different from the average estimate of 1.10.

In model 3, indicators of economic activity status and de facto marital status are introduced along with their interactions with the incentive treatment. There is a suggestion that the effect of the incentive may be stronger for those whose main activity is looking after the family or home ($P = 0.07$) and those with an 'other' economic status ($P = 0.06$), relative to those in employment. The effect may be weaker for students ($P = 0.09$). With respect to marital status, there is a suggestion that the effect of the incentive may be greater amongst the small group who are separated from partners ($P = 0.11$) and those who have never been married ($P = 0.11$). A six-category indicator of level of education was also tested but was dropped from the model as none of the interaction terms even approached significance, providing no evidence that the effect of the incentive level varies by the level of education of the sample member.

In summary, the overall message seems to be that increasing the value of the incentive was effective at improving response rate across all the demographic groups tested, to broadly similar degrees. Where there were differences, the effect tended to be stronger amongst groups with relatively low response rates, notably 16–24-year-olds, but this was not a clear pattern. In so far as this is true, increasing the value of an incentive during the course of a panel survey could help to reduce nonresponse bias, though we have not assessed this directly here. It should be noted that the apparently smaller number of significant associations in the logistic regression models, compared to the descriptive analysis of Table 12.2, could partly be a result of the definition of the dependent variable. Table 12.2 shows that a reduction in the proportion of nonresponding cases is often associated with a reduction in the number of telephone or proxy responses too, whereas the modelling treats telephone and proxy responses in the same way as full responses.

12.6 CONCLUSION

Several aspects of the use of respondent incentives are shared between cross-sectional and longitudinal surveys. In both contexts, an incentive sent unconditionally in advance of the interview appears to be most effective in increasing response rate, cash incentives are more effective than gifts in-kind, a higher incentive amount tends to produce a higher response rate and there is some, but not consistent, evidence that incentives are more effective for those who are least likely to respond to the survey.

However, some aspects are distinct in the case of longitudinal surveys. Effects on long-term retention rates are perhaps more important than effects on wave-specific response rates and the evidence suggests that the effect of a repeated incentive can become more pronounced the more waves are involved. Retention rates are important in their own right as sample size cannot easily be manipulated for a long-term panel survey as it can for a cross-sectional survey. But arguably of more importance is attrition bias. Here, the effect of incentives is less clear. There is some evidence that incentives act disproportionately on sample members with low retention propensities, suggesting that they have the potential to reduce bias. But some studies failed to show any effect at all of incentives on sample composition.

Longitudinal surveys often adjust their practices with regard to incentives over time, typically increasing the value of the incentive to keep it in line with the general cost of living and in some cases varying the type of incentives used. Our own study, reported in Section 12.5.2 above, suggests that even a small increase in the value of an incentive on a mature panel can bring a significant improvement in response rates. This might suggest that regular small increases in the value of the incentive could be more effective than an occasional larger increase. Alternatively, the finding could be specific to the context in which the value of the incentive had remained unaltered for eight annual waves. Further research is needed on this point.

Even though most longitudinal surveys do not use differential incentive amounts for respondents based on past response behaviour or predicted response propensities, most have some circumstances under which a household or individual may receive more than the standard amount. Variation in incentive amounts within a sample is therefore already accepted, albeit perhaps implicitly only in order to reflect differences in the burden of participation. The great potential of longitudinal surveys to allow incentives to be tailored to the sample member's individual circumstances has not yet been realised. Tailoring could take many forms. The value of the incentive is just one dimension that could be varied between sample members. Others might include the nature of the incentive (e.g. different kinds of vouchers), the way it is partitioned over time or waves (e.g. two small incentives, perhaps one with a between-wave mailing and one with the advance letter, versus a single larger incentive per wave), the timing of administration relative to data collection waves, and so on. Longitudinal surveys have thus far demonstrated very little willingness to experiment with targeted treatments, so evidence on effective strategies is thin.

Overall, it seems clear that the use of respondent incentives is an important element of the strategy to minimise attrition for many longitudinal surveys. The evidence suggests consistently that attrition rates would be higher in the absence of incentives, but we have limited knowledge of what the optimum strategies are for any given design of longitudinal survey and whether or how incentive strategies translate into improvements in the accuracy of estimation over the longer term. In particular, we still know relatively little about the effect of changing incentive amounts or delivery methods during a longitudinal survey, targeting particular groups based on demographic characteristics or previous response history, the use of differential incentive amounts for different cases or circumstances and the longer term effect of incentives on attrition and bias. We urgently need to extend the research knowledge base if we are to be able to use survey budgets effectively and wisely when choosing respondent incentive strategies for longitudinal surveys.

ACKNOWLEDGEMENTS

We would like to thank our colleagues on other longitudinal surveys for their invaluable assistance in providing details of the incentive strategies used on their studies. Our thanks to: Kate McGonagle (PSID, University of Michigan); Nicole Watson (HILDA, University of Melbourne); Kymn Kochanek (NLSY/NORC, University of Chicago); Jürgen Schupp (SOEP, Berlin); Sylvie Michaud (NLSCY, Statistics Canada).

REFERENCES

Balakrishnan, P. V., Chawla, S. K., Smith, M. F. and Micholski, B. P. (1992). Mail survey response rates using a lottery prize giveaway incentive. *Journal of Direct Marketing*, 6, 54–59.

Baumgartner, R. and Rathbun, P. (1997). Prepaid monetary incentives and mail survey response rates. Paper presented at the *Annual Conference of the American Association for Public Opinion Research*, Norfolk, Virginia.

Brüderl, J., Castiglioni, L., Krieger, U. and Pforr, K. (2008). The effects of incentives on attrition bias: results of an experimental study. *Paper presented at the 1st International Panel Survey Methods Workshop*, July, University of Essex, UK.

Burton, J., Laurie, H. and Moon, N. (1999). "Don't ask me nothin' about nothin', I might just tell you the truth". The Interaction between unit non-response and item non-response. Paper presented at the *International Conference on Survey Response*, October, Portland, Oregon.

Church, A. H. (1993). Estimating the effect of incentives on mail survey response rates: a meta-analysis. *Public Opinion Quarterly*, 57, 62–79.

Couper, M. P., Ryu, E. and Marans, R. W. (2006). Survey incentives: cash vs in-kind; face-to-face vs mail; response rate vs nonresponse error. *International Journal of Public Opinion Research*, 18, 89–106.

Groves, R. M., Singer, E. and Corning, A. (2000). Leverage-saliency theory of survey participation: description and illustration. *Public Opinion Quarterly*, 64, 299–308.

Halpern, S. D. and Asch, D. A. (2003). Improving response rates to mailed surveys: what do we learn from randomized controlled trials? *International Journal of Epidemiology*, 32, 637–638.

Hill, D. H. and Willis, R. J. (2001). Reducing panel attrition: a search for effective policy instruments. *Journal of Human Resources*, 36, 416–438.

Jäckle, A. and Lynn, P. (2004). In the long run: lessons from a panel survey respondent incentives experiment. In *Proceedings of the Survey Research Methods Section of the American Statistical Association 2004* (pp. 4794–4801). Retrieved September 26, 2008, from www.amstat.org/sections/srms/Proceedings/

Jäckle, A. and Lynn, P. (2008). Respondent incentives in a multi-mode panel survey: cumulative effects on nonresponse and bias. *Survey Methodology*, 34, 105–117.

James, J. M. and Bolstein, R. (1990). The effect of monetary incentives and follow-up mailings on the response rate and response quality in mail surveys. *Public Opinion Quarterly*, 54, 346–361.

James, J. M. and Bolstein, R. (1992). Large monetary incentives and their effect on mail survey response rates. *Public Opinion Quarterly*, 56, 442–453.

James, T. L. (1997). Results of the wave 1 incentive experiment in the 1996 Survey of Income and Program Participation. *1997 Proceedings of the Survey Research Methods Section of the American Statistical Association* (pp. 834–839). Washington, DC: American Statistical Association.

Juster, F. T. and Suzman, R. M. (1995). An overview of the Health and Retirement Study. *Journal of Human Resources*, 30 (Suppl.), S7–S56.

Kim, C. D., Dant, S., Lee, C. C. and Whang, Y.-O. (2001). Increasing response rate in industrial mail surveys: the effect of respondent involvement in sweepstakes incentive. *Journal of the Academy of Marketing Studies*, 5, 49–56.

Laurie, H. (2007). The effect of increasing financial incentives in a panel survey: an experiment on the British Household Panel Survey, wave 14. *ISER Working Paper*, No. 2007-05. Colchester: University of Essex. Retrieved September 26, 2008, from www.iser.essex.ac.uk/pubs/workpaps/pdf/2007-05.pdf

Lengacher, J. E., Sullivan, C. M., Couper, M. P. and Groves, R. M. (1995). *Once Reluctant, Always Reluctant? Effects of Differential Incentives on Later Survey Participation in a Longitudinal Study*. Survey Research Centre, University of Michigan.

Lynn, P., Buck, N., Burton, J., Jäckle, A. and Laurie, H. (2005). A review of methodological research pertinent to longitudinal survey design and data collection. *ISER Working Paper*, No. 2005–29. Colchester: University of Essex.

Lynn, P. and Sturgis, P. (1997). Boosting survey response through a monetary incentive and fieldwork procedures: an experiment. *Survey Methods Centre Newsletter*, 17, 18–22.

Lynn, P., Taylor, B. and Brook, L. (1997). Incentives, information and number of contacts: testing the effects of these factors on response to a panel survey. *Survey Methods Centre Newsletter*, 17, 7–12.

Lynn, P., Thomson, K. and Brook, L. (1998). An experiment with incentives on the British Social Attitudes Survey. *Survey Methods Centre Newsletter*, 18, 12–14.

McCool, S. F. (1991). Using probabilistic incentives to increase response rates to mail-return highway intercept diaries. *Journal of Travel Research*, 30, 17–19.

Mack, S., Huggins, V., Keathley, D. and Sundukchi, M. (1998). Do monetary incentives improve response rates in the Survey of Income and Program Participation? In *Proceedings of the American Statistical Association, Survey Research Methods Section* (pp. 529–534). Washington, DC: American Statistical Association.

Martin E., Abreu, D. and Winters, F. (2001). Money and motive: effects of incentives on panel attrition in the Survey of Income and Program Participation. *Journal of Official Statistics*, 17, 267–284.

Rodgers, W. (2002). Size of incentive effects in a longitudinal study. Presented at the *2002 American Association for Public Research Conference*, mimeo, Survey Research Centre, University of Michigan, Ann Arbor.

Scherpenzeel, A., Zimmermann, E., Budowski, M., Tillmann, R., Wernli, B. and Gabadinho, A. (2002). Experimental pre-test of the biographical questionnaire, *Working Paper*, No. 5-02. Neuchatel: Swiss Household Panel. Retrieved September 26, 2008, from http://aresoas.unil.ch/workingpapers/WP5_02.pdf

Schrapler, J. P. (2003). Respondent behaviour in panel studies – a case study for income-nonresponse by means of the British Household Panel Study (BHPS), *ISER Working Paper*, No. 2003-08. Colchester: University of Essex.

Shettle, C. and Mooney, G. (1999). Monetary incentives in government surveys, *Journal of Official Statistics*, 15, 231–250. Retrieved October 22, 2008, from www.jos.nu/Articles/abstract.asp?article=152231

Singer, E. (2000). The use of incentives to reduce nonresponse in household surveys. *Working Paper*, No. 051, Survey Methodology Program, Institute for Social Research, University of Michigan.

Singer, E. (2002). The use of incentives to reduce nonresponse in household surveys. In R. M. Groves, D. A. Dillman, J. L. Eltinge and R. J. A. Little (Eds), *Survey Nonresponse* (pp. 163–177). New York: John Wiley & Sons, Inc.

Singer, E., Groves, R. M. and Corning, A. D. (1999b). Differential incentives: beliefs about practices, perceptions of equity, and effects on survey participation. *Public Opinion Quarterly*, 63, 251–260.

Singer, E. and Kulka, R. A. (2000). Paying respondents for survey participation. *Survey Methodology Program Working Paper*, No. 092, Survey Research Center, Institute for Social Research, University of Michigan, Ann Arbour.

Singer, E., Van Hoewyk, J. and Gebler, N. (1999a). The effect of incentives on response rates in interviewer-mediated surveys. *Journal of Official Statistics*, 15, 217–230. Retrieved October 22, 2008, from www.jos.nu/Articles/abstract.asp?article=152217

Singer, E., Van Hoewyk, J. and Maher, P. (1998). Does the payment of incentives create expectation effects? *Public Opinion Quarterly*, 62, 152–164.

Singer, E., Van Hoewyk, J. and Maher, M. P. (2000). Experiments with incentives in telephone surveys. *Public Opinion Quarterly*, 64, 171–188.

Tzamourani, P. (2000). An experiment with promised contribution to charity as a respondent incentive on a face-to-face survey. *Survey Methods Newsletter*, 20, 13–15.

Tzamourani, P. and Lynn, P. (2000). Do respondent incentives affect data quality? Evidence from an experiment. *Survey Methods Newsletter*, 20, 3– 7.

Warriner, K., Goyder, J., Gjertsen, H., Hohner, P. and McSpurren, K. (1996). Charities, no; lotteries, no; cash, yes. Main effects and interactions in a Canadian incentive experiment. *Public Opinion Quarterly*, 60, 542–562.

Weinberg, D. H. (2002). The Survey of Income and Program Participation – recent history and future developments, *Working Paper*, No. 232. Washington, DC: US Census Bureau.

Willimack, D. K., Schuman, H., Pennell, B-E. and Lepkowski, J. M. (1995). Effects of a prepaid nonmonetary incentive on response rates and response quality in a face-to-face survey. *Public Opinion Quarterly*, 59, 78–92.

Zimmerman, E., Budowski, M., Gabadinho, A., Scherpenzeel, A., Tillmann, R. and Wernli, B. (2003). The Swiss Household Panel Survey 2004–2007. Proposal submitted to the Swiss National Science Foundation, *Working Paper*, No. 2_03. Neuchatel: Swiss Household Panel. Retrieved September 26, 2008, from http://aresoas.unil.ch/workingpapers/WP2_03.pdf

CHAPTER 13

Attrition in Consumer Panels

Robert D. Tortora

The Gallup Organization

13.1 INTRODUCTION

This chapter examines attrition in the Gallup Poll Panel (GPP). The GPP is a consumer panel (Sudman and Wansink, 2002) that surveys members on a wide variety of topics, including market, opinion and social research. After providing an outline of consumer panels the paper describes the GPP. Then the remainder of the paper focuses on attrition in the GPP and the results of three experiments that were conducted with the goal of finding treatments that would minimize or reduce attrition.

Gallup decided to build a panel for several reasons. First and foremost is the declining response rate in Random Digit Dial (RDD) surveys in the USA (Tuckel and O'Neill, 2001; Steeh *et al.*, 2001), Tortora (2004) documents a decline in response rate for a large quarterly RDD survey conducted by Gallup[1] from 37.0 % to 20.4 % from December 1997 through to September 2003. This decline occurred while cooperation rates declined from 49.0 % to 36.9 %, answering machines went from 3.5 % to 18.3 % and 'ring, no answers' increased from 18.3 % to 22.5 %. These factors not only increase the cost of RDD surveys, since more effort has to be expended to maintain response rates, but quality questions also arise. In addition, coverage is becoming more problematic as the percentage of cell/mobile telephone-only users rapidly increases in the USA (Tortora *et al.*, 2007). These telephone numbers are typically excluded in traditional RDD surveys in the USA.

Thus increasing costs and possible decreasing quality indicate the need to investigate alternatives to current RDD surveys. One such alternative is to build a panel.[2]

---

[1] A quarterly sample size of 15 000 using a minimum of five call attempts to resolve the status: that is, obtain an interview, a refusal and an out-of-scope number such as a business, etc. for each telephone number.

[2] Gallup decided to build this panel using RDD recruiting, but knowing that in the future this method may not be viable.

Methodology of Longitudinal Surveys P. Lynn

However, due to the high cost of managing and maintaining a panel and possible bias, attrition is a crucial issue. Excessive attrition can cause recruiting costs to rise to unacceptable levels, especially with the Gallup approach of using RDD for recruiting. High levels of attrition are accompanied by the potential for biased results, particularly if attrition is not at random but among certain panel (demographic) groups.

Consumer panels can be separated into two general types: those where panel members report on essentially the same topic over time[3] and those where panel members report on a variety of topics over an often irregular time period (called a discontinuous panel). Sudman and Wansink (2002, pp. 14 and 15) identify over 50 panels in 10 countries worldwide. Thus while some panels may be focused on a particular topic such as food consumption or media use, discontinuous panels survey members on a variety of topics that can include such topics as evaluations of new products, marketing and advertising experiments and opinion polls. The time period between surveys of panel members can vary, with some panels contacting members on a regular basis, for example to measure restaurant food consumption, and other panels requesting respondents to provide data on an irregular basis, when the need arises. Göritz (2002) defines access panels as a group of people who have agreed to take part in surveys on a repeated basis. This definition does not distinguish between panels that use random versus nonrandom methods of recruitment.

Consumer panels vary in how they are constructed or how they recruit members. Many panels use nonprobability methods to recruit members. Recruitment can occur through telephone calls from lists (not necessarily representative of a population), placing advertisements in various media outlets, recruiting on line and through the use of large mailing lists. Especially in panels that are constructed for market research purposes, some kind of incentive is generally used to attract members. The incentive can take many forms, such as cash, gifts and entry into lotteries (see Göritz (2004) for a description of several incentive experiments in online panels). One common way in which panels implement incentives is by rewarding respondents when they join as well as accumulating points for gifts by completing additional surveys.

As Sudman and Wansink note, many modes of data collection can be used for these panels. Over the last few years online discontinuous panels are appearing on a regular basis. In these panels almost all of the data are collected on line. With these panels in particular the universe of coverage is often unknown so other techniques such as quota sampling and models (Lee, 2006) must be used. It is interesting to note that many panels tout their size as an important feature versus whether they represent any particular universe.

There are several other potential sources of error that consumer panels have in common with longitudinal surveys: conditioning of respondents; nonresponse bias, either item or unit, in any particular survey of the panel (analogous to a wave of a longitudinal survey); and bias due to attrition caused by people dropping out of the panel.

A conditioning bias (Clinton, 2001; Sturgis *et al.*, Chapter 7 of this volume) can occur if membership in a panel influences respondent responses or behaviours. For example, if membership in the GPP causes members to become more interested in politics, then these members may change their opinions on surveys related to political actions. However this paper does not examine conditioning bias.

---

[3] Often called a longitudinal survey.

We distinguish between nonresponse bias, which can occur on any given survey when a panel member does not respond to a particular survey request, and attrition bias, which may occur when a panel member is no longer a member of the panel and hence would not respond to any further survey requests. Attrition bias can occur when those who drop out of a panel are systematically different to those who continue to participate. Note that in the case of potential nonresponse bias a panel member may not respond to a particular survey request, but will/does respond to a later request(s). Neither does this chapter address item nonresponse, although Lee *et al.* (2004) found that item nonresponse to an initial survey indicates some disposition to attrition during subsequent diary surveys.

The next section describes the Gallup Poll Panel in more detail.

13.2 THE GALLUP POLL PANEL

Why the Gallup Poll Panel? As noted above, RDD surveys are becoming problematic in the USA. The problems of potential nonresponse and coverage bias associated with RDD surveys result in increased cost in attempts to maintain quality. Besides perhaps offering a replacement for traditional RDD surveys, panels have other advantages: speed of data collection, especially if using telephone interviewing; the ability to target subpopulations without extensive screening; and finally, for any given survey of the panel, lower one-time data collection costs even when including a fee for panel maintenance.

The GPP uses a RDD telephone survey to recruit members.[4] A random adult is not selected in the recruit interview, rather any adult who answers the telephone or comes to the telephone is recruited, since the goal of the panel is to recruit up to three other members of the household aged 13 and above. Recruiting covers the continental USA, Alaska and Hawaii. Before making the request to join the panel the recruit survey asks[5] five questions about current topics of interest in the USA. Topics might include Presidential Approval Rating, the economy, the war in Iraq, etc. No incentives are offered for panel participation. The message used to encourage joining the panel revolves around panelists being able to influence government and business decisions. During the telephone recruit various household-level demographics are obtained, in addition to some individual demographics of the person on the phone.[6]

Recruiting was rather uneven over the period ending in November 2005. The two major causes of this were client demand on the Gallup call centres and the Gallup learning process in building a panel. In several instances over the initial year of the panel client demand became so heavy that Gallup discontinued panel recruitment, sometimes for as little as a week and once for over two months. At other times while client demand was high a small amount of recruiting went on. A major part of the learning process

[4] For the time period covering the data in this paper only English language recruiting was used.

[5] Initially the recruit questionnaire did not include these questions but a small pilot found that a high level of resistance to joining the panel was reduced if a set of questions about topics of interest were asked first. The author believes these questions legitimize the survey and the request to join the panel.

[6] A 1 in 20 sample of persons who refuse the recruit survey are asked for person-level demographics. At the time of this analysis there appears to be no differences between those agreeing to join the panel and those refusing. However, note that this comparison is based on the initial household respondent, who is not a randomly selected adult.

revolved around the need to build up questionnaires on a set of different topics to administer to the panelists, which contributed to this unevenness of the recruiting.

Upon agreement to join, the potential panel member is sent a 'Welcome Package'. This package includes a letter from George Gallup, Jr. explaining the purpose of the panel, a fact sheet on 'Why Should I do This?', a small 'token of appreciation' like a refrigerator magnet of the GPP and an intake (booklet) questionnaire where up to three other members of the household, aged 13 plus, can join the panel. Since there is evidence (see Hoogendoorn and Sikkel, 1998) that attrition increases as the difficulty and length of tasks panel members are asked to perform increase, the burden on members of the Gallup Poll Panel is intentionally reduced. Those that elect to join the panel are committing to two or three surveys per month, with a typical survey lasting between 5 and 15 minutes, although Gallup has asked members to complete longer surveys.

Up to five call attempts are made to contact and interview an adult in each household selected in the RDD sample. Slightly more than 63 % of those contacted agreed to join the panel, and 55 % of those who agreed to join the panel actually returned the Welcome Questionnaire. The overall CASRO or AAPOR3 response rate was 27 %.

Each person who agrees to join becomes a panel member if they provide the following set of minimal demographics; gender, birth date,[7] race, ethnicity and education. Otherwise, they are administered a telephone demographic callback survey to fill in these details.

Gallup provides token gifts and a quarterly magazine to panel members. Gallup does not offer cash, prizes or lotteries for members to join or to complete surveys. Gallup uses this approach to control panel costs and as a way to minimize attracting the professional survey respondent: a person who joins a panel for the money or gifts. Panel member engagement is encouraged by providing members with token gifts (generally costing around $1, such as memberships cards, calendars and magnets). In addition panel households receive the quarterly magazine, organized around a specific theme that attempts to appeal to a wide variety of panel members.

After returning the intake questionnaire, panel members are asked to complete surveys on a variety of topics. Some of the surveys are about specific activities of panel members, such as the use of banking and financial institutions, travel and hotel usage, eating out, automobile ownership or leasing. Others are on a variety of political and social survey topics. The former types of surveys are used to measure overall customer engagement with a business (the particular bank, hotel, restaurant chain, make and mode of car) and to give Gallup the ability to target panel members for future surveys, especially those aimed at population subgroups with low incidence. Table 13.1 summarizes the 32 GPP surveys conducted in chronological order up to November 2005. It shows the month and year the survey was fielded, the sample size (number invited to take part), mode(s) of data collection used and the subpopulation sampled. The first survey in Table 13.1, Clothes Shopping 1, was conducted in September 2004, even though recruiting started in May.[8] Many of the surveys were to build up a panel database to allow Gallup to efficiently target consumers for a variety of client topics. Any survey with an asterisk (*) falls in this category. These surveys collected data on panel members'

[7] Birth date, rather than age, is important to collect so that panel members, correct age can be used for sample selection based on age or to avoid asking age on each panel survey.

[8] Surveying started once a sufficient number of panel members had returned the Welcome Package.

Table 13.1 GPP surveys from September 2004 to November 2005 in chronological order, with sample size, mode of data collection and sampled population.

Survey topic	Month and Year	Sample size	Mode(s)	Population
Clothes Shopping 1*	9/04	973	Mail/Telephone/Web	Age 13 +
Automobile Ownership and Leasing 1*	10/04	4076	Mail/Web	Age 18 +
Retail Department Store Shopping Habits*	10/04	6635	Mail/Telephone/Web	Age 13 +
Election Opinion Poll	11/04	8610	Mail/Telephone/Web	Age 18 +
Bank Usage and Habits 1*	12/04	7847	Mail/Web	Age 18 +
Shopping for Electronics and Usage*	1/05	8112	Mail/Web	Age 18 +
Education	1/05	9048	Mail/Telephone/Web	Age 13 +
Health*	1/05	8082	Mail/Web	Age 18 +
Valentine Holidays	2/05	2000	Telephone	Age 18 +
Internal Research	2/05	2330	Web	Age 18 +
Alcohol Purchase and Use*	2/05	10 370	Mail/Web	Age 21 +
Teens	2/05	1647	Telephone	Age 13–17
Clothes Shopping 2*	2/05	10 754	Mail/Telephone/Web	Age 13 +
Dining Out*	2/05	15 832	Mail/Web	Age 18 +
Do It Yourself Shopping*	3/05	19 701	Mail/Telephone/Web	Age 18 +
Well-Being	3/05	720	Web	Age 18 +
Lodging: Business and Leisure Travel*	3/05	18 231	Mail/Web	Age 18 +
Wedding Preparation	3/05	1500	Telephone	Age 18 +, Females
Internal Research	4/05	5999	Telephone	Age 18 +
Bank Usage and Habits 2*	4/05	22 554	Mail/Telephone/Web	Age 18 +
Grocery Shopping Habits*	5/05	25 338	Mail/Web	Age 13 +
Internal Research	5/05	1010	Telephone	Age 18 +
Teenager Freedom	5/05	1364	Mail/Web	Age 13–17
Credit Card Usage*	5/05	13 864	Mail/Web	Age 18 +
Electronic Services	5/05	3923	Web	Age 18 +
Internal Research	6/05	900	Web	Age 18 +, Employed
Personal Finance*	6/05	20 309	Mail/Telephone/Web	Age 18 +
Charitable Services*	7/05	22 938	Mail/Web	Age 18 +
Client Test: Incentives	7/05	519	Web	Age 18 +
Teen School Issues	8/05	1336	Telephone/Web	Age 13–17
Social Networks	8/05	22 836	Mail/Web	Age 18 +
Internet Shopping	9/05	2500	Web	Age 18 +
Wage and Discrimination*	9/05	21 635	Mail/Telephone/Web	Age 18 +

recent activities in these fields and on customer engagement. Typical questions also included where panel members shopped and whether they purchased, owned or used particular consumer brands. Other surveys included Gallup Poll surveys such as the Election Opinion Survey, social surveys such as any of the teen population surveys and the Social Networks Survey, as well as surveys for Gallup clients such as the Valentine

Holiday Survey and the Internet Shopping Survey. Side-by-side testing of the Gallup Poll using the GPP and RDD are part of the process of evaluating the quality of the panel. The Election Opinion Survey is also an example of this testing. In addition, Gallup uses the panel for various internal tests and evaluations. While the goal was to send two surveys per month, the surveys were not administered at regular intervals but rather as client needs dictated and as survey questionnaires for building up the panel database became available. On average, panelists did 10.7 surveys and were classified as active members for 10.1 months.

In summary, the advantages of building a panel are: as a possible replacement for an apparently failing data collection method (RDD); the ability to target populations such as shoppers at Do It Yourself retail outlets; the ability to target rare populations such as luxury car owners or 13–17-year-olds; the ability to reduce the field period for data collection; the flexibility to choose the most appropriate mode(s) of data collection; and for any given survey to reduce (client) costs when compared to RDD. In the future, if RDD continues to fail, an important quality issue that will have to be addressed is whether some form of self-selected or opt-in panel will work and which statistical models provide the ability to make inferences to a population of interest.

At this point in time, however, a major issue with the GPP is attrition. A person is said to attrite from the GPP if they request no longer to receive invitations to do GPP surveys. With a discontinuous consumer panel, attrition has the potential to bias the results of any particular survey of the panel. In addition, if attrition is large, the panel costs increase for recruiting to replace the attrited members. This can be compounded if the attrition is among particular subpopulations, such as young adults, males, etc. If attrition occurs among a particular group it may be difficult for a RDD-recruited panel to identify acceptable ways of finding replacements. For example, targeting young male adults in an RDD survey is difficult. However, for some subpopulations RDD can work by targeting groups of telephone numbers with approximately known demographics, such as Blacks or Hispanics in the USA. Thus, many panels use gifts (examples of gifts might include electronic products, vacation trips, etc.), lotteries or cash payments to minimize attrition. These incentives can be expensive and also require building an incentive management system to ensure that panelists get the right 'gift' on time. In discussion with several panel managers about gifts and incentives one point that came up several times, with of course no scientific support, was that making a mistake in delivering a 'gift' to a panel member[9] was a sure cause of attrition. Thus, Gallup choose for the GPP to use token gifts and a quarterly magazine targeted for panel members rather than larger gifts, with the idea that the Gallup brand will engage members and constrain attrition to an acceptable level.

The analysis in Section 13.3 below covers panel membership over the same time period, from the initiation of the panel in May 2004 when recruiting started until November 2005. To be included in this analysis a person must have been sampled for at least one survey during the time period. On 21 November 2005 the Panel consisted of 21 404 active

[9] In fact, any mistake, like having the wrong name on a mail or web questionnaire, or sending a male a questionnaire obviously for a female, is a quick way to get someone to drop out of a panel. Among panel managers the consensus for why this happens is that panel members realize that the panel (management) knows a lot about each member and mistakes are thought to be a sign of sloppy work or uncaring personnel.

members and 7330 inactive individual members. Active panel members are defined as those people still in the pool who can be asked to do a survey and in fact have been asked to do at least one survey. Active panel members have been in the panel for between 1 and 14 months. The active panel members were in 18 809 households. Panelists become inactive by requesting to be removed from the panel. They can opt out individually or for all household members in the GPP. Early in 2005 Gallup considered introducing a rule designating as inactive those panelists who did not respond to three consecutive surveys. However, a study of around 1500 panel members showed that around one-third of those who did not complete three consecutive surveys did eventually complete another survey. As a result, this rule for assigning nonresponders as inactive was never made operational.

There are two other differences between the GPP and many longitudinal panels. First the GPP makes no attempt to follow movers. If someone moves and does not notify Gallup of a new address or telephone number they attrite (unless they are an Internet user and their e-mail address does not change). Second, longitudinal surveys are usually interested in estimating change, and this is not the case with the GPP. The latter is generally making a series of cross-sectional survey estimates.

To summarize, the GPP differs from many discontinuous panels. First, it is a panel of randomly recruited members. In the USA this author knows of only the Knowledge Networks Panel (Pineau and Slotwiner, 2003) that also recruits using RDD sampling. Second, the GPP does not rely on gifts (monetary or otherwise) or lotteries in an effort to retain members. Third, the GPP always starts any interview using Dr. George Gallup's idea that any survey conducted by Gallup should start out with 'political' questions such as the Presidential Approval Rating. Finally, a member of the GPP can be interviewed by telephone, mail or web. In the next section the three experiments that Gallup ran are described.

13.3 ATTRITION ON THE GALLUP POLL PANEL

13.3.1 Descriptive Analysis

This section describes attrition by comparing the demographics of active and inactive panel members. Again, an active member is defined as any person in the panel eligible to receive a future panel survey. An inactive member is one who has requested to no longer receive survey invitations. The demographics studied include gender, age group, education, race (white versus all others) and marital status. As noted above, a total of 28 734 panel members are studied, of which 74.2 % were active as of November 2005.

Table 13.2 shows the association of attrition with each of five demographic variables. Attrition is greater amongst men than amongst women. Amongst adults, attrition is highest for the youngest age group (18–24) and decreases with increasing age. However, attrition is much lower amongst teenagers (13–17) than amongst any adult group. This may reflect the particular salience of the teen surveys and the relative infrequency with which teens were asked to participate in surveys (see Table 13.1). College graduates are less likely to attrite than those with lower levels of education. There is a particularly strong association of attrition with race: Whites attrite at about half the rate of non-Whites. Finally, an association is also observed with marital status. Divorced panel members had the lowest levels of attrition and those living as married or never married had the highest levels.

Table 13.2 Bivariate associations between attrition rate and demographics.

Variable	Category	% Inactive	P^a
Total		25.7	
Gender	Male	28.0	<0.001
	Female	23.8	
Age	13–17	11.0	<0.001
	18–24	46.5	
	25–34	31.6	
	35–44	27.8	
	45–54	24.4	
	55 plus	22.8	
Education	Less than high school	28.2	<0.001
	High school graduate	29.3	
	Some college	26.2	
	Trade/technical/vocational training	29.9	
	College graduate	21.4	
	Postgraduate	19.4	
Race	White	23.1	<0.001
	Other	43.4	
Marital status	Never married	28.1	<0.001
	Married	25.2	
	Living as married	31.8	
	Separated	24.3	
	Divorced	20.0	

[a] *P* values are based on chi-square tests ignoring the clustering of panel members within households.

Table 13.3 shows the relationship between length of panel membership, in months, and attrition. Length of membership indicates how long ago a panel member was recruited. Attrition is very low amongst those recruited in the four months prior to November 2005. It then increases with increasing length of membership, reaching 46.0 % for those in the panel for 12 months. Attrition is lower for those in the panel over a year.

It is also interesting to observe the association between attrition and total number of survey requests. Though number of survey requests is highly correlated with length of membership (0.787, $P < 0.01$), it shows a somewhat different relationship with attrition (Table 13.4). Attrition rates are similar – between 20 % and 30 % – for panel members with 5–11 requests, but then increase sharply with number of requests, reaching 63 % amongst those with 14 requests but then decreasing again. The high rate of attrition (70 %) amongst those receiving just two requests is also striking. Given that the analysis is restricted to panel members who participated in at least one survey, this suggests that there is a sizeable subset of people who respond once and then resign from the panel. There also seems to be a tendency to drop out after completing 12–15 surveys.

In summary, attrition in the GPP looks much like nonresponse in many cross-sectional surveys (Keeter *et al.*, 2006). Males, younger adults, less educated and those of races other than Caucasians attrite at higher rates. Just looking at length of panel membership, we see a jump in the percentage of inactive members at 11 and 12 months and then a drop after that. The Kaplan–Meier survival plots indicate a mean survival

Table 13.3 Association between attrition rate and length of panel membership.

Length of membership (months)	% Inactive	P^a
2	0.0	<0.001
3	3.8	
4	0.0	
5	12.3	
6	24.4	
7	20.6	
8	19.6	
9	22.7	
10	24.0	
11	40.0	
12	46.0	
13	35.8	
14	33.8	

[a] *P* value is based on a chi-square test ignoring the clustering of panel members within households.

Table 13.4 Association between attrition rate and number of survey requests.

Number of survey requests	% Inactive	P^a
2	69.9	<0.001
3	21.1	
4	45.3	
5	24.6	
6	28.7	
7	22.1	
8	21.4	
9	28.7	
10	23.6	
11	23.2	
12	30.0	
13	47.8	
14	62.6	
15	43.1	
16	27.0	
17	19.2	
18	16.1	
19	12.1	
20	11.6	
21	12.8	
22	50.0	

[a] *P* value is based on a chi-square test ignoring the clustering of panel members within households.

time of 11 survey requests. The overall level of attrition, at 26 % by November 2005, may not seem excessively high but it must be remembered that this is in the context of the 27 % recruitment rate reported earlier. Aside from the costs of further recruitment associated with attrition, attrition can potentially introduce a nonresponse bias (though see Clinton (2001) for a discussion of the Knowledge Networks panel, where he found no evidence of a nonresponse bias due to attrition). For these reasons, we are interested in finding ways of reducing rates of attrition in consumer panels.

13.3.2 Experiments

In an attempt to find ways of reducing attrition, three experiments were conducted. The first experiment tested alternative interviewer introductions at the RDD recruitment stage, the second related to respondent commitment and the third related to survey mode. Here we describe the experiments and their effect on attrition rates.

Experiment 1: Interviewer Introductions

As noted by Meegama and Blair (1999): 'The goal of an effective introduction appears to not only provide minimum information, but the right kind of information – eliciting cooperation of the potential respondent and, secondarily, enhancing rapport during the interview.' For panel surveys, introductions may also influence the attrition rate. Different studies of introductions have found mixed results with respect to refusal rates. For example, when introductions provide more information, studies have found that refusals to participate in surveys were more frequent (Blumberg *et al.*, 1974; Kearney *et al.*, 1983), less frequent (Hauck and Cox, 1974) or about the same (Leuptow *et al.*, 1977; Singer, 1978; Sobal, 1982).

Meegama and Blair (1999) found no difference in refusal rates when they tested a 'salient' introduction versus an 'overcoming objections' introduction. Similar to the GPP experiment, Houtkoop-Steenstra and van den Bergh (2000) compared three scripted introductions to an agenda-based introduction. Interviewers were given more freedom to craft their introductions; their only requirement was the use of a few key words. The agenda-based introduction produced higher response rates and higher appointment rates, and therefore lower refusal rates.

Two introductions to the panel were used in our experiment. Each was randomly assigned to potential panel members. In the first case the interviewers were instructed to read a 'longer' introduction verbatim. This introduction covered all of the points associated with belonging to the GPP. For example, it included mentioning that panel members would be surveyed about twice a month and that each survey would last between 5 and 15 minutes. This introduction also mentioned that Gallup was recruiting other family members aged 13 or more. The second introduction was shorter and gave the interviewers latitude to tailor the introduction as they went along. This introduction did not mention the number of surveys or the length of a survey. Recruits were only told that other family members could join but with no mention of age. If respondents asked about such issues, interviewers would answer. Both the long and short introduction had the five poll questions.

This experiment developed after an initial recruitment test where the interviewers began with the long introduction, that is, they immediately jumped into a request to join the GPP. This proved difficult, as respondents were very wary of an immediate request to join a panel. Also, interviewers did not like the longer introduction, saying they thought they could be just as effective with a shorter statement that gave them some flexibility to respond to questions. Therefore the recruitment process was modified to: precede the request to join by asking five Gallup Poll-like questions; and to test interviewer claims that using a shorter introduction to the panel would be just as effective as the longer verbatim introduction. A discussion with interviewers after the modifications indicated that both the asking of the Poll-like questions and the freedom to tailor the request to join the panel were perceived to improve the recruitment process. Thus the experiment described above was implemented.

Two research questions are evaluated: did either introduction prove more effective at gaining initial agreement to join the panel; and did either introduction result in higher attrition over time. The rate of agreement to join the panel did not differ between the short (the 95 % confidence interval is 64 ± 2 %) and long (the 95 % confidence interval is 62 ± 2 %) introductions. Neither did the attrition rate differ, being 25.5 % with the short introduction and 26.1 % with the long introduction ($P > 0.05$).

Experiment 2: Respondent Commitment

The second experiment involved asking those who completed the intake survey, a self-completion booklet questionnaire, to sign and date the questionnaire. A random half of those households agreeing on the telephone to join the GPP was sent an intake questionnaire with a signature and date block.

Each person in the household who joined was asked to provide their signature and the date. The other half received the same intake questionnaire without the signature and date block. The signature and date block appeared as follows:

> Thank you for agreeing to join The Gallup Poll Panel. We appreciate your signing below – your signature indicates you've answered these questions and will consider answering questions on future Gallup Polls.
>
> Signature: ____________________________
>
> Today's Date: __ - __ - __ (MM-DD-YR)

This signature block was designed to see if those who sign the questionnaire are more committed to the panel than those who are not asked to sign. This treatment is in the spirit of Charles Cannell, who sometimes asked interviewees to commit to certain processes (Singer, 1978, personal communication 2005). Also, in many face-to-face surveys organizations ask interviewers to sign a completed questionnaire that attests to the fact that the interview took place.

An analysis of the differences in return of intake questionnaires found no significant difference on the return rates of the signature version (55 %) versus the no signature version (54 %). Again, there was also no difference in the observed attrition rates, these being 25.2 % with a signature request and 26.5 % without ($P > 0.05$).

Experiment 3: Data Collection Modes

This experiment involved using Gallup telephone interviewers for some surveys of the panel – an approach with major cost implications. One group of panel members was randomly assigned to a treatment where the first survey each month was done by telephone and the second was self-administered. Those in the other treatment only received self-administered surveys. In either treatment those that had self-administered surveys received an Internet survey if they are frequent web users (use the web at least twice a week) while others received a mail questionnaire. Panel members were randomly split into two groups, with 70 % of the panel members assigned to the telephone group and 30 % assigned to the self-administered treatment. The overall attrition rate was 24.1 % amongst the telephone group and 29.9 % amongst the group that only received self-administered surveys. The difference is highly significant ($\chi^2 = 104.4$, $P < 0.001$) though it should be noted that the lower attrition rate is associated with much higher data collection costs.

13.3.3 Logistic Regression

Here we present a logistic regression model of attrition in which the independent variable is an indicator of whether the panel member is active (= 1) or inactive (= 0).

All of the variables in Tables 13.2 and 13.3 are included in the model. The total number of survey requests is excluded from the model since, as noted above, it has a correlation of 0.787 ($P < 0.001$) with length of panel membership. In addition, length of membership is collapsed into nine categories because of the relatively small number who had been panel members for five months or fewer. For each variable, the last category presented in Table 13.2 or 13.3 is treated as the reference category. Indicators of treatment group for each of the three experiments reported in Section 13.3.2 are also included in the model. The Hosmer–Lemeshow goodness-of-fit test is not significant, with $F = 4.853$ (8 d.f.) and $P = 0.773$. Table 13.5 summarizes the model.

The largest odds ratio in Table 13.5 is for those with membership less than six months relative to those with membership of 14 months. It is hardly surprising that attrition is more likely to occur the longer a member is in the panel. Other significant odds ratios are all consistent with the bivariate findings reported in Section 13.3.1, with just one exception. Consistent with the bivariate analysis, we see that: the odds of retention for males are only 78 % of those of females; the odds for Whites are 177 % of those for non-Whites; the odds for teens are four times those aged 55 and older; and the odds for 28-24-year-olds are only 34 % of those aged 55 and older. However, while the bivariate analysis showed that persons living as married or never married attrite at higher rates than others, this is no longer the case once age is taken into account. The effects for the indicators of the three experiments are as in the bivariate analysis, as one would expect (as allocation was at random): only the telephone interviewing protocol was significant.

13.3.4 A Serendipitous Finding: The Relationship Between Type of Survey and Attrition

Anecdotal evidence from the records of calls made by GPP members to the Gallup Hotline to request that their membership be terminated led to an interesting observation. The Hotline calls indicated that some individuals were quitting because

Table 13.5 Logistic regression of attrition.

	B	SE	Wald	d.f.	<*P*	Exp(*B*)
Gender (Ref. Female)	0.249	0.031	65.40	1	<0.001	0.780
Age group (Ref. 55+)			384.94	5	<0.001	
13–17	1.392	0.179	60.72	1	<0.001	4.022
18–24	−1.086	0.077	198.99	1	<0.001	0.337
25–34	−0.526	0.052	102.41	1	<0.001	0.591
35–44	−0.334	0.043	59.50	1	<0.001	0.716
45–54	−0.14	0.041	11.55	1	0.001	0.869
Education (Ref. Postgraduate)			266.64	5	<0.001	
Less than high school	−0.882	0.083	111.56	1	<0.001	0.414
High school graduate	−0.571	0.049	135.80	1	<0.001	0.565
Some college	−0.351	0.048	54.58	1	<0.001	0.704
Trade/technical/vocational Training	−0.585	0.071	67.60	1	<0.001	0.557
College graduate	−0.073	0.047	2.46	1	0.117	0.93
Marital status (Ref. Divorced)			12.57	4	0.014	
Never married	−0.14	0.09	2.42	1	0.12	0.87
Married	−0.225	0.079	8.21	1	0.004	0.798
Living as married	−0.282	0.165	2.91	1	0.088	0.755
Separated	−0.264	0.091	8.45	1	0.004	0.768
Race (Ref. Non-White)	0.569	0.046	149.73	1	<0.001	1.766
Membership (Ref. 14 mths)			353.43	9	<0.001	
Less than 6 months	2.977	0.726	16.81	1	<0.001	19.621
Six months	1.724	0.218	62.48	1	<0.001	5.609
Seven months	0.586	0.077	57.15	1	<0.001	1.796
Eight months	0.522	0.048	119.47	1	<0.001	1.686
Nine months	0.335	0.05	45.02	1	<0.001	1.398
Ten months	0.292	0.055	28.31	1	<0.001	1.338
Eleven months	−0.398	0.245	2.64	1	0.104	0.672
Twelve months	−0.475	0.107	19.53	1	<0.001	0.622
Thirteen months	−0.103	0.055	3.51	1	0.061	0.903
Short introduction (Expt. 1)	0.054	0.03	3.22	1	0.073	1.056
No signature on intake (Expt. 2)	−0.052	0.03	2.90	1	0.089	0.95
Telephone contact (Expt. 3)	0.371	0.032	134.30	1	<0.001	1.449
Constant	0.428	0.116	13.73	1	<0.001	1.534

they had been expecting to be doing only Poll-like surveys and not market research surveys. To examine this possible issue as a cause for attrition an index variable was constructed that summarized the mix of types of survey that each individual completed. Each GPP survey was classified as a poll-like/social survey or a market research survey. For each poll-like/social survey the index was incremented by 1, and for each market research survey the index was decreased by 1. Table 13.6 shows the association between this variable and attrition. A chi-square test is highly significant ($P < 0.001$). The attrition rate is much lower for panel members with positive scores on the index, that is, for those who completed relatively more poll-like or social surveys.

Table 13.6 Association between attrition rate and survey-type index.

Survey-type index	% Inactive	n
−4	38.5	26
−3	34.9	430
−2	33.7	2920
−1	40.5	8212
0	38.9	5887
1	4.9	7661
2	2.5	2217
3	1.3	476
4	0.0	18

13.4 SUMMARY

The Gallup Poll Panel is a consumer panel that asks members to participate in a variety of opinion, social and market research surveys. Attrition is a serious concern for consumer panels as it can introduce nonresponse bias and it can introduce extra recruitment costs. The analysis presented here suggests that the demographic characteristics associated with attrition in the GPP are similar to the characteristics associated with nonresponse in many RDD surveys. Males, younger adults, those with lower education and those of a race other than White drop out at higher rates than others. In addition, length of membership and number of survey requests have an impact on attrition, with a median survival time of 11 months.

The effects of three experiments on attrition were also studied. The experiments consisted of a short versus long telephone recruit introduction, a request for signature on the intake mail questionnaire and receipt of only self-administered questionnaires or of one telephone interview per month. Only the last of these experiments produced a significant effect on attrition in both the bivariate analysis and the logistic regression. Attrition was found to reduce when the first survey each month was carried out by telephone.

Finally, the mix of surveys in which panel members took part appears to have an impact on attrition. Panel members who took part in more social and political surveys, as opposed to market research surveys, were less likely to drop out of the panel.

REFERENCES

Blumberg, H. H., Fuller, C. and Hare, A. P. (1974). Response rates in postal surveys. *Public Opinion Quarterly*, 38, 113–123.

Clinton, J. D. (2001). *Panel Bias from Attrition and Conditioning: A Case Study of the Knowledge Networks Panel*, Department of Political Science Technical Report. California: Stanford University.

Göritz, A. S. (2002). *Web-based mood induction*. Unpublished Doctoral Dissertation, University of Erlangen-Nürnberg, Germany.

Göritz, A. S. (2004). The impact of material incentives on response quantity, response quality, sample composition, survey outcome, and cost in online access panels. *International Journal of Market Research*, 46, 327–345.

Hauck, M. and Cox, M. (1974). Locating a sample by random digit dialing *Public Opinion Quarterly*, 38, 253–256.

Hoogendoorn, A. W. and Sikkel, D. (1998). Response burden and panel attrition. *Journal of Official Statistics*, 14, 189–205.

Houtkoop-Steenstra, H. and van den Bergh, H. (2000). Effects of introductions in large scale telephone survey interviews. *Sociological Methods and Research*, 28, 281–300.

Kearney, K., Hopkins, R. A., Mauss, A. L. and Weisheit, K. A. (1983). Sample bias resulting from a requirement for parental consent. *Public Opinion Quarterly*, 47, 96–102.

Keeter, S., Kennedy, C., Dimock, M., Best, J. and Craighill, P. (2006). Gauging the impact of growing nonresponse on estimates from a national RDD telephone survey. *Public Opinion Quarterly*, 70, 759–779.

Lee, E., Hu, M. and Toh, R. (2004). Respondent non-cooperation in surveys and diaries: an analysis of item non-response and panel attrition. *International Journal of Market Research* 46, 311–326.

Lee, S. (2006). Propensity score adjustment as a weighting scheme for volunteer panel web surveys. *Journal of Official Statistics*, 22, 329–349.

Leuptow, L., Mueller, S. A., Hammes, R. R. and Master, L. R. (1977). The impact of informed consent regulations on response rate response bias. *Sociological Methods and Research*, 6, 183–204.

Meegama, N. and Blair, J. (1999). The effects of telephone introductions on cooperation: an experimental comparison. In *Proceedings of the Survey Research Methods Section* (pp. 394–397). Washington, DC: American Statistical Association.

Pineau, V. and Slotwiner, D. (2003). Probability samples vs. volunteer respondents in internet research: defining potential effects on data and decision-making in marketing applications. *Technical Paper*. California, USA: Knowledge Networks.

Singer, E. (1978). Informed consent: consequences for response rate and response quality in social surveys. *American Sociological Review*, 66, 348–361.

Sobal, J. (1982). Disclosing information in interview introductions and methodological consequences of informed consent. *Sociological and Social Research*, 66, 348–61.

Steeh, C., Kirgis, N., Cannon, B. and DeWitt, J. (2001). Are they really as bad as they seem? nonresponse rates at the end of the twentieth century. *Journal of Official Statistics*, 17, 227–247.

Sudman, S. and Wansink, B. (2002). *Consumer Panels* (2nd edn). Chicago, IL: American Marketing Association.

Tortora, R. (2004). Response trends in a national random digit dial survey, *Advances in Methodology and Statistics*, 1, 21–32.

Tortora, R., Groves, R. and Peytcheva, E. (2007). Multiplicity-based sampling for the mobile telephone population: coverage, nonresponse and measurement issues. In J. M. Lepkowski, *et al.* (Eds), *Advances in Telephone Survey Methodology* (pp. 133–148). New York: John Wiley and Sons, Inc.

Tuckel, P. S. and O'Neill, H. W. (2001). The vanishing respondent in telephone surveys. Presented at the *Annual Conference of the American Association for Public Opinion Research*, Montreal, Canada.

CHAPTER 14

Joint Treatment of Nonignorable Dropout and Informative Sampling for Longitudinal Survey Data

Abdulhakeem A. H. Eideh
Al-Quds University, Palestine

Gad Nathan
Hebrew University of Jerusalem, Israel

14.1 INTRODUCTION

In this chapter we study, within a modelling framework, the joint treatment of nonignorable dropout and informative sampling for longitudinal survey data, by specifying the probability distribution of the observed measurements when the sampling design is informative. This might be a useful approach for longitudinal surveys with initial nonresponse (which can be thought of as a form of informative sampling) and subsequent nonignorable attrition. The sample distribution of the observed measurements model is extracted from the population distribution model, assumed to be multivariate normal. The sample distribution is derived first by identifying and estimating the conditional expectations of first-order sample inclusion probabilities (assuming complete response at the first time period), given the study variable, based on a variety of models such as linear, exponential, logit and probit. Next, we consider a logistic model for the informative dropout process. The proposed method combines two methodologies used in the analysis of sample surveys: for the treatment of informative sampling and of informative dropout. One incorporates the dependence of the first-order inclusion probabilities at the initial time period on the study variable (see Eideh and Nathan, 2006), while the other incorporates the dependence of the probability of nonresponse on unobserved or missing observations (see Diggle and Kenward, 1994). An empirical example based on data from the British Labour Force Survey illustrates the methods proposed.

Methodology of Longitudinal Surveys P. Lynn

Data collected by sample surveys, and in particular by longitudinal surveys, are used extensively to make inferences on assumed population models. Often, survey design features (clustering, stratification, unequal probability selection, etc.) are ignored and the longitudinal sample data are then analysed using classical methods based on simple random sampling. This approach can, however, lead to erroneous inference because of sample selection bias implied by informative sampling – the sample selection probabilities depend on the values of the model outcome variable (or the model outcome variable is correlated with design variables not included in the model). For example, if the sample design is clustered, with primary sampling units selected with probabilities proportional to size (e.g. size of locality), and the dependent variable (e.g. income) is related to the size of the locality, ignoring the effects of this dependence can cause bias in the estimation of regressions coefficients. In theory, the effect of the sample selection can be controlled for by including among the model all the design variables used for the sample selection. However, this possibility is often not operational because there may be too many of them or because they are not of substantive interest. Initial nonresponse in a longitudinal survey may also be considered as a form of informative sampling.

To overcome the difficulties associated with the use of classical inference procedures for cross-sectional survey data, Pfeffermann, Krieger and Rinott (1998) proposed the use of the sample distribution induced by assumed population models, under informative sampling, and developed expressions for its calculation. Similarly, Eideh and Nathan (2006) fitted time series models for longitudinal survey data under informative sampling.

In addition to the effect of complex sample design, one of the major problems in the analysis of longitudinal data is that of missing values. In longitudinal surveys we intend to take a predetermined sequence of measurement on a sample of units. Missing values occur when measurements are unavailable for one or more time points, either intermittently or from some point onwards (attrition).

The literature dealing with the treatment of longitudinal data considers three major areas of research:

(1) *Analysis of complete nonsurvey longitudinal data (without nonresponse)*. The predominant method of analysis for longitudinal data has long been based on the application of generalized linear models (GLMs) (McCullagh and Nelder, 1999) to repeated measures and the use of generalized estimating equations (GEEs) to estimate the model parameters (see Diggle *et al.*, 1994). The generalized linear model describes the conditional distribution of the outcome, given its past, where the distribution parameters may vary across time and across subjects as a stochastic process, according to a mixing distribution. Two different approaches to longitudinal analysis are dealt with by means of similar GLMs. In the "subject-specific" approach, sometimes referred to as the *random effects model*, the heterogeneity between subjects is explicitly modelled, while in the "population-averaged" approach, sometimes referred to as the *marginal model*, the average response is modelled as a function of the covariates, without explicitly accounting for subject heterogeneity.

Frequently longitudinal sample surveys deal with hierarchical populations, such as individuals within households or employees within establishments, for which multilevel modelling is appropriate. In another approach, Goldstein, Healy and Rasbash

(1994) consider the analysis of repeated measurements using a two-level hierarchical model, with individuals as second level and the repeated measurements as the first level.

Path analysis, which has long been a preferred method of modelling complex relationships between large numbers of variables in cross-sectional analysis of structured data-sets in the social sciences, has been generalized to modelling longitudinal data primarily by means of *Graphical Chain Modelling* (GCM) and *Structural Equation Modelling* (SEM) – see Wermuth and Lauritzen (1990), Mohamed *et al.* (1998) and Smith *et al.* (Chapter 22 of this volume).

Other models used for the analysis of longitudinal data include *Antedependence Models* designed to deal with nonstationarity. Zimmerman and Nunez-Anton (2000) propose structured and unstructured antedependence models for longitudinal data, primarily in the context of growth analysis.

(2) *Treatment of nonresponse in longitudinal data in the nonsurvey context.* The analysis of longitudinal data with nonignorable missing values has received serious attention in the last 20 years. For example, Little (1993) explores pattern-mixture models and pattern-set mixture models for multivariate incomplete data. Diggle and Kenward (1994) propose a likelihood-based method for longitudinal data subject to informative and monotone missingness. Little (1995) discusses methods that simultaneously model the data and the dropout process via two broad classes of models – random-coefficient selection models and random-coefficient pattern-mixture models, based on likelihood inference methods. Little and Wang (1996) fit pattern-mixture models for multivariate incomplete data with covariates, via maximum likelihood and Bayesian methods, using the Expectation–Maximization (EM) algorithm. Troxel, Harrington and Lipsitz (1998) use full likelihood methods to analyse continuous longitudinal data with nonignorable (informative) missing values and nonmonotone patterns. Lipsitz, Ibrahim and Molenberghs (2000) use a Box-Cox transformation for the analysis of longitudinal data with incomplete responses, while Ibrahim, Chen and Lipsitz (2001) estimate the parameters in the generalized linear mixed models with nonignorable missing response data and with nonmonotone patterns of missing data in the response variable. For further discussion see Schafer (1995) and Little and Rubin (2002).

(3) *Treatment of effects of complex sample design and of nonresponse in longitudinal surveys.* Some recent work has considered the use of the sample distribution under informative sampling. Longitudinal survey data may be viewed as the outcome of two processes: the process that generates the values of units in the finite population, often referred as the superpopulation model, and the process of selecting the sample units from the finite population, known as the sample selection mechanism. Analytical inference from longitudinal survey data refers to the superpopulation model. When the sample selection probabilities depend on the values of the model response variable at the first time period, even after conditioning on auxiliary variables, the sampling mechanism becomes informative and the selection effects need to be accounted for in the inference process. Pfeffermann, Krieger and Rinott (1998) propose a general method of inference on the population distribution (model) under informative sampling, which consists of approximating the parametric distribution of the sample

measurements. The sample distribution is defined as the distribution of the sample measurements, given the selected sample. Under informative sampling, this distribution is different from the corresponding population distribution, although for several examples the two distributions are shown to belong to the same family and only differ in some or all of the parameters. Several authors discuss and illustrate a general approach to the approximation of the marginal sample distribution for a given population distribution and of the first-order sample selection probabilities. Pfeffermann and Sverchkov (1999) propose two new classes (parametric and semiparametric) of estimators for regression models fitted to survey data. Sverchkov and Pfeffermann (2004) use the sample distribution for predicting finite population totals under informative sampling. Pfeffermann, Moura and Silva (2001) propose a model-dependent approach for two-level modelling that accounts for informative sampling. Pfeffermann and Sverchkov (2003) consider four different approaches to defining parameter-estimating equations for generalized linear models, under informative probability sampling design, utilizing the sample distribution. Chambers, Dorfman and Sverchkov (2003) describe a framework for applying a common exploratory data analysis procedure – nonparametric regression – to sample survey data. Eideh (2007) uses the sample distribution for deriving best linear unbiased predictors of the area means for areas in the sample and for areas not in the sample. For more discussion see Eideh and Nathan (2003, 2006) and Nathan and Eideh (2004).

The joint treatment of the effects of complex sample design and of nonresponse in longitudinal surveys has been considered by several authors. Feder, Nathan and Pfeffermann (2000) develop models and methods of estimation for longitudinal analysis of hierarchically structured data, taking unequal sample selection probabilities into account. The main feature of the proposed approach is that the model is fitted at the individual level but contains common higher level random effects that change stochastically over time. The model allows the prediction of higher and lower level random effects, using data for all the time points with observations. This should enhance model-based inference from complex survey data. The authors introduce a two-stage procedure for estimation of the parameters of the model proposed. At the first stage, a separate two-level model is fitted for each time point, thus yielding estimators for the fixed effects and for the variances. At the second stage, the time series likelihood is maximized only with respect to the time series model parameters. This two-stage procedure has the further advantage of permitting appropriate first- and second-level weighting to account for possible informative sampling effects. Pfeffermann and Nathan (2001) use time series structures with hierarchical modelling for imputation for wave nonresponse. Skinner and Holmes (2003) consider a model for longitudinal observations that consists of a permanent random effect at the individual level and autocorrelated transitory random effects corresponding to different waves of investigation. Eideh and Nathan (2006) fit time series models for longitudinal survey data under informative sampling via the sample likelihood approach and pseudo maximum likelihood methods and introduce a new test of sampling ignorability based on the Kullback–Leibler information measure.

None of the above studies consider simultaneously the problem of informative sampling and the problem of informative dropout when analysing longitudinal survey data. In this paper we study, within a modelling framework, the joint treatment of nonignorable dropout and informative sampling for longitudinal survey data, by specifying the probability distribution of the observed measurements, when the sampling design is informative. This is the most general situation in longitudinal surveys, and other combinations of sampling informativeness and response informativeness can be considered as special cases. The sample distribution of the observed measurements model is extracted from a population distribution model, such as the multivariate normal distribution. The sample distribution is derived first by identifying and estimating the conditional expectations of first-order (complete response) sample inclusion probabilities, given the study variable, based on a variety of models such as linear, exponential, logit and probit. Next, we consider a logistic model for the informative dropout process. The proposed method combines two methodologies used in the analysis of sample surveys for the treatment of informative sampling and informative dropout. One incorporates the dependence of the first-order inclusion probabilities on the study variable (see Pfeffermann *et al.*, 1998), while the other incorporates the dependence of the probability of nonresponse on unobserved or missing observations (see Diggle and Kenward, 1994). This is possible in longitudinal surveys by using models based on observations in previous rounds.

The main purpose here is to consider how to account for the joint effects of informative sampling designs and of informative dropout in fitting general linear models for longitudinal survey data with correlated errors.

14.2 POPULATION MODEL

Let y_{it} $(i = 1, \ldots N; t = 1, \ldots T)$ be the measurement on the *ith* subject at time $t = 1, \ldots T$. Associated with each y_{it} are the (known) values x_{itk} $(k = 1, \ldots p)$ of p explanatory variables. We assume that the y_{it} follow the regression model:

$$y_{it} = \beta_1 x_{it1} + \cdots + \beta_p x_{itp} + \varepsilon_{it}, \tag{14.1}$$

where the ε_{it} are a random sequence of length T associated with each of the N subjects. In our context, the longitudinal structure of the data means that we expect the ε_{it} to be correlated within subjects.

Let $\mathbf{y}_i = (y_{i1}, \ldots y_{iT})'$ and $\mathbf{x}_{it} = (x_{it1}, \ldots x_{itp})'$ and let $\boldsymbol{\beta} = (\beta_1, \ldots \beta_p)'$ be the vector of unknown regression coefficients. The general linear model for longitudinal survey data treats the random vectors $\mathbf{y}_i$ $(i = 1, \ldots N)$ as independent multivariate normal variables, that is

$$\mathbf{y}_i \underset{p}{\sim} MVN(\mathbf{x}'_i\boldsymbol{\beta}, \mathbf{V}), \tag{14.2}$$

where $\mathbf{x}_i$ is the matrix of size T by p of explanatory variables for subject i, and $\mathbf{V}$ has the (jk)th element $v_{jk} = \mathrm{cov}_p\,(y_{ij}, y_{ik})$, where $j,k = 1, \ldots T$ (see Diggle *et al.*, 1994).

14.3 SAMPLING DESIGN AND SAMPLE DISTRIBUTION

We assume a single-stage informative sampling design, where the sample is a panel sample selected at time $t = 1$ and all units remain in the sample until time $t = T$. Examples of longitudinal surveys, some of which are based on complex sample designs, and of the issues involved in their design and analysis can be found in Friedlander *et al.* (2002), Herriot and Kasprzyk (1984) and Nathan (1999). In many of the cases described in these papers, a sample is selected for the first round and continues to serve for several rounds. Then it is intuitively reasonable to assume that the first-order inclusion probabilities, π_i (which are the product of the sample selection probabilities and the probabilities of survey participation at time $t = 1$), depend on the population values of the response variable, y_{i1}, at the first occasion only, and on $\mathbf{x}_{i1} = (x_{i11}, \ldots x_{i1p})'$.

14.3.1 Theorem 1

Let $\mathbf{y}_i \underset{p}{\sim} f_p(\mathbf{y}_i|\mathbf{x}_i, \theta)$ be the population distribution of $\mathbf{y}_i$ given $\mathbf{x}_i$ and $\mathbf{y}_{i,T-1} = (y_{i2}, y_{i3}, \ldots y_{iT})'$. If we assume that π_i depends only on y_{i1} and on $\mathbf{x}_{i1}$, then the (marginal) sample distribution of $\mathbf{y}_i$ given $\mathbf{x}_i$ is given by:

$$f_s(\mathbf{y}_i \,|\, \mathbf{x}_i, \boldsymbol{\theta}) = f_s(y_{i1} \,|\, \mathbf{x}_{i1}, \boldsymbol{\theta}) f_p(\mathbf{y}_{i,T-1} \,|\, y_{i1}, \mathbf{x}_i; \boldsymbol{\theta}), \tag{14.3}$$

where

$$f_s(y_{i1} \,|\, \mathbf{x}_{i1}, \boldsymbol{\theta}) = \frac{E_p(\pi_i \,|\, y_{i1}, \mathbf{x}_{i1})}{E_p(\pi_i \,|\, \mathbf{x}_{i1}, \boldsymbol{\theta})} f_p(y_{i1} \,|\, \mathbf{x}_{i1}, \boldsymbol{\theta}) \tag{14.4}$$

is the sample distribution of y_{i1} given $\mathbf{x}_{i1}$ and

$$E_p(\pi_i|\mathbf{x}_{i1}, \boldsymbol{\theta}) = \int E_p(\pi_i \,|\, y_{i1}, \mathbf{x}_{i1}) f_p(y_{i1} \,|\, \mathbf{x}_{i1}, \boldsymbol{\theta}) \mathrm{d}y_{i1}. \tag{14.5}$$

Proof: See Eideh and Nathan (2006).

Assuming independence of the population measurements, Pfeffermann, Krieger and Rinott (1998) establish an asymptotic independence of the sample measurements, with respect to the sample distribution, under commonly used sampling schemes for selection with unequal probabilities. Thus the use of the sample distribution permits the use of standard efficient inference procedures as likelihood-based inference.

Note that, for a given population distribution, $f_p(y_{i1} \,|\, \mathbf{x}_{i1}, \boldsymbol{\theta})$, the sample distribution, $f_s(y_{i1} \,|\, \mathbf{x}_{i1}, \boldsymbol{\theta}, \gamma)$, in Equation (14.4) is completely determined by $E_p(\pi_i \,|\, y_{i1}, \mathbf{x}_{i1}; \gamma)$.

Consider the following approximation models for the population conditional expectation; $E_p(\pi_i \,|\, y_{i1}, \mathbf{x}_{i1}; \gamma)$, proposed by Pfeffermann *et al.* (1998) and Skinner (1994):

Exponential Inclusion Probability Model:

$$E_p(\pi_i|y_{i1}, \mathbf{x}_{i1}) = \exp(a_0^* + a_0 y_{i1} + a_1 x_{i11} + a_2 x_{i12} + \cdots + a_p x_{i1p}) \tag{14.6}$$

Linear Inclusion Probability Model:

$$E_p(\pi_i|y_{i1}, \mathbf{x}_{i1}) = b_0^* + b_0 y_{i1} + b_1 x_{i11} + b_2 x_{i12} + \cdots + b_p x_{i1p} \tag{14.7}$$

Nathan and Eideh (2004) consider additional approximations for the population conditional expectation, namely the logit and probit models.

14.3.2 Theorem 2

We assume the multivariate normal population distribution of $\mathbf{y}_i$ given $\mathbf{x}_i$, defined by Equation (14.2). Then under the exponential inclusion probability model given by Equation (14.6) it can be shown that the sample distribution of $\mathbf{y}_i$, given $\mathbf{x}_i$,θ,a_0, is multivariate normal:

$$\mathbf{y}_i \mid \mathbf{x}_i, \boldsymbol{\theta}, a_0 \underset{S}{\sim} MVN(\boldsymbol{\mu}^*, \mathbf{V}), \tag{14.8}$$

where $\boldsymbol{\mu}^* = (\mathbf{x}'_{i1}\boldsymbol{\beta} + a_0 v_{11}, \mathbf{x}'_{i2}\boldsymbol{\beta}, \ldots, \mathbf{x}'_{iT}\boldsymbol{\beta})'$. Thus, only the mean of the sample distribution (Equation 14.8) differs from that of the population distribution (Equation 14.2), whereas the variance matrix, $\mathbf{V}$, remains the same. The distributions coincide when $a_0 = 0$, that is, when the sampling design is noninformative.

Similar results are obtained for the linear, logit and probit inclusion probability models.

14.4 SAMPLE DISTRIBUTION UNDER INFORMATIVE SAMPLING AND INFORMATIVE DROPOUT

Missing values arise in the analysis of longitudinal data whenever one or more of the sequence of measurements from units within the study is incomplete, in the sense that intended measurements are not taken, or lost, or otherwise unavailable. For example, firms are born and die, plants open and close, individuals enter the survey and exit and animals may die during the course of the experiment. We follow much of the literature on the treatment of missing values in longitudinal data in restricting ourselves to dropout (or attrition), that is, to patterns in which missing values are only followed by missing values.

Suppose we intend to take a sequence of measurements, $y_{i1}, \ldots y_{iT}$, on the ith sampled unit. Missing values are defined as dropout if whenever y_{i_j} is missing so are y_{i_k} for all $k \geq j$. One important issue that then arises is whether the dropout process is related to the measurement process itself. Following the terminology in Rubin (1976) and Little and Rubin (2002), a useful classification of dropout processes is:

(1) Completely random dropout (CRD): the dropout process and measurement processes are independent, that is, the missingness is independent of both observed and unobserved data.
(2) Random dropout (RD): the dropout process depends only on the observed measurements, that is, those preceding dropout.
(3) Informative dropout (ID): the dropout process depends both on the observed and on the unobserved measurements, that is, those that would have been observed if the unit had not dropped out.

Following Diggle and Kenward (1994), assume that a complete set of measurements on a sample unit $i = 1, \ldots n$ could potentially be taken at all times: $t = 1, \ldots T$.

Let $\mathbf{y}_i^* = (y_{i1}^*, y_{i2}^*, ...y_{iT}^*)'$ denote the complete vector of intended measurements and $\mathbf{y}_i = (y_{i1}, y_{i2}, \ldots y_{i,d_i-1})'$ denote the vector of observed measurements, where d_i denotes the time of dropout, with $d_i = T + 1$ if no dropout occurs for unit i. We assume that $\mathbf{y}_i^*$ and $\mathbf{y}_i$ coincide for all time periods during which the ith unit remains in the study, that is, $y_{it} = y_{it}^*$ if $t < d_i$.

We define D_i as the random variable $2 \leq D_i \leq T + 1$, which takes the value of the dropout time d_i of the ith unit, $i = 1, 2, \ldots n$.

Let $H_{it} = \{y_{i1}, y_{i2}, \ldots y_{i,t-1}, \mathbf{x}_i\}$. Then under the exponential inclusion probability model (Equation 14.6), according to Theorem 2, the sample pdf of the complete series, $\mathbf{y}_i^* = (y_{i1}^*, y_{i2}^*, ...y_{iT}^*)'$, is multivariate normal as defined by Equation (14.8).

The general model for the dropout process assumes that the probability of dropout at time $t = d_i$ depends on H_{id_i} and on $y_{id_i}^*$. Then for $d_i \leq T$, Diggle and Kenward (1994) propose the following *logistic model* for an informative dropout process with dropout at time d_i:

$$\text{logit}[P_{d_i}(H_{id_i}, y_{id_i}^*; \boldsymbol{\varphi})] = \log \frac{P_{d_i}(H_{id_i}, y_{id_i}^*; \boldsymbol{\varphi})}{1 - P_{d_i}(H_{id_i}, y_{id_i}^*; \boldsymbol{\varphi})} = \phi_1 y_{i1} + \cdots + \phi_{d_i-1} y_{i,d_i-1} + \phi_{d_i} y_{id_i}^*, \tag{14.9}$$

where $\boldsymbol{\varphi}$ is a vector of unknown parameters.

Once a model for the dropout process has been specified, we can derive the joint distribution of the observed random vector $\mathbf{y}_i$, under the assumed multivariate normal sample distribution of $\mathbf{y}_i^*$, via the sequence of conditional sample pdfs of y_{it} given $H_{it} - f_{ts}(y \mid H_{it}, \mathbf{a}^*, \boldsymbol{\beta}, \mathbf{V}_0, \boldsymbol{\varphi})$ – and the sequence of conditional sample pdfs of y_{it}^* given $H_{it} - f_{ts}^*(y|H_{it}, \mathbf{a}^*, \boldsymbol{\beta}, \mathbf{V}_0)$.

For an incomplete sequence $\mathbf{y}_i = (y_{i1}, y_{i2}, \ldots y_{i,d_i} - 1)'$ with dropout at time d_i, the joint sample distribution is given by:

$$f_s(\mathbf{y}_i|\mathbf{x}_i) = f_{s,d_i-1}^*\{\mathbf{y}_{i,d_i-1}|\mathbf{x}_i\} * \left\{ \prod_{t=2}^{d_i-1} [1 - P_t(H_{it}, y_{it}; \boldsymbol{\varphi})] \right\} P(D_i = d_i|H_{id_i}), \tag{14.10}$$

where $f_{s,d_i-1*}\{\mathbf{y}_{i,d_i-1}|\mathbf{x}_i\} = f_s^*(y_{i1}|\mathbf{x}_{i1}) f_p^*(y_{i2}, ..., y_{i,d_i-1}|y_{i1}, \mathbf{x}_i)$ (see Theorem 2), and

$$P(D_i = d_i|H_{id_i}) = \int P_t(H_{it}, y_{it}; \boldsymbol{\varphi}) f_{tp}^*(y_{it}|H_{it}, \boldsymbol{\beta}, \mathbf{V}_0) dy_{it}. \tag{14.11}$$

Note that Equations (14.10) and (14.11) take into account the effect of informative sampling and informative dropout.

14.5 SAMPLE LIKELIHOOD AND ESTIMATION

In this section, we extend the methods of estimation for the analysis of longitudinal survey data under informative sampling (Eideh and Nathan, 2006), to take into account the effects of attrition or dropout, according to the model proposed by Diggle and Kenward (1994). We propose two alternative methods, based on the results of the previous section on the sample distribution of the observed measurements, in the presence of informative sampling and informative dropout.

14.5.1 Two-Step Estimation

The two sets of parameters in Equation (14.10) that need to be estimated are those on which the population distribution depends, $\boldsymbol{\theta} = (\boldsymbol{\beta}, \sigma^2, v_{jk}), j,k = 1, \ldots T$, and the parameters on which the sample distribution of observed measurements depends, $\boldsymbol{\theta}^* = (\boldsymbol{\theta}, a_0, \boldsymbol{\varphi})$, where a_0 is the parameter indexing the informative sampling process and $\boldsymbol{\varphi}$ is the parameter indexing the dropout process; see Equations (14.6) and (14.10). Thus the parameters of the sample distribution of the observed measurements, $\boldsymbol{\theta}^*$, include the parameters of the sample and population distributions. The parameters of the population distribution can be estimated using the sample distribution of observed measurements and using a two-step method:

(1) Estimation of a_0. According to Pfeffermann and Sverchkov (1999), the following relationship holds:

$$E_s(w_i \mid y_{i1}) = \frac{1}{E_p(\pi_i | y_{i1})}. \tag{14.12}$$

Thus, we can estimate a_0 by regressing $-\log(w_i)$ against $y_{i1}, x_{i11}, \ldots x_{i1p}$ $(i \in s)$.

(2) Substitution of the ordinary least-squares estimator, $\tilde{a}_0$, of a_0 in the sample distribution of observed measurements. The contribution to the log-likelihood function for the observed measurements of the ith sampled unit can be written as:

$$\begin{aligned} L_i(\boldsymbol{\beta}, \sigma^2, v_{jk}, \boldsymbol{\varphi}) &= \log f_s(\mathbf{y}_i | \mathbf{x}_i, \boldsymbol{\beta}, \sigma^2, v_{jk}, \tilde{a}_0) \\ &= \log f^*_{s,d_i-1}\{\mathbf{y}_{i,d_i-1} | \mathbf{x}_i, \boldsymbol{\beta}, \sigma^2, v_{jk}, \tilde{a}_0\} \\ &\quad + \log \left\{ \prod_{t=2}^{d_i-1} [1 - P_t(H_{it}, y_{it}; \boldsymbol{\varphi})] \right\} \\ &\quad + \log P(D_i = d_i | H_{id_i}, \boldsymbol{\beta}, \sigma^2, v_{jk}, \boldsymbol{\varphi}). \end{aligned} \tag{14.13}$$

Thus the full sample log-likelihood function for the observed measurements to be maximized with respect to $(\boldsymbol{\beta}, \sigma^2, v_{jk}, \boldsymbol{\varphi})$ is given by:

$$L_{rs}(\boldsymbol{\beta}, \sigma^2, v_{jk}, \boldsymbol{\varphi}) = \sum_{i=1}^{n} L_i(\boldsymbol{\beta}, \sigma^2, v_{jk}, \boldsymbol{\varphi}) = L_1(\boldsymbol{\beta}, \sigma^2, v_{jk} | \tilde{a}_0) + L_2(\boldsymbol{\varphi}) + L_3(\boldsymbol{\beta}, \sigma^2, v_{jk}, \boldsymbol{\varphi}), \tag{14.14}$$

where the three components of Equation (14.14) correspond to the three terms of Equation (14.13). The explicit form of $L_1(\boldsymbol{\beta}, \sigma^2, v_{jk} | \tilde{a}_0)$ is obtained from Equation (14.8), replacing T by $d_i - 1$. $L_2(\boldsymbol{\varphi})$ is determined by Equation (14.9) The last term $L_3(\boldsymbol{\beta}, \sigma^2, v_{jk}, \boldsymbol{\varphi})$ is determined by Equation (14.10), which requires the distribution $f^*_{tp}(y_{it} | H_{it}, \boldsymbol{\beta}, \mathbf{V}_0)$.

14.5.2 Pseudo Likelihood Approach

This approach is based on solving the estimated census maximum likelihood equations. The census maximum likelihood estimator of $\boldsymbol{\theta} = (\boldsymbol{\beta}, \sigma^2, v_{jk})$ solves the census likelihood equations, which in our case are:

$$U(\boldsymbol{\theta}) = \frac{\partial}{\partial \boldsymbol{\theta}} \log L_p(\boldsymbol{\theta}) = \sum_{i=1}^{N} \frac{\partial}{\partial \boldsymbol{\theta}} \log L_{ip}(\boldsymbol{\beta}, \sigma^2, v_{jk}). \tag{14.15}$$

Thus the pseudo maximal likelihood (PML) estimator of $\boldsymbol{\theta} = (\boldsymbol{\beta}, \sigma^2, v_{jk})$ is the solution of:

$$\begin{aligned} \frac{\partial L_{ws}(\boldsymbol{\beta}, \sigma^2, v_{jk})}{\partial \boldsymbol{\theta}} &= \sum_{i=1}^{n} w_{i1} \frac{\partial \log f_p^*(y_{i1} | \mathbf{x}_{i1}, \boldsymbol{\beta}, \sigma^2, v_{11})}{\partial \boldsymbol{\theta}} \\ &+ \sum_{i=1}^{n} w_i^* \frac{\partial f_p^*(y_{i2}, y_{i3}, \ldots y_{iT} | y_{i1}, \mathbf{x}_i, \boldsymbol{\beta}, \sigma^2, v_{jk})}{\partial \boldsymbol{\theta}}, \end{aligned} \tag{14.16}$$

where we can take $w_i^* = w_i$ or $w_i^* = N/n$. For more details, see Binder (1983) and Eideh and Nathan (2006).

14.6 EMPIRICAL EXAMPLE: BRITISH LABOUR FORCE SURVEY

The British Labour Force Survey (LFS) is a household survey, gathering information on a wide range of labour force characteristics and related topics. Since 1992 it has been conducted on a quarterly basis, with each sample household retained for five consecutive quarters, and a fifth of the sample replaced each quarter. The survey was designed to produce cross-sectional data, but in recent years it has been recognized that linking together data on each individual across quarters would produce a rich source of longitudinal data that could be exploited to give estimates of gross change over time (see Tate, 1999). Labour force gross flows are typically defined as transitions over time between the major labour force states: "employed", "unemployed" and "not in labour force" (or economically inactive). Quarter-to-quarter gross flows show how individuals with each labour force state or classification in one quarter are classified in the following quarter. Gross flows provide estimates of the number of individuals who went from employed in one quarter to employed in the next quarter, employed to unemployed, employed to not in labour force, and so forth. Estimates of labour force flows are useful for answering questions such as: how much of the net increase in unemployment is due to individuals losing jobs and how much is due to individuals formerly not in the labour force starting to look for jobs; and how many unemployed individuals become discouraged and leave the labour force? A number of problems are associated with the estimation of gross flows. Some of these problems are: quarter to quarter nonresponse; measurement errors or response errors; sample rotation; and complex sample design effects. In this numerical example we consider only the quarter-to-quarter nonresponse. The problem of handling quarter-to-quarter nonresponse was discussed and studied by Clarke and Chambers (1998) and Clarke and Tate (1999).

In order to accommodate the differences between the assumptions of Sections 14.2 to 14.5 and those required for the present application, primarily due to the fact that the LFS data relate to categorical rather than to continuous variables, the following modifications were made.

Table 14.1 Gross flow estimates (percentages).

Flow[a]	Unweighted	Weighted	Exponential SMLE
EE	70.62	69.32	69.78
EU	1.08	1.25	1.17
EN	1.53	1.71	1.61
UE	1.61	1.62	1.61
UU	3.78	4.41	4.16
UN	1.00	1.12	1.07
NE	1.40	1.29	1.35
NU	1.06	1.14	1.12
NN	17.92	18.12	18.13

[a] E = employment; U = unemployment; N = nonactive.

Let $n_h(a,b)$ be the number of individuals in household h with labour force flow (a,b) $(a,b = 1,2,3)$ and let $\omega(a,b) > 0$ be the probability of an individual having labour force flow (a,b).

We assume that nonresponse is of whole households, so that responses for all individuals are obtained if the household responds and none are obtained if the household fails to respond. This closely approximates to the situation in most household labour force surveys. Let S_{11} denote the subset of households who responded in both quarters, i.e. the subset representing the longitudinal-linked data on the same persons.

The estimates of labour force gross flows are shown in Table 14.1. The methods of estimation used are as follows:

(1) *Unweighted method.* The second column of Table 14.1 gives estimates from the unweighted data, obtained by maximizing the sample likelihood,

$$\hat{\omega}_U(a,b) = \frac{\sum_{h \in S_{11}} n_h(a,b)}{n_{11}}; a,b = 1,2,3, \tag{14.17}$$

where $n_{11} = \sum_{h \in S_{11}} \sum_{a,b} n_h(a,b)$ is the total number of persons in the households of subset S_{11}

(2) *Weighted method.* The third column of Table 14.1 gives estimates from the weighted data at the household level, computed as:

$$\hat{\omega}_h(a,b) = \frac{\sum_{h \in S_{11}} w_h n_h(a,b)}{\sum_{h \in S_{11}} w_h}; a,b = 1,2,3, \tag{14.18}$$

where w_h is the longitudinal weight of household h.

(3) *The sample likelihood method.* The sample likelihood was derived under the assumptions of the exponential model, for the household weights as a function of the labour force flow frequencies, defined by Equation (14.6) and on the basis of the relationships between the population likelihood and that of the respondent sample, defined by Equations (14.8) and (14.10).

The fourth column in Table 14.1 gives the estimates (SMLE), based on the sample log-likelihood under the exponential model, which in this case is given by:

$$E\left(w_h^{-1}|\{n_h(a,b)\},\{\alpha_h(a,b)\}\right) = \exp\left(\alpha(0,0) + \sum_{a,b}\alpha(a,b)n_h(a,b)\right), \qquad (14.19)$$

where $\alpha(0,0),\alpha(1,1),\ldots\alpha(3,3)$ are parameters to be estimated.

The main findings from Table 14.1 are:

- There are small differences between unweighted and weighted gross flow estimates.
- There are only small differences between gross flow estimates based on household-level weighting and those obtained based on sample likelihoods, under the exponential models. The household-level weighted estimates use the calibrated longitudinal weights, while the sample likelihood method uses the predicted weights based on modelling. Also the calibrated weights as constructed by the Office for National Statistics are functions of auxiliary variables (age, tenure, marital status) and do not depend on the labour force frequencies. Thus these calibrated weights might be considered as ignorable because they depend only on auxiliary variables and do not depend on the labour force status. The fact that the differences between them are small implies that the estimates based on the sample likelihoods are basically just reconstructing the present weights (possibly with some smoothing) and may not reflect the full effects of informative nonresponse.
- Both the household-level weighting and sample likelihood procedures for estimating the labour force gross flows seem to reduce at least part of the effects of nonresponse, compared to the unweighted method. Based on their simulation study, Clarke and Tate (2002) recommend similarly that weighting should be used to produce flow estimates that offer a considerable improvement in bias over unadjusted estimates. Although the sample likelihood estimates cannot be shown to be better than the weighted estimators, their similarity to the weighted estimates indicates that they also are an improvement over the unweighted estimates.

14.7 CONCLUSIONS

We have introduced an alternative method of obtaining weighted estimates of gross flows, taking into account informative nonresponse. We have presented empirical estimates obtained with this method and compared them with estimates obtained using classical methods. Our alternative method is based on extracting the response labour force sample likelihood as a function of the population labour force likelihood and of the response probabilities, based on the reciprocals of the adjusted calibrated weights. The proposed method is model based while the classical method is based on the adjusted weights. Thus we think that the new method is more efficient than the weighted method, although no hard evidence for this is available. However the two methods, sample likelihood and weighting, give approximately the same estimates of labour force gross flows when the propensity scores are based on the reciprocals of the adjusted calibrated weights.

Initially we considered that the estimates of gross flows based on the response sample likelihood might explain the nonignorable nonresponse. The similarity of the results of the weighted and response likelihood methods is not surprising, since the calibrated weights used in both methods are only a function of auxiliary variables and do not depend on the labour force status. The interesting result is that if we have sample data that contain the response variable and the sampling weights and for nonresponse the calibrated adjusted weights, then basing inference using the classical weighted method and the proposed method based on the response likelihood may give similar results.

REFERENCES

Binder, D. A. (1983). On the variances of asymptotically normal estimators from complex surveys. *International Statistical Review*, 51, 279–292.

Chambers, R. L., Dorfman, A. and Sverchkov, M. (2003). Nonparametric regression with complex survey data. In R. Chambers and C. Skinner (Eds), *Analysis of Survey Data* (pp. 151–173). New York: John Wiley & Sons, Inc.

Clarke, P. S. and Chambers, R. L. (1998). Estimating labour force gross flows from surveys subject to household level nonignorable nonresponse. *Survey Methodology*, 24, 123–129.

Clarke, P. S. and Tate, P. F. (1999). Methodological issues in the production and analysis of longitudinal data from the Labour Force Survey. *GSS Methodology Series*, No. 17. London: Office for National Statistics.

Clarke, P. S. and Tate, P. F. (2002). An application of non-ignorable non-response models for gross flows estimation in the British Labour Force Survey. *Australian and New Zealand Journal of Statistics*, 4, 413–425.

Diggle, P. and Kenward, M. G. (1994). Informative drop-out in longitudinal data analysis, with discussion. *Applied Statistics*, 43, 49–73.

Diggle, P. J., Liang, K. Y. and Zeger, S. L. (1994). *Analysis of Longitudinal Data*. Oxford: Oxford University Press.

Eideh, A. H. (2007). A correction note on small area estimation. *International Statistical Review*, 75, 122–123.

Eideh, A. H. and Nathan, G. (2003). Model-based analysis of Labour Force Survey gross flow data under informative nonresponse. *Contributed Paper for the 54th Biennial Session of the International Statistical Institute*, August 13–20, Berlin, Germany.

Eideh, A. H. and Nathan, G. (2006). Fitting time series models for longitudinal survey data under informative sampling. *Journal of Statistical Planning and Inference*, 136, 3052–3069.

Feder, M. F., Nathan, G. and Pfeffermann, D. (2000). Time series multilevel modeling of complex survey longitudinal data. *Survey Methodology*, 26, 53–65.

Friedlander, D., Okun, B. S., Eisenbach, Z. and Elmakias, L. L. (2002). Immigration, social change and assimilation: educational attainment among Jewish ethnic groups in Israel. *Research Reports*. Jerusalem: Israel Central Bureau of Statistics.

Goldstein, H., Healy, M. J. R. and Rasbash, J. (1994). Multilevel time series models with applications to repeated measures data. *Statistics in Medicine*, 13, 1643–1655.

Herriot, R. A. and Kasprzyk, D. (1984). The Survey of Income and Program Participation. *American Statistical Association, Proceedings of the Social Statistics Section* (pp. 107–116). Washington, DC: American Statistical Association.

Ibrahim, J. G., Chen, M. H. and Lipsitz, S. R.(2001). Missing responses in generalized linear mixed models when the missing data mechanism is nonignorable. *Biometrika*, 88, 551–564.

Lipsitz, S. R., Ibrahim, J. and Molenberghs, G. (2000). Using a Box-Cox transformation in the analysis of longitudinal data with incomplete responses. *Applied Statistics*, 439, 287–296.

Little, R. J. A. (1993). Pattern-mixture models for multivariate incomplete data. *Journal of the American Statistical Association*, 88, 125–134.

Little, R. J. A. (1995). Modeling the drop-out mechanism in repeated-measures studies. *Journal of the American Statistical Association*, 90, 1112–1121.

Little, R. J. A. and Rubin, D. B. (2002). *Statistical Analysis with Missing Data*. New York: John Wiley & Sons, Inc.

Little, R. J. A. and Wang, Y. (1996). Pattern-mixture models for multivariate incomplete data with covariates, *Biometrics*, 52, 98–111.

McCullagh, P. and Nelder, J. A. (1999). *Generalized Linear Models*. London: Chapman & Hall.

Mohamed, W. N., Diamond, I. and Smith, P. W. F. (1998). The determinants of infant mortality in Malaysia: a graphical chain modelling approach. *Journal of the Royal Statistical Society A*, 161, 349–366.

Nathan, G. (1999). A review of sample attrition and representativeness in three longitudinal surveys. *GSS Methodology Series*, No. 13. London: Office for National Statistics.

Nathan, G. and Eideh, A. H. (2004). L'analyse des données issues des enquêtes longitudinales sous un plan de sondage informatif. In P. Ardilly (Ed.), *Échantillonage et Méthodes d'Enquêtes* (pp. 227–240). Paris: Dunod.

Pfeffermann, D., Krieger, A. M. and Rinott, Y. (1998). Parametric distributions of complex survey data under informative probability sampling. *Statistica Sinica*, 8, 1087–1114.

Pfeffermann, D., Moura, F. A. S. and Silva, N. P. L. (2001). Multilevel modeling under informative sampling. *Proceedings of the Invited Paper Sessions for the 53rd session of the International Statistical Institute* (pp. 505–532). The Hague, The Netherlands: International Statistical Institute.

Pfeffermann, D. and Nathan, G. (2001). Imputation for wave nonresponse – existing methods and a time series approach. In R. Groves, D. Dillman, J. Eltinge and R. Little (Eds), *Survey Nonresponse* (pp. 417–429). New York: John Wiley & Sons, Inc.

Pfeffermann, D. and Sverchkov, M. (1999). Parametric and semi-parametric estimation of regression models fitted to survey data. *Sankhya*, 61 B, 66–186.

Pfeffermann, D. and Sverchkov, M. (2003). Fitting generalized linear models under informative probability sampling. In R. Chambers and C. J. Skinner (Eds), *Analysis of Survey Data* (pp. 175–194). New York: John Wiley & Sons, Inc.

Rubin, D. B. (1976). Inference and missing data. *Biometrika*, 63, 581–592.

Schafer, J. L. (1995). *Analysis of Incomplete Multivariate Data by Simulation*. London: Chapman & Hall.

Skinner, C. J. (1994). Sample models and weights. *American Statistical Association, Proceedings of the Section on Survey Research Methods* (pp. 133–142). Alexandria, VA: American Statistical Association.

Skinner, C. J. and Holmes, D. (2003). Random effects models for longitudinal data. In R. Chambers and C. Skinner (Eds), *Analysis of Survey Data* (pp. 205–218). New York: John Wiley & Sons, Inc.

Sverchkov, M. and Pfeffermann, D. (2004). Prediction of finite population totals based on the sample distribution. *Survey Methodology*, 30, 79–92.

Tate, P. F. (1999). Utilising longitudinally linked data from the British Labour Force Survey. *Survey Methodology*, 25, 99–103.

Troxel, A. B., Harrington, D. P. and Lipsitz, S. R. (1998). Analysis of longitudinal data with non-ignorable non-monotone missing values. *Applied Statistics*, 47, 425–438.

Wermuth, N. and Lauritzen, S. L. (1990). On substantive research hypotheses, conditional independence graphs and graphical chain models. *Journal of the Royal Statistical Society B*, 52, 21–50.

Zimmerman, D. L. and Nunez-Anton, V. (2000). Modeling nonstationary longitudinal data. *Biometrics*, 56, 699–705.

CHAPTER 15

Weighting and Calibration for Household Panels

Ulrich Rendtel and Torsten Harms
Freie Universität Berlin, Germany

15.1 INTRODUCTION

A household panel is a special type of panel survey that aims to obtain information not only about the sampled persons in the panel, but also about their household context. This is typically necessary in order to assess important political and sociological questions such as poverty or effects of parenthood that are defined at the household level and not (solely) at the personal level. To obtain full information about the household, a household panel design typically involves all persons in the household being interviewed. The resulting household variables are then constructed by the agency that conducts the survey.

At the estimation stage it would however be inefficient to discard the individual information provided by the nonsample household members. Thus ways have to be found to include these persons in panel estimates. Subsequently, one can also follow these cohabiters and thus interpret a household panel as a kind of sampling scheme that is induced by the household composition at the various waves. This rather elaborate scheme is further complicated by the fact that households are not stable entities. Examples might be a household split if a child leaves the parental household – or the opposite effect, a person who does not belong to the initial sample moving into the household. Such events require clear rules on how to deal with them during field work, so-called 'follow-up rules'. These rules define the circumstances in which persons can

Methodology of Longitudinal Surveys P. Lynn

enter or leave the panel. Subsequently, one has to deal with such events at the estimation stage. In this chapter, we will present different possible follow-up rules as they occur in some of the largest ongoing panel studies and will show that many follow-up procedures can be incorporated into the estimator of totals as some sort of weighting of the individuals in the households.

In practice, dealing with follow-up is not the only challenge. As with other surveys, panel studies usually suffer from selective nonresponse and attrition (see Watson and Wooden, Chapter 10 of this volume). Since follow-up rules can be incorporated into weighting estimators, it seems natural to take one step further to adjust the weights in order to lessen the impact of nonresponse and attrition. This can be done using a calibration approach (Deville and Särndal, 1992). We will discuss the implementation and effects of adjustments for nonresponse (Section 15.5) and calibration in panel surveys in general (Section 15.4).

15.2 FOLLOW-UP RULES

As mentioned in the introduction, household panel studies usually follow each sampled person until he or she eventually drops out of the target population (e.g. by death or emigration). The sampled person retains an initial sampling probability that forms the basis on which to construct suitable estimators. The information of cohabiters is primarily collected to retrieve the household information for the initially sampled persons. Consequently one could omit the individual information of these cohabiters when estimating population quantities for individuals.

Given the high maintenance costs of panel studies, it is however inefficient to use only the data of the initially sampled persons. As we will later see, some panels also contact all persons who were interviewed in the last wave. This includes persons who were not initially sampled but interviewed due to the fact that they were living together with a sampled person. The motivation behind this approach is that persons who were already contacted before are relatively easy to track and thus are less costly for the fieldwork than, for example, drawing refreshment samples. During the course of a panel there are thus multiple possibilities for persons who were not initially sampled to enter the panel. It should be noted that this topic does not only apply to cohabiters in households but also to other groups that are sampled in panel surveys, such as staff panels in workforce studies.

Another motivation to follow more persons than just the initial sample over time is to balance out the negative effects of panel attrition and mortality on the sample size by including cohabiters permanently in the panel. This has been treated in detail by Rendtel (1995). Furthermore, one could make provisions for newborns to enter the panel via their sampled parents. This allows the collection of information about early childhood and adolescence. Most ongoing household panels have follow-up rules that allow entrance to the panel via some sort of association with already sampled persons.

15.2.1 Population Definitions

Let U^0 denote the target population at time 0, the start of the panel. We denote by the subscripts $i,i' \in U^0$ persons who are part of this population. The aim of a panel survey is to follow these individuals over time. However in a longitudinal setting it is nontrivial to define the study population U^t at time t. Some definitions might be:

- *Initial population*. This means that an initial population U^0 is followed over time. Newborns do not belong to U^t and dropouts due to death or moving out of the population frame are discarded in the subsequent estimation. As a consequence this initial population usually differs from the cross-sectional population at a latter wave t. The advantage of this static population concept is that the estimation of longitudinal phenomena becomes relatively easy. This concept can be further refined, for example, to only consider adults of the initial population.
- *Initial population plus cohabiters*. Here U^t consists of all persons in U^0 plus their current cohabiters in the household at t. As we have stated before, for many social and economic phenomena the household is the relevant unit, so it might be natural to include in U^t current cohabiters of the members of U^0.
- *Initial population plus their offspring*. Another common extension of the initial-population concept is to include U^0 plus their offspring in U^t. As a consequence, U^t can be viewed as a demographic development of the initial population U^0. Such a concept is, for example, applied in the US Panel of Income Dynamics (PSID), the British Household Panel Survey (BHPS) and the European Community Household Panel (ECHP).
- *Contagion models based on household association*. Under a first-degree contagion model, a person k at wave t belongs to U^t if he or she lives or has lived in a household that also contains a member of the initial population U^0. Higher degrees of contagion could also be defined. An example might be the infinite-degree contagion model, where a person k is defined to belong to U^t if there exists a chain of households with individuals living together that ultimately lead to a person $i \in U^0$. This type of longitudinal population has, for example, been adapted for the German Socio-Economic Panel (SOEP). The aim of such a contagion model is to make the population U^t more representative of the cross-sectional population at time t.
- *All persons ever part of the cross-sectional populations up to* t. Here the population U^t is the unification of all cross-sectional populations that existed between wave 0 and t. As not all persons were present at all times, the resulting missing values have to be treated properly when it comes to describe features of this population. This is not a trivial problem, especially when it comes to estimates that measure changes between waves.

At wave t the initial population has led to a population U^t. These persons are denoted in the following by subscripts $k,k' \in U^t$. In all cases, a person who dies or drops out of the population for other reasons (e.g. emigration) is excluded from the population. This is consistent with the description of longitudinal populations as already noted by Kish (1965). It is important to note that U^t encompasses only those individuals who either belonged to U^0 at wave 0 or entered the population U^t according to the above population definitions.

In any case, the above-defined population U^t should not be mistaken for the current population at time t, which, for example, includes households consisting purely of immigrants. In this chapter we will not treat immigration, which usually requires the use of additional refreshment samples. Still, one should keep in mind that for most panel studies the population U^t is different from the cross-sectional population at time t (see Elliot *et al.*, Chapter 2 of this volume). We will however allow for the inclusion of newborns into an ongoing panel. The difference between immigration and population gains via birth is that newborns are naturally assigned to an existing household within the population, while this is generally not the case for immigrants.

15.2.2 Samples and Follow-up

From an initial population U^0 described in the previous part, a sample $s^0 \subset U^0$ is drawn according to a specified sampling plan and the selected persons $i \in s^0$ will be interviewed. Over time this sample evolves into a set of persons eligible to be interviewed at time t, denoted by s^t. This set might be different from s^0 according to the follow-up rules that are applied to the initial sample. These rules specify which persons are interviewed in subsequent waves, given a realized sample in the previous wave:

- *Follow only initial sample.* This is the simplest case, meaning that only the persons $i \in s^0$ are interviewed in the subsequent waves. This type of follow-up rule corresponds to a population concept, where U^t is identical to the initial population U^0.
- *Follow only initial sample plus current cohabitors.* This extension includes the current cohabiters of the initial population into U^t. By doing so, one can increase the efficiency by using the data that are automatically collected for these cohabiters. However after leaving the household these cohabiters are no longer followed.
- *Follow initial population plus their offspring.* This type of follow-up rule is applied in surveys such as the Panel Socio-Économique (PSELL) in Luxembourg and the BHPS, where children of members of s^0 are also interviewed as part of s^t even after moving out of the household of their parents.
- *Follow* n*-degree cohabiters.* Depending on n, this follow-up rule corresponds to a population that is defined via an n-degree contagion model. In the case of $n = 1$, all cohabiters of sampled persons are interviewed and followed if they leave the household. In the case of $n = \infty$, every person who has ever been interviewed will be followed from wave to wave. It should be noted that this type of follow-up rule might have the tendency to yield exploding sample sizes. However, in practice this effect is balanced by panel attrition, so that, for example, for the case of the German Socio-Economic Panel (SOEP) even following all cohabiters leads to declining sample sizes over time.

Given different follow-up rules, the sample represents different population concepts that we described in the previous subsection. It is important to note that this $s^t \subset U^t$ is not a sample drawn from U^t but rather a consequence of the development of the household composition over time amongst the sample $s^0 \subset U^0$. It is important to note that s^t can include persons who were not initially sampled but entered the panel via the follow-up rules. In the design-based context that we will follow, this development pattern of the households is considered as fixed beforehand and thus the only source of randomness lies in the selection of the sample $s^0 \subset U^0$, which is then transferred onto the sample s^t.

15.3 DESIGN-BASED ESTIMATION

Given the data collected for a sample s^t, one wishes to make estimates concerning the population U^t. We will restrict ourselves to the estimation of totals or means of variables y^t that are measured at time t and could be of the following types:

- *Cross-sectional variables*. This might be a very common case where y_k^t measures age, income or other variables of individual $k \in U^t$.
- *Longitudinal variables*. Such variables might be a change of income or a binary variable that indicates a change, such as if the person has moved. In the latter case y_k^t would be 0 if the person did not move and 1 if the person moved. In order to analyse mobility, we would thus be interested in the total or the mean of y^t, which would be the number or proportion of persons who moved during the analysed period.

It should be pointed out that the variable y_k^t might also require information that not only comes directly from person k but measures a more complex phenomenon, such as the household income of the household of person k. Generally the population total is then given by $T_{y^t} = \Sigma_{k \in U^t} y_k^t$. Other measures, such as correlations or proportions, can be expressed as functions of such totals and estimated similarly. A detailed description of this approach can be found in Särndal *et al.* (1992).

15.3.1 The Horvitz–Thompson Estimator

At the beginning of a panel study, individuals $i \in U^0$ are sampled according to a specific sampling design. In this design, π_i is the probability of person i to be selected, and $\pi_{ii'}$ is the joint sampling probability of individuals i and i'. The sampling process assigns each person $i \in U^0$ the indicator I_i, which is 1 if the person has been selected or 0 otherwise. Thus s^0 is the set of individuals $i \in U^0$ with $I_i = 1$.

Assume that we are interested in the total of a variable y_i^0, for example the income of person i at time 0. Then we could use the common Horvitz–Thompson (HT) estimator to estimate the population total of y^0 at wave 0, $T_{y^0} = \Sigma_{i \in U^0} y_i^0$:

$$\hat{T}_{y^0}^{HT} = \sum_{i \in s^0} d_i y_i^0$$

with $d_i = 1/\pi_i$ being the so-called design weight. This means that each individual in the sample is weighted with the inverse of his or her selection probability. It can be shown that the above estimator is unbiased in the design-based sense (see Horvitz and Thompson, 1952).

Generally, household panels initially select households and all persons in the selected households are sampled at wave 0. This corresponds to a typical cluster sampling scheme. Thus, initially, each individual inherits the sampling probability of his or her associated household.

Now what happens if we wish to estimate T_{y^t} at latter wave? In this case the usual HT estimator would be:

$$\hat{T}_{y^t}^{HT} = \sum_{k \in s^t} d_k y_k^t$$

with d_k being the inverse selection probability for a person $k \in U^t$ to enter s^t. Again this estimator would be unbiased, however the difficulty lies in the determination of d_k. As s^t is the longitudinal evolution of s^0, one should consider each individual $i \in U^0$ with their corresponding selection probability and check if selection of i at wave 0 would have led to k being in s^t. Often this is not feasible as the future development is only known for the selected persons ($I_i = 1$) and not for the remainder of U^0 ($I_i = 0$).

15.3.2 Link Functions

There is a different method to obtain unbiased estimates while at the same time not discarding the information of the persons who were not initially sampled but interviewed due to the fact that they live in the household of a sampled person. The principal idea is redistributing the initial weights d_i of the sampled persons $i \in s^0$ onto the persons $k \in s^t$. We will follow the idea of Kalton and Brick (1995) that is based on concepts presented by Ernst (1989), and use link functions that describe the transfer of weights between persons over time. The household mobility defines a link between persons in wave 0 and wave t. A function l_{ik} that maps each pair of individuals $i \in U^0$, $k \in U^t$ to a real value is called a link function.

Given such a link function l_{ik}, the weight w_k of a person $k \in U^t$ is then the sum of the weights d_i of the initially sampled person redistributed according to the link function:

$$w_k = \sum_{i \in s^0} l_{ik} d_i .$$

Note that this redistribution is applied regardless of whether the person $k \in U^t$ was also part of the initial sample or not. The above representation illustrates the principal idea, namely that the initially sampled persons share their design weights with the interviewed persons who were not initially sampled.

The following list gives some examples of follow-up schemes and link functions that are applied in practice. Note that typically the link functions state how to redistribute the weights from one wave to the next. Thus in all the following examples the link functions l_{jk} describe links between persons j and k from the populations U^{t-1} and U^t:

- *No weight share.* This might be the most straightforward idea that can be followed in a panel survey: only those individuals $i \in U^0$ who were initially sampled are included in estimation over time. Thus each individual simply retains his or her original weight with no modification due to the household context. For $j \in U^{t-1}$ and $k \in U^t$ we have:

$$l_{jk} = \begin{cases} 1 & \text{if individual } j \text{ is identical to } k \\ 0 & \text{otherwise.} \end{cases}$$

  This type of weighting also relates to the idea of cohort studies. While being very simple, panel mortality will inevitably lead to an ever-shrinking sample. In addition such a weighting scheme does waste information if all individuals living in the same household are interviewed.
- *Weight share.* If one follows all persons of the household, the following link function can be utilized, as described by Lavallée (1995). Consider the individuals $j \in U^{t-1}$ and $k \in U^t$ where k lives in household h at time t. Let N_h be the size of this household, excluding newborns.

$$l_{jk} = \begin{cases} \frac{1}{N_h} & \text{if individual } j \text{ also lives in household } h \\ 0 & \text{otherwise.} \end{cases}$$

Thus the initial sample weights are redistributed evenly among all household members at wave t. By excluding newborns in the denominator, this effectively means that the total weight of a household with a newborn rises, as the new child inherits the average weight of the other household members. Population gains due to birth are thus adequately reflected by the weights. This linkage function is treated by Taylor and Buck (1994) for British Household Panel Survey (BHPS).

- *Equal household share.* Consider again individuals $j \in U^{t-1}$ and $k \in U^t$ where k lives in household h at time t. Furthermore let H_h be the number of different households at wave $t-1$ that contained individuals who can now be found in household h. And let C_{jh} be the number of persons who lived together with individual j at time $t-1$ and can now be found in household h.

$$l_{jk} = \begin{cases} \frac{1}{H_h C_{jh}} & \text{if individual } j \text{ also lives in household } h \\ 0 & \text{otherwise.} \end{cases}.$$

Note that by the definition of C_{jh} we do not have to make any special provisions for newborns.

- *Weight share of adults.* This is the same as the common weight share approach, but here not all persons are counted, only the adults. This is a suitable method for the case where the longitudinal population consists only of the adults in the initial population at time $t-1$. Thus in this case the weights are redistributed among the adults only instead of among all persons in the household. Note that since age is not a fixed variable over time one has to fix the attribute 'adult' to one of the two time-points in order to get well-defined links. An application for this method is given in the Panel Socio-Économique (PSELL) in Luxembourg.
- *Restricted weight share of adults.* This is a modification of the previous linkage function that is applied in the US Panel Study of Income Dynamics (PSID). Here the weights are not redistributed among all persons in a household but are averaged only between the head of the household and his or her spouse. Children born in such a household then inherit this weight. Other persons who belong to the household keep their initial weight. Thus for part of the sample (married couples) the weight share approach is applied, whereas for other parts of the sample no weight share is done.
- *Weighted links proportional to initial selection probability.* Consider the link function l_{jk} of the equal household share method proposed above. This link function can be modified so that each link leading from individual j to k does not have identical weight but is proportional to the initial selection probability π_j:

$$\tilde{l}_{jk} = \begin{cases} \frac{\pi_j l_{jk}}{\sum_{j' \in s^0} \pi_{j'} l_{j'k}} & \text{if individual } j \text{ also lives in household } h. \\ 0 & \text{otherwise.} \end{cases}$$

In special cases of independent household sampling, Rendtel (1999) has shown that the above link $\tilde{l}_{jk}$ yields the minimal variance for the resulting estimator, which led to an application of this scheme in the German Socio-Economic Panel (SOEP).

We are mostly interested in the association of individuals at wave t with those at the beginning of the panel. Thus for $i \in U^0$ and $k \in U^t$ we can leave out the intermediate waves and directly link the persons at wave t with those at wave 0 via a link function l_{ik}. This link function can be calculated in a recursive fashion. Given a link function l_{ij} that links the individuals $i \in U^t$ and $j \in U^{t-1}$, we have:

$$l_{ik} = \sum\nolimits_{j \in U^{t-1}} l_{jk} \sum\nolimits_{i \in U^0} l_{ij}$$

where l_{jk} is the usual one-period link function that links $j \in U^{t-1}$ with $k \in U^t$. Lavallée and Deville (2006) discuss this topic in more detail. Generally it can be stated that since all follow-up rules used in practice ensure that there are no breaks in the individual history, the link function l_{ik} can always be constructed. The only possible difficulties arise if a person temporarily drops out of the observed panel due to fieldwork or refusal. In this case the link function is missing. We will discuss treatment of these cases later on.

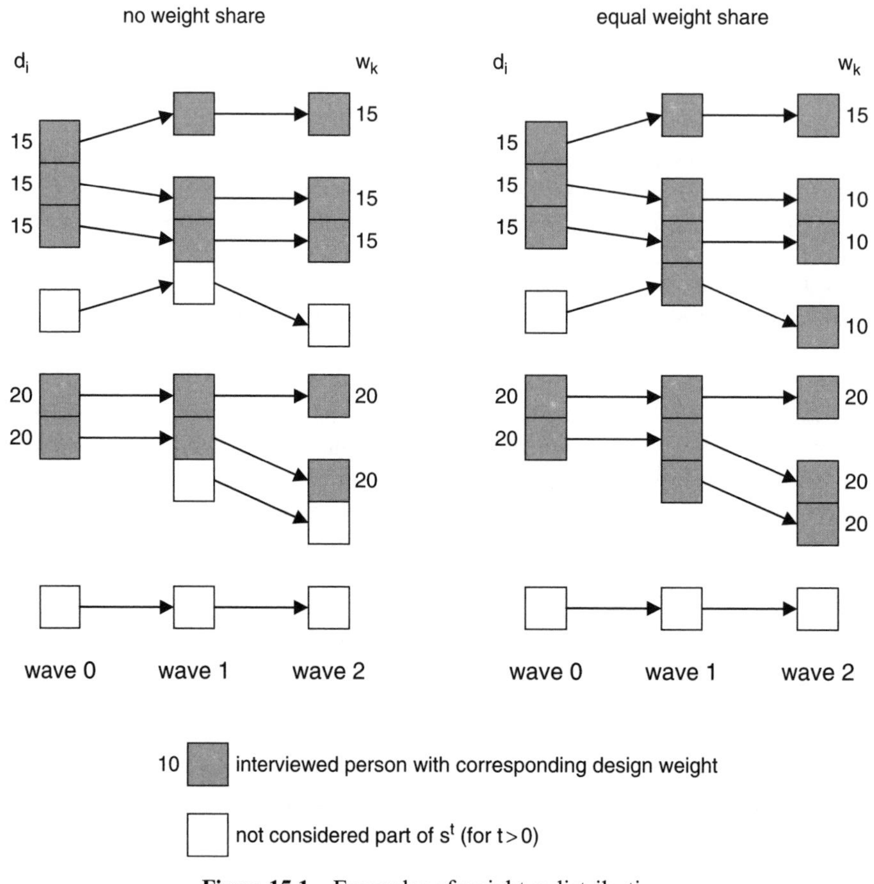

Figure 15.1 Examples of weight redistribution.

The weights of the various follow-up rules are illustrated in Figure 15.1. The resulting weighted estimator $\hat{T}_{y^t}$ for the total $T_{y^t} = \sum_{i \in U^t} y_i^t$ of a variable y^t at wave t is then given by:

$$\hat{T}_{y^t} = \sum_{k \in s^t} w_k y_k^t = \sum_{k \in s^t} \left(y_k^t \sum_{i \in s^0} l_{ik} d_i \right).$$

As we can see, we do not need to know the link functions l_{ik} for all $i \in U^0$ and $k \in U^t$ but it suffices to know the l_{ik} that relate the persons i and k contained in s^0 and s^t.

Under more complex survey designs, such as overlapping panels, or in the presence of immigration and refreshment sampling the weighting scheme becomes considerably more elaborate. A good coverage of topics that arise in these situations can be found in Lavallée (1995) and Merkouris (2001).

15.3.3 Convexity and Variance of the Weighted Estimator

In order to ensure that there is no design bias under general conditions, we note once more that all variables y_k^t as well as l_{ik} are – given the follow-up rules – fixed properties of the population for which we do not impose any model. The only source of randomness is due to the sampling indicators I_i for $i \in U^0$ at the beginning of the panel.

Thus we have:

$$E(\hat{T}_{y^t}) = \sum_{k \in U^t} \left(\sum_{i \in U^0} l_{ik} d_i \pi_i y_k^t \right) = \sum_{k \in U^t} y_k^t \left(\sum_{i \in U^0} l_{ik} \right).$$

Clearly this estimator is only unbiased for $T_{y^t} = \sum_{k \in U^t} y_k^t$ if, for all $k \in U^t$, $\sum_{i \in U^0} l_{ik} = 1$.

If a sampling scheme fulfils this requirement, we call it convex. Thus for all convex weighting schemes the estimator is unbiased. Note that all of the link functions described above are convex by construction.

Turning back to our two examples we can see that usually more than one convex weighting scheme exists. Often, the weighting scheme is determined by the survey agency. In the general case, however, one could choose the weighting scheme that yields minimum variance. For some simple situations, Kalton and Brick (1995) are able to determine the optimal allocation of weights.

For the variance, we can first rewrite our estimator as follows:

$$\hat{T}_{y^t} = \sum_{i \in U^0} d_i I_i \left(\sum_{k \in U^t} l_{ik} y_k^t \right) = \sum_{i \in s^0} d_i \tilde{y}_i^t$$

where $\tilde{y}_i^t = \sum_{k \in U^t} l_{ik} y_k^t$ can be seen as the 'future' contribution y^t of person $i \in U^0$ to the estimation of the total of T_{y^t} under the fixed weight distribution scheme.

Again, noting that only the I_i are random with covariance $C[I_i, I_{i'}] = \pi_{ii'} - \pi_i \pi_{i'}$, the variance is thus:

$$V(\hat{T}_{y^t}) = \sum_{i \in U^0} \sum_{i' \in U^0} (\pi_{ii'} - \pi_i \pi_{i'}) d_i d_{i'} \tilde{y}_i^t \tilde{y}_{i'}^t$$

and an unbiased estimate is given by

$$\hat{V}(\hat{T}_{y^t}) = \sum_{i \in s^0} \sum_{i' \in s^0} (\pi_i^{-1} \pi_{i'}^{-1} - \pi_{ii'}^{-1}) \tilde{y}_i^t \tilde{y}_{i'}^t.$$

Note that both the estimator for the total as well as the variance estimator can be readily calculated because they only require knowledge about the sampled persons and their follow-up behaviour.

Since one can determine the variance of the estimator, one could in theory even choose the convex link functions l_{ik} that minimize the variance while at the same time yielding unbiased estimates (see Lavallée and Deville, 2006). In practice however the follow-up rule is fixed beforehand by constraints of the fieldwork.

15.4 CALIBRATION

Often, one wishes to adjust the weights of individuals in order to reproduce exact estimates for known totals of covariates x. The motivation behind such an adjustment is manifold, as it can be used for the reduction of a possible bias but also for variance reduction. Up to now, we have not allowed for any frame imperfections or nonresponse problems. In the above setting without nonresponse, our weighted estimators are already unbiased. We will thus focus more on the aspect of variance reduction that can be achieved via calibration. However, we will revisit the merits of calibration for biased estimators in the context of nonresponse in the next section.

Apart from variance reduction, calibration is also a tool for survey homogenization. By calibrating different surveys onto a variable x, the calibrated estimate for the variable x is identical for each of the different surveys. This is often a highly desired feature in official statistics in order to obtain at least identical age–gender or geographical classifications (see Schulte-Nordholt *et al.*, 2004; Ehling and Rendtel, 2004).

In other cases, the rich data available from panel studies might be used in order to provide cross-sectional estimates either at the individual level or the household level. As we have seen in Sections 15.2.1 and 15.2.2, panels are generally not defined to provide valid estimates for the cross-sectional population. Thus by calibrating the population to reproduce at least some known totals for key variables x of the cross-sectional population, one hopes to reduce the potential bias due to nonresponse and immigration from using the panel data.

The aim of calibration is to adjust the weights of sampled persons in order to produce exact estimates for known totals T_x of some covariate x. Typical choices for x are sociodemographic key variables such as geographic indicators or age–gender groups. In our example of an age–gender indicator, x would be of the form $(0, \ldots 0,1,0, \ldots 0)'$ where the 1 at the zth position indicates that the individual belongs to group z of the age–gender cross-tabulation. Usual calibration, such as in Deville and Särndal (1992), aims at adjusting each individual weight by factor g, so that the g-adjusted estimator yields exactly the known population total T_x.

Using the panel to assess cross-sectional household characteristics, calibration at the level of individuals yields a potentially undesirable feature, namely that persons in the same household usually obtain different adjustment factors. Thus after calibration the weights of individuals within the household are no longer homogeneous. Following LeMaître and Dufour (1987) or Zieschang (1990), this might not be compatible with the understanding of households as clusters, where each individual has a weight that is equal to the inverse selection probability of the household.

15.4.1 Types of Calibration within Panels

For the described calibration within panels, there are three possible ways:

(1) *Initial calibration (IC)*. At the beginning of the panel, the weights d_i of the selected persons $i \in U^0$ are adjusted by factors g_i^0 that have to fulfil:

$$\sum_{i \in s^0} d_i g_i^0 x_i^0 = T_{x^0}$$

for a known total $T_x^0 = \Sigma_{i \in U^0} x_h^0$. The calibrated estimator for y^0 would then be:

$$\hat{T}_{y^0}^{IC} = \sum_{i \in s^0} d_i g_i^0 y_i^0.$$

This is typically done to make the sample representative at least for key socio-demographic variables x^0 at the start of the panel. Obviously, these modified weights $d_i g_i^0$ can then again be redistributed within the newly formed households in the subsequent waves as described by the follow-up rules: $w_k^C = \Sigma_{i \in s^0} l_{ik} d_i g_i^0$. This would lead to an estimator at wave t:

$$\hat{T}_{y^t}^{IC} = \sum_{k \in s^t} y_k^t w_k^C.$$

(2) *Final calibration (FC)*. At wave t the redistributed weights w_k of the individuals $k \in s^t$ are adjusted by a factor g_k^t that is as close as possible to the constant 1, so that:

$$\sum_{k \in s^t} w_k g_k^t x_k^t = T_{x^t}$$

for a known total $T_{x^t} = \Sigma_{i \in s^t} x_k^t$.

This adjustment can be applied if one wishes to use the panel at wave t to make estimates pertaining to the current cross-sectional population. Note that a panel is usually a longitudinal study with undercoverage of newborns and immigrants. By fitting the weights to some demographic data x, the panel is at least without bias regarding the variable x^t.

(3) *Initial and subsequent final calibration (IFC)*. Here, both initial and final calibration are carried out. At the second stage the weights w_k^C, as defined in (1) above, are further adjusted by a factor g_k^t that has to fulfil:

$$\sum_{k \in s^t} w_k{}^C g_k^t x_k^t = T_{x^t}.$$

This approach, which is carried out sequentially, is applied by Eurostat for the European Community Household Panel (ECHP). For an integrated way to perform both calibrations simultaneously, see Lavallée (2002). The aim is to obtain a weighted sample for the initial population as well as a weighted sample for the population at time t that can both be regarded as representative with respect to x. The resulting weights can then be used to provide cross-sectional estimates, which is often desirable given the rich data contained within many panels. However, care should be taken to clearly label the resulting weights as 'cross-sectional', as is also done in the ECHP.

Since the above equations usually do not determine unique adjustment factors g, one usually chooses, from all admissible combinations of g, those that are closest to the constant adjustment by a factor of 1. Thus, one minimizes

$$\sum\nolimits_{i \in s^0} d_i (g_i^0 - 1)^2 \quad or \quad \sum\nolimits_{k \in s^t} w_k (g_k^t - 1)^2$$

depending on the type of calibration one carries out. In those cases the resulting estimator can be expressed explicitly, as we will show in the next subsection.

15.4.2 Bias and Variance

In this section we discuss the bias and variance of the calibrated estimators described above. We mostly present only the results, typically derived via Taylor linearization. For more details, the reader is referred to the existing literature, such as Deville and Särndal (1992).

For all calibration estimators, it can be shown that for large case numbers the bias induced by calibration can be neglected. However the variance of the calibration estimator is usually different from the expression for the weighted estimator. Similar to the usual variance estimator for $\hat{T}_{y^t}$, we will use the shorthand:

$$\tilde{y}_i^t = \sum\nolimits_{k \in s^t} l_{ik} y_k^t \quad \text{and} \quad \tilde{x}_i^t = \sum\nolimits_{k \in s^t} l_{ik} x_k^t$$

for the future contributions towards y^t and x^t that come from person $i \in U^0$.

Initial Calibration (IC)

If one carries out initial calibration at wave 0 and estimates at the same wave, the resulting estimator $\hat{T}_{y^0}^{IC}$ is identical to the common calibration estimator as described by Särndal *et al.* (1992), whose standard manipulations will be adhered throughout this section.

More interesting is the other case, namely the use of calibrated weights in order to derive an estimator at wave t. Here one has to take into account that these weights $d_i g_i^0$ for $i \in s^0$ are redistributed onto persons $k \in s^t$ according to the follow-up rule $w_k^C = \sum_{i \in s^0} l_{ik} d_i g_i^0$. Thus we have

$$\hat{T}_{y^t}^{IC} = \sum\nolimits_{k \in S^t} y_k^t w_k^C.$$

Taylor linearization leads to an alternative expression:

$$\hat{T}_{y^t}^{IC} = T_{x^0}' \hat{B}^{IC} + \sum\nolimits_{i \in S^0} d_i \tilde{e}_i^{IC}$$

with

$$\tilde{e}_i^{IC} = \tilde{y}_i^t - x_i^{0\prime} \hat{B}^{IC},$$
$$\hat{B}^{IC} = \left(\sum\nolimits_{i \in s^0} d_i x_i^0 x_{i'}^0\right)^{-1} \left(\sum\nolimits_{i \in s^0} d_i x_i^0 \tilde{y}_i^t\right).$$

The above estimator can thus be interpreted as a regression of the future contributions $\tilde{y}_i^t$ versus the values x_i^0 at wave 0. The corresponding variance estimator, following a suggestion of Deville and Särndal (1992), would then be:

$$\hat{V}(\hat{T}_{y^t}^{IC}) = \sum\nolimits_{i \in s^0} \sum\nolimits_{i' \in s^0} (\pi_i^{-1} \pi_{i'}^{-1} - \pi_{ii'}^{-1}) g_i^0 g_{i'}^0 \tilde{e}_i^{IC} \tilde{e}_{i'}^{IC}.$$

Final Calibration (FC)

In this case, calibration is carried out at wave t while the only source of randomness is the sample selection process at wave 0. Thus we obtain different variance formulas than in the case of initial calibration.

Standard manipulations similar to the case of initial calibration (IC) lead to a closed form of the calibrated estimator:

$$\hat{T}_{y^t}^{FC} = T_{x^t}{}'\hat{B}^{FC} + \sum\nolimits_{i\in s^0} d_i \tilde{e}_i^{FC}$$

with

$$\tilde{e}_i^{FC} = \left(\tilde{y}_i^t - \tilde{x}_i^{t\prime}\hat{B}^{FC}\right)\left(\sum\nolimits_{k\in s^t} l_{ik} g_k^t\right),$$

$$\hat{B}^{FC} = \left(\sum\nolimits_{k\in s^t} w_k x_k^t x_k^{t\prime}\right)^{-1}\left(\sum\nolimits_{k\in s^t} w_k x_k^t y_k^t\right).$$

This is also some sort of regression estimation, but now it concerns the future contributions towards x^t and y^t of the persons $i\in s^0$. Again, the variance can then be estimated as a double sum:

$$\hat{V}(\hat{T}_{y^0}^{FC}) = \sum\nolimits_{i\in s^0}\sum\nolimits_{i'\in s^0}(\pi_i^{-1}\pi_{i'}^{-1} - \pi_{ii'}^{-1})\tilde{e}_i^{FC}\tilde{e}_{i'}^{FC}.$$

Initial and Final Calibration (IFC)

Here we have two adjustment factors g_i^0 for $i \in U^0$ and g_k^t for $k \in U^t$ that are determined as described above. With $w_k^C = \sum_{i\in s^0} l_{ik} d_i g_i^0$ being the calibrated weight that is determined via the follow-up rules, the estimator for T_{y^t} is then:

$$\hat{T}_{y^t}^{IFC} = \sum\nolimits_{k\in U^t} y_k^t g_k^t w_k^C.$$

Due to the nonlinearity of the double calibration, the resulting estimator can only be expressed approximately in terms of residuals:

$$\hat{T}_{y^t}^{IFC} \approx T_{x^t}{}'\hat{B}^t + T_{x^0}{}'\hat{B}^0 + \sum\nolimits_{i\in s^0} d_i e_i^{IFC},$$

and thus the variance estimator would be

$$\hat{V}(\hat{T}_{y^t}^{IFC}) = \sum\nolimits_{i\in s^0}\sum\nolimits_{i'\in s^0}(\pi_i^{-1}\pi_{i'}^{-1} - \pi_{ii'}^{-1})g_i^0 g_{i'}^0 e_i^{IFC} e_{i'}^{IFC}$$

with

$$e_i^{IFC} = -x_{i'}^0\hat{B}^0 + \sum\nolimits_{k\in U^t} l_{ik} g_k^t (y_k^t - x_{k'}^t\hat{B}^t),$$

$$\hat{B}^t = \left(\sum\nolimits_{k\in s^t} w_k x_k^t x_{k'}^t\right)^{-1}\left(\sum\nolimits_{k\in s^t} w_k x_k^t y_k^t\right),$$

$$\hat{B}^0 = \left(\sum\nolimits_{i\in s^0} d_i x_i^0 x_{i'}^0\right)^{-1}\left[\sum\nolimits_{i\in s^0} d_i x_i^0 \sum\nolimits_{k\in s^t} l_{ik}(y_k^t - x_k^t\hat{B}^t)\right],$$

which again indicates some form of double regression of x^t vs. y_i^t and the residuals of this regression vs. x^0. The variance formulas for such simultaneous calibration can be found in Lavallée (2002).

Application for More than Two Waves
The Taylor linearization technique might be useful to derive the variance estimator if the number of waves where calibration takes place is limited. In the case of calibration in each wave and longer running panels, this approach might be less attractive. In such a situation, one may change to resampling techniques like bootstrap or jackknife. However, one should bear in mind that these approaches require a replication of the initial sampling process and the calibration procedure at each iteration, which may be a computational burden.

15.5 NONRESPONSE AND ATTRITION

Up to this point, we have not considered the treatment of nonresponse or attrition in the context of the weight share approach. However these topics are important when dealing with panel surveys in real life. We use attrition to mean nonresponse or dropout of persons who were successfully interviewed in previous waves. Nonresponse refers to data that could not be obtained from persons who were contacted for the first time. Immigration and frame imperfections are other topics that are related to the problem of nonresponse but will not be covered here. In all cases the problem is the same, namely that information is missing for a part of the sample.

With either kind of nonresponse, estimates based on the available data usually yield biased estimates, since nonresponse can depend on the variable of interest y. The weight share approach seems to aggravate this problem, since it redistributes the effects of nonresponse or attrition onto other members of the panel according to the follow-up rules. Fortunately, this topic can be readily analysed using link functions.

Consider our longitudinal sample s^t. At wave t we define $r^t \subseteq s^t$ as those persons who respond to the interview at wave t. The problem with attrition within longitudinal surveys is that we wish to predict T_{y^t} but, due to attrition, the unbiased weighted estimator:

$$\hat{T}_{y^t} = \sum\nolimits_{k \in s^t} w_k y_k^t = \sum\nolimits_{k \in r^t} w_k y_k^t + \sum\nolimits_{k \in s^t \setminus r^t} w_k y_k^t$$

cannot be calculated, since the second sum is unknown. This lack of knowledge usually makes it impossible to construct an unbiased estimator and thus without any additional information one cannot hope to obtain unbiased estimates.

15.5.1 Empirical Evidence Regarding Nonresponse and Attrition

Before we present the treatment of nonresponse and attrition under the linkage approach, it is important to have a clear picture of the various factors that cause nonresponse and attrition and their specific effects:

- *General analysis of nonresponse and attrition.* There exists a variety of models that explain nonresponse and attrition over the course of a panel. A recent example is the general sequential model by Lepkowski and Couper (2002), which can be divided into the following stages: initial nonresponse, failure to trace persons from wave to wave, noncontact with the successfully traced persons and noncooperation of the contacted persons. Empirical results tend to show that noncooperation of the contacted persons seems to be more important than difficulties in keeping track of the panel sample from wave to wave (Nicoletti and Buck, 2004). Over the course of the panel the effects of nonresponse and attrition for each wave are then accumulated. Sisto and Rendtel (2004) found that the consequences of cumulative attrition seem to be less severe than the effect of the initial nonresponse.

- *Effects of fieldwork on nonresponse and attrition.* This cause might be the one that is the easiest to control by the statistical agency. Typical events that increase attrition are interviewer changes, which can be a consequence of fluctuations among interviewer staff but also could be related to other variables such as moves. This has been analysed in detail by Behr *et al.* (2005). Other changes in the survey implementation, such as a change in mode of data collection, can also affect attrition.
- *Impact of the questionnaire design.* Generally a stable questionnaire increases the chances of cooperation. This might be due to the fact that it is easier to fill out repeating questionnaires, but also due to the fact that participants might feel less invasion into their privacy by repeatedly responding to a questionnaire they once accepted. A typical example illustrating this effect is the German Socio-Economic Panel (SOEP), which asked in one single wave for the total value of assets of the interviewed persons, causing a spike in nonresponse and attrition.
- *Choice of variables for modeling response.* In order to model response the appropriate variables have to be selected. Apart from personal attributes such as age, income or gender, the above discussion might suggest some useful variables to include into the modelling approach, such as interview mode, or a dummy variable indicating if the person has already been interviewed successfully. Generally modelling the response probability might be possible with similar variables for different surveys, which is why we will include a short analysis here. Table 15.1 summarizes the results of a logistic analysis of the participation in a panel after already having completed one questionnaire successfully, using European Community Household Panel (ECHP) data where all countries used the same questionnaire (Behr *et al.*, 2005). The significant positive (+) or negative (−) effects of the different variables are shown.

  As can be seen, the only clear effect is the negative impact of an interviewer change on the response rate. It thus seems advisable for each survey to evaluate separately the effect of covariates on nonresponse or attrition. As a general statement, one should note that it is possible to conduct such analyses either from wave 0 to t or sequentially for each wave. The latter method seems advisable, as the individual history of the respondents might provide powerful predictors of the response propensity. Examples include Nicoletti and Buck (2004), who use the item nonresponse rate in the past as a predictor for attrition in the current wave, and Uhrig (2008), who uses time-varying covariates in pooled models.

The focus on item nonresponse as a special case of nonresponse circumvents to some extent an intrinsic problem of nonresponse analysis, namely that the explanatory variables must be known for every member of the sample. In the case that attrition depends on changes that occurred between the current and the previous wave, such a variable can usually not be observed for all attriters. If no other source of information (e.g. registers) is available, such change variables have to be excluded from the analysis of nonresponse.

Before coming to the analytical treatment of nonresponse or attrition, we would also point out that such an analysis of the causes of nonresponse might not only be used to provide models to correct for nonresponse, but might also be a helpful tool to reduce the amount of nonresponse and attrition in the future.

Table 15.1 Logit results predicting response by country.

Country	Income (I)			Age		Household moved	Marital status		Female	Higest level of education		Main activity status		Interviewer change
	I	I^2	I^3	< 30	>64		Separated or divorced	Widowed or never married		Third level	Less than second stage	Unemployed	Inactive	
Germany	+	–	+			–	–	–			–			–
Germany (SOEP)	+			–	–	–	–	–					+	–
United Kingdom		–	+					–		+			+	–
United Kingdom (BHPS)	+		+						+				+	
Finland	–	+	–				+	–	+					
Dennmark					–	–	–	–		+	–	–		NA
Ireland	–	+		–	+	–		–		+	+			–
The Netherlands		+		–		+		–	+	+	–			–
Belgium			+	–		–			+	+				–
Luxembourg	+	–	+	–	+	+	+	–	+	+	–	–	+	–
France	+	–	+		–					+		–		–
Spain					+			–				+	+	NA
Portugal	–					–		–	+	–	+	–	–	NA
Austria				–										–
Italy	–	+	–			–	–	–						–
Greece	–			–				–		+				NA

15.5.2 Treatment via Model-Based Prediction

There are two different approaches to cope with nonresponse. The first model-based approach tries to predict variables that are not observed due to nonresponse. The second approach, which will be covered in the next subsection, tries to model the nonresponse process itself.

If there exists a linear relationship of the form $y_k^t = x_k^{t\prime}B$ for all $k \in U^t$ and the total T_x^t is known, then the calibrated estimators $\hat{T}_{y^t}^{FC}$ and $\hat{T}_{y^t}^{IFC}$ are asymptotically unbiased. A typical example is the spending of y on certain goods, which is roughly proportional to the overall income x. In this case the total income in the population might be well known from tax registers or at least very efficient estimates might exist. Such situations allow for the unbiased estimation of T_y even under nonresponse. We will illustrate this effect for the final calibration (FC) estimator: Note that this estimator can be written as:

$$\hat{T}_{y^t}^{FC} = T_{x^t}{}'\hat{B}^{FC} + \sum\nolimits_{i \in s^0} d_i \sum\nolimits_{k \in U^t} l_{ik} (y_k^t - x_k^{t\prime}\hat{B}^{FC}).$$

For large sample sizes one can show, similarly to Deville and Särndal (1992), that $\hat{B}^{FC} \approx B$ and thus the residual $y_k^t - x_k^{t\prime}\hat{B}^{FC}$ goes to zero, whereas $T_x^{t\prime}\hat{B}^{FC} \approx T_{x^t}{}'B = T_{y^t}$. The same arguments hold for the initial and final calibration (IFC) estimator, which also requires $y_k^t = x_i^{t\prime}B$ to hold in order to be asymptotically unbiased regardless of selective nonresponse. In the case of initial calibration (IC) we would need a slightly different model to hold, namely $\tilde{y}_i t = x_i^{0\prime}B$: that is, the future contribution to y^t of person i is a linear function of x^0. Due to the different type of calibration, in this case it is knowledge of T_{x^0} rather than T_{x^t} that is required.

Note however that unbiased (and even exact) estimates are only obtained if the linear relationship holds for all persons, regardless of whether they belong to the responding group r^t or the attrition group $s^t \setminus r^t$. In practice this constraint will not strictly hold but generally one can hope that calibration on explanatory variables x not only reduces the variance but also the attrition bias, (see Lundström and Särndal, 2005).

A further generalization would be to allow for other models that yield estimates $\hat{y}_k^t$. This can then be incorporated into a design-based estimator, as has been shown by Lehtonen and Veijanen (1998) for discrete measures, or be used to simply impute the dataset and apply the estimator as if no nonresponse had occurred (see Rubin (2004) for the more elaborate multiple imputation). Again, only a perfect prediction would be able to fully correct the nonresponse bias but even for less than perfect predictors it can generally be hoped that bias will be reduced.

15.5.3 Treatment via Estimation of Response Probabilities

A second approach is to model the response probability for each person $k \in s^t$ and then reweigh the responding persons with the inverse of this estimated probability. Rizzo, Kalton and Brick (1996) as well as Clarke and Tate (2002) have studied ways of implementing such a procedure. Again, one can show that the resulting estimator is unbiased if the predicted response probability matches the true response probability. Estevao and Särndal (2000) have shown that this weight adjustment can also be interpreted as a modified calibration approach whose properties are similar to the usual or linear calibration that we presented in the previous section.

The advantage of this approach is that the resulting weights can be used regardless of the variable of interest. On the contrary, prediction of y requires a specific model for each variable of interest. On the downside, weighting with the inverse estimated response probability may inflate the variance of the weights and the resulting estimates. An explanatory variable (e.g. interviewer change) for the response that is independent of the variable y has no effect on the bias of the resulting estimator, whether it is included in the response model or not. However, the variance of the resulting estimator increases by including such a variable into the response model. In many cases the aim of nonresponse treatment is to reduce the nonresponse bias. Regarding this target only, all important variables for nonresponse should be included in the model.

We will present here the logistic response propensity model that is widely used in practice and incorporates the simple expansion estimator as well as response homogeneity group (RHG) estimators. A good overview of this topic can be found in Lundström and Särndal (2005).

For modelling purposes, we will now consider three substantially different types of nonresponse.

Initial Nonresponse (INR)

This means that a sampled person $i \in s^0$ does not participate in the panel. This is modelled via a response indicator R_i^0, which is 1 if i responds and 0 otherwise. Similarly θ_i indicates the response probability that we will have to estimate. We consider the simple binomial response model, namely that R_i^0 is independent of $R_{i'}^0$ if $i \neq i'$. This might be a reasonable approach in the case of panel studies that sample individuals. In the case of sampling full households, nonresponse is typically clustered at the household level. In such a case, it might be more reasonable to model the response probability at the household level.

We assume that θ_i depends on a covariate x_i^0 via the logistic form:

$$\theta_i = \frac{e^{x_i^{0\prime}\lambda^0}}{1 + e^{x_i^{0\prime}\lambda^0}}.$$

Our aim is to find an estimate $\hat{\theta}_i$, given some knowledge about x^0. We assume that we know x_i^0 for all $i \in r^0$ and that the total $\sum_{i \in s^0} x_i^0$ is known. Thus it is sufficient to find an estimator $\hat{\lambda}^0$ and then set:

$$\hat{\theta}_i = \frac{e^{x_i^{0\prime}\hat{\lambda}^0}}{1 + e^{x_i^{0\prime}\hat{\lambda}^0}}.$$

With $g_i^0 = 1/\hat{\theta}_i = (1 + e^{-x_i^0\hat{\lambda}^0})$, λ^0 can be estimated implicitly by setting:

$$\sum_{i \in r^0} x_i^0 g_i^0 = \sum_{i \in s^0} x_i^0.$$

Note that we estimate by reweighting the respondents with the inverse response probability. Another way would be by weighting the full sample with the response probability. We prefer the approach above as it might be more likely to know the individual value x_i for the respondents than for the full sample, for which only the total of x might be available from other sources.

The resulting estimator at waves 0 and t can then be found by reweighting each element $i \in r^0$ with g_i^0, which is the inverse of the estimated response probability:

$$\hat{T}_{y^0}^{INR} = \sum\nolimits_{i \in r^0} d_i g_i^0 y_i^0,$$

$$\hat{T}_{y^t}^{INR} = \sum\nolimits_{k \in U^t} y_k^t \sum\nolimits_{i \in r^0} l_{ik} d_i g_i^0.$$

We can easily see that this type also covers the simple expansion estimator and the response homogeneity group (RHG) approach. For the expansion estimator, consider x_i to be constantly equal to 1. In this case $\hat{\lambda}^0$ is uniquely determined by the constraint resulting in $\hat{\theta}$ that corresponds to the response fraction. The same is true for the RHG estimator, which requires x variables that encode membership in the response homogeneity group in a binary fashion (e.g. membership to the qth RHG is encoded in a vector x that is 0 everywhere except at the qth place where it is 1).

These estimators correspond to the initial calibration (IC) estimator. The only differences are that the adjustment factor g_i^0 now has a different functional form than in the case of the linear calibration estimator in the previous section. In addition, the calibration constraint is now different. Thus again Taylor linearization can be used in order to assess bias and variance of this estimator. We will only give results for the estimator at time t, $\hat{T}_{y^t}^{INR}$. The results for the estimator at time 0 can be found in Kott (2004).

The linearized version of $\hat{T}_{y^t}^{INR}$ is:

$$\hat{T}_{y^t}^{INR} = \sum\nolimits_{i \in r^0} d_i g_i^0 \tilde{y}_i{}^t + \left(\sum\nolimits_{i \in s^0} x_i^0 - \sum\nolimits_{i \in r^0} x_i^0 g_i^0\right) \hat{B}^{INR}$$

where:

$$\hat{B}^{INR} = \left[\sum\nolimits_{i \in r^0} x_i^0 x_i^0 (g_i^0 - 1)\right]^{-1} \left[\sum\nolimits_{i \in r^0} l_{ik} d_i x_i^0 (g_i^0 - 1) \sum\nolimits_{k \in s^t} y_k^t\right]$$

For the subsequent analysis, note that we now have two sources of randomness that affect our estimators: the sampling process, as well as the outcome for the response variable R_i^0. If the response model holds, that is, R_i^0 is the outcome of independent Bernoulli trials with success probability θ_i for each trial that can be described via the previous logistic model, then it can be shown that $\hat{T}_{y^t}^{INR}$ is asymptotically unbiased with a combined expectation under the sampling and response process:

$$E(\hat{T}_{y^t}^{INR}) = E\left[E(\hat{T}_{y^t}^{INR} | s^0)\right] = T_y^t.$$

The variance can be decomposed into its corresponding sampling and response variance:

$$V(\hat{T}_{y^t}^{INR}) = V\left[E(\hat{T}_{y^t}^{INR} | s^0)\right] + E\left[V(\hat{T}_{y^t}^{INR} | s^0)\right] = V_1(\hat{T}_{y^t}^{INR} | s^0) + V_2(\hat{T}_{y^t}^{INR} | s^0).$$

These terms can be estimated via

$$\hat{V}_1(\hat{T}_{y^t}^{INR}) = \sum\nolimits_{i \in r^0} \sum\nolimits_{i' \in r^0} (\pi_i^{-1} \pi_{i'}^{-1} - \pi_{ii'}^{-1}) \hat{\theta}_{ii'}^{-1} \tilde{y}_i^t \tilde{y}_{i'}^t,$$

where $\hat{\theta}_{ii'}$ estimates the joint response probability of the elements $i,i' \in s^0$ via $\hat{\theta}_{ii} = \hat{\theta}_i$ and $\hat{\theta}_{ii'} = \hat{\theta}_i \hat{\theta}_{i'}$ in the case of $i \neq i'$. For the second variance term note:

$$\hat{V}_2(\hat{T}_{y^t}^{INR}) = \sum_{i \in r^0} \hat{\theta}_i (1 - \hat{\theta}_i) z_i^2$$

where $z_i = d_i g_i \tilde{y}_i t - x_i^0 g_i \hat{B}^{INR}$.

Attrition (ATT)

Here a sample person that responded at time 0 drops out during the course of the panel. This could be associated with difficulties in the fieldwork but also could be due to a reluctance to continue participation in the panel. Viewed from the estimation problem at wave t, this type of nonresponse can be treated similarly to the initial nonresponse (INR). The only difference is that the nonresponse occurs at a later point of time, thus usually more information is available in order to model the response process.

Final Nonresponse (FNR)

After having established a sample s^t according to the follow-up rules, it usually occurs that the desired data cannot be collected for all contacted persons $k \in s^t$. This would also be the type of nonresponse that one has to deal with in a cross-sectional context. We denote the response indicator by R_k^t and the corresponding response probability by ϕ_k.

With the common logistic response model with $\phi_k = e^{x_k^{t\prime} \lambda^t} / (1 + e^{x_k^{t\prime} \lambda^t})$ and thus an adjustment factor $g_k^t = 1/\hat{\phi}_k = (1 + e^{-x_k^t \hat{\lambda}^t})$, where $\hat{\lambda}^t$ is determined via $\sum_{k \in r^t} w_k x_k^t g_k^t = \sum_{k \in s^t} w_k x_k$, the resulting estimator is given as:

$$\hat{T}_{y^t}^{FNR} = \sum_{k \in r^t} w_k y_k^t g_k^t.$$

Again, the variance can be found via Taylor approximation. Combining both yields:

$$\hat{T}_{y^t}^{FNR} = \sum_{k \in r^t} y_k^t g_k^t \sum_{i \in s^0} l_{ik} d_i + \hat{B}\left(\sum_{k \in s^t} x_k \sum_{i \in s^0} l_{ik} d_i - \sum_{k \in r^t} x_k^t g_k^t \sum_{i \in s^0} l_{ik} d_i \right),$$

with

$$\hat{B} = \left[\sum_{k \in r^t} x_k^t x_k^t (g_k^t - 1) \sum_{i \in s^0} l_{ik} d_i \right]^{-1} \left[\sum_{k \in r^t} y_k^t x_k^t (g_k^t - 1) \sum_{i \in s^0} l_{ik} d_i \right],$$

which is again unbiased if the response model holds and allows for similar variance estimation to that shown above in the case of initial nonresponse.

The Case of Missing Link Functions

In this chapter we have dealt with each form of nonresponse or attrition separately. It should be noted that a combined treatment of the different forms of attrition and nonresponse is possible, analogous to the use of calibration as outlined in Section 15.4.1. Note that in all cases we assume that the link functions l_{ik} are known if i and k are both respondents. This might not always be the case, as knowledge of the link function is usually a result of the fieldwork. An example might be that the interviewer fails to register the household association for an interviewed person. We will not cover the case of missing link functions, however a simple treatment would be to consider persons $k \in U^t$ with l_{ik} unknown as nonrespondents. Alternatively, these link functions can be estimated, as has been shown by Lavallée (2002).

15.6 SUMMARY

In this chapter, we formalized the concept of making use of the information of non-sample household members interviewed during the course of a household panel survey. For gains in efficiency these other household members are usually followed and also interviewed at subsequent waves. Since households are not stable units, panel surveys have to make provisions on how to deal with such developments of the household composition over time. We showed that these follow-up rules can be incorporated relatively easily into the estimation stage, allowing for both unbiased estimates as well as valid variance estimators. In order to improve the estimation, one could also adjust the resulting weights using the calibration approach. Due to the longitudinal nature of panel surveys there are several possible adjustments, however for each of them the resulting estimator and variance formulas can be obtained even in the case of complex follow-up rules or nonresponse.

REFERENCES

Behr, A., Bellgardt, E. and Rendtel, U. (2005). Extent and determinants of panel attrition in the European Community Household Panel. *European Sociological Review*, 21, 489–512.

Clarke, P. S. and Tate, P. F. (2002). An application of non-ignorable non-response models for gross-flows estimations in the British Labour Force Survey. *Australian and New Zealand Journal of Statistics*, 44, 413–425.

Deville, J. C. and Särndal, C. E. (1992). Calibration estimators in survey sampling. *Journal of the American Statistical Association*, 82, 376–381.

Ehling, M. and Rendtel, U. (2004). *Harmonisation of Panel Surveys and Data Quality*. Wiesbaden: Statistisches Bundesamt.

Ernst, l. (1989). Weighting issues for longitudinal household and family estimates. In D. Kasprzyk *et al.* (Eds), *Panel Surveys* (pp. 139–159). Hoboken, NJ: John Wiley & Sons, Inc.

Estevao, V. M. and Särndal, C. E. (2000). A functional form approach to calibration. *Journal of Official Statistics*, 16, 379–399.

Horvitz, D. G. and Thompson, D. J. (1952). A generalization of sampling without replacement from a finite universe. *Journal of the American Statistical Association*, 47, 663–685.

Kalton, G. and Brick, J. M. (1995). Weighting schemes for household panel surveys. *Survey Methodology*, 21, 33–44.

Kish, L. (1965). *Survey Sampling*. Hoboken: John Wiley & Sons, Inc.

Kott, P. (2004). Using calibration weighting to adjust for nonresponse and coverage errors. *Proceedings of the Mini Meeting on Current Trends in Survey Sampling and Official Statistics*, Calcutta. http://www.nass.usda.gov/research/reports/calcutta-for-web.pdf

Lavallée, P. (1995). Cross-sectional weighting of longitudinal surveys of individual households using the weight share method. *Survey Methodology*, 21, 25–32.

Lavallée, P. (2002). *Le Sondage Indirect*. Brussels: Éditions de l'Université de Bruxelles.

Lavallée, P. and Deville, J. C. (2006). Indirect sampling: the foundations of the generalized weight share method. *Survey Methodology*, 32, 165–176.

Lehtonen, R. and Veijanen, A. (1998). Logistic generalized regression estimators. *Survey Methodology* 24, 51–55.

LeMaître, G. and Dufour, J. (1987). An integrated method for weighting persons and families. *Survey Methodology*, 13, 199–207.

Lepkowski, J. and Couper, M. (2002). Nonresponse in the second wave of longitudinal household surveys. In Groves *et al.* (Eds), *Survey Nonresponse* (pp. 259–274). Hoboken, NJ: John Wiley & Sons, Inc.

Lundström, S. and Särndal, C. E. (2005). *Estimation in Surveys with Nonresponse*. Hoboken: John Wiley & Sons, Inc.

Merkouris, T. (2001). Cross-sectional estimation in multiple-panel household surveys. *Survey Methodology* 27, 171–181.

Nicoletti, C. and Buck, N. (2004). Explaining interviewee contact and co-operation in the British and German Household Panels. In M. Ehling and U. Rendtel (Eds), *Harmonisation of Panel Surveys and Data Quality* (pp. 143–166). Wiesbaden: Statistisches Bundesamt.

Rendtel, U. (1995). *Panelausfälle und Panelrepräsentativität*. Frankfurt: Campus Verlag.

Rendtel, U. (1999). *The application of the convex weighting estimator to household panel surveys*. Unpublished manuscript, Freie Universität Berlin, Berlin.

Rizzo, L., Kalton, G. and Brick, J. M. (1996). A comparison of some weighting adjustment methods for panel nonresponse. *Survey Methodology*, 22, 43–53.

Rubin, D. (2004). *Multiple Imputation*. Hoboken: John Wiley & Sons, Inc.

Särndal, C. E., Swensson, B. and Wretman, J. (1992). *Model Assisted Survey Sampling*. New York: Springer Verlag.

Schulte-Nordholt, E., Hartgers, M. and Gircour, R. (2004). *The Dutch Virtual Census 2001*. Voorburg/Heerlen: Statistics Netherlands.

Sisto, J. and Rendtel, U. (2004). Nonresponse and attrition effect on design-based estimates of household income. In M. Ehling and U. Rendtel (Eds), *Harmonisation of Panel Surveys and Data Quality* (pp. 210–219). Wiesbaden: Statistisches Bundesamt.

Taylor, A. and Buck, N. (1994). Weighting adjustments in the BHPS wave one – wave two weights. In A. Taylor, *et al.* (Eds), *British Household Panel Survey – User Manual Volume A* (pp. A5- 1–A5-12). Colchester: University of Essex.

Uhrig, S. C. N. (2008). The nature and causes of attrition in the British Household Panel Survey. *ISER Working Paper* 2008–05. Colchester: University of Essex. Retrieved June 14, 2008, from www.iser.essex.ac.uk/pubs/workpaps/pdf/2008-05.pdf

Zieschang, K. D. (1990). Sample weighting methods and estimation of totals in the Consumer Expenditure Survey. *Journal of the American Statistical Association*, 85, 986–1001.

CHAPTER 16

Statistical Modelling for Structured Longitudinal Designs

Ian Plewis

University of Manchester, UK

16.1 INTRODUCTION

Repeatedly measuring individuals or cases – the defining characteristic of longitudinal surveys – generates datasets that are hierarchically structured. We can regard the repeated measurements as a cluster of correlated measures within each individual, the correlations arising from the shared individual characteristics. This way of thinking about longitudinal data is useful whenever we repeatedly measure the same underlying variable over time, even though we might use different instruments at different measurement occasions. For example, when we repeatedly measure children in order to understand more about the way they develop we expect to find systematic relations with age: children grow, learn more and become more skilled at dealing with a range of social situations as they get older. These changes do not, however, all take place in the same way or at the same rate for every child. The same sorts of ideas can also be applied in other contexts. We might, for example, analyse changes in political attitudes during adulthood, changes in physical and mental functioning for older people and changes in income across the lifecourse.

In many situations, each subject has their own 'growth' function and their own growth parameters, where the term 'growth' is used to capture *any* systematic changes with age or time. At their simplest, these functions will be straight lines and the parameters of most interest will be the slopes of these lines. We would expect to find systematic variation in these parameters across subjects – for example, differences between boys and girls in educational progress, different rates of decline in physical functioning for smokers and nonsmokers and different income profiles for men and women. We can investigate these variations by specifying a statistical model that properly represents

Methodology of Longitudinal Surveys P. Lynn

both the underlying growth function and the variation in the growth parameters across subjects' characteristics.

Growth curves have been used for many years to study children's physical development (Goldstein, 1979a) but only more recently with educational, psychological and economic measures. The impetus for extending the approach can be attributed to the increasing availability of data from large-scale longitudinal surveys, and to the development of statistical methods, known variously as multilevel models, hierarchical linear models and random coefficient models, that enable soundly based inferences about patterns and correlates of growth to be made. The focus of the example used in this chapter is on data from cohort studies of children and particularly on their educational development but, as indicated above, the approach is a general one that can be applied to a variety of research questions about change.

To get a full description of all the individual and contextual influences on growth we would need to address the following kinds of questions:

(1) How does a particular variable (e.g. reading attainment) develop with age and how does this development vary with fixed characteristics of the child, such as their sex, and also with time-varying explanatory variables, such as their family's income?
(2) Does does this description generalize to more than one cohort?
(3) How do two or more variables (e.g. physical and mental functioning) change together with age (and across cohorts)?
(4) Do members of the same household show similar patterns of growth? In other words, are the characteristics of the household that are shared by its members – living in poor housing conditions, for example – associated with each person's growth? Children also share their parents' development and so the association between development or growth across generations is also potentially interesting.
(5) Is there a spatial or institutional dimension to growth? For example, is educational progress related to the neighbourhood(s) the child is brought up in or the schools they attend, or both?

As we shall see, we can represent these different influences on growth in terms of different levels, which in turn can be represented as components of a statistical model.

The rest of the chapter is ordered as follows. First, the fact that data are structured in particular ways is used to set out a general statistical model for their analysis that represents different influences on growth. This is then followed by a series of analyses that illustrate the range of models that can be fitted to existing data. The measurement difficulties posed for this kind of analytical approach by the absence of fixed scales are discussed at this point. The chapter concludes with discussions of extensions of the models to other kinds of longitudinal data, and other methodological issues.

16.2 METHODOLOGICAL FRAMEWORK

Suppose we have a long series of cohort studies, for example, conducted every 2 years over a period of 50 years. Index this series by k ($k = 1 \dots K$). For each cohort, we might select a different sample of areas or neighbourhoods, j, where $j = 1 \dots J_k$. The cohort members (often, but not necessarily, children), i, are sampled from these areas ($i = 1 \dots I_{jk}$). Each cohort member is, in principle, measured on a number of occasions (t) and on more than one outcome (l) so that $t = 1 \ldots T_{ijk}$ and $l = 1 \ldots L_{tijk}$. In other words, we would have a five-level nested hierarchy with outcomes at the bottom of the hierarchy (which we call level 1) and cohorts at the top (level 5). Note that the number of outcomes is allowed to vary across occasions and that the number of measurement occasions can vary (usually unintentionally) across cohort members. Note also that the ordering of the subscripts is the one generally adopted in the multilevel literature but it is not necessarily consistent with the ordering used in the sample survey literature for multistage samples.

We will assume that the outcomes **y** have a multivariate normal distribution. A growth model that encompasses all these sources of variation can be written as:

$$y_{ltijk} = \sum_{q=0}^{Q_l} b_{lqijk} a_{tijk}^{q} + e_{ltijk}. \qquad (16.1)$$

This is a polynomial in age (a) of order Q_l that can vary with l, the order determined from the data but with Q_l usually less than $\max(T_{ijk})$. Using polynomials is a flexible and convenient way of representing nonlinear change with age. Spline functions are another possibility, as described by Snijders and Bosker (1999, Chapter 12). To reduce the difficulties created by the correlations between higher order polynomials, it is usual to centre age around its overall mean. This centring also helps with the interpretation of the estimates from the model. Note that all the coefficients – the intercept terms b_0 and the growth parameters b_1, b_2, etc. – can vary from response to response (l), from cohort member to cohort member (i), from area to area (j) and from cohort to cohort (k). In other words, we treat them as random effects and focus on the estimation of their variances and, within levels, their covariances. We assume that random effects are uncorrelated across levels and also that:

$$\begin{aligned} E(e_{ltijk}) &= \sigma^2_{(l)e} \qquad l = l',\ t = t',\ i = i',\ j = j',\ k = k' \\ &= \sigma_{(ll')e} \qquad l \neq l',\ t = t',\ i = i',\ j = j',\ k = k' \\ &= 0, \text{ otherwise.} \end{aligned}$$

In other words, we assume that the variances of the within-individual residuals do not vary across occasions (i.e. with age) and that the residuals for the responses are correlated at each occasion but that there is no correlation across occasions, either within or between responses (i.e. no autocorrelation). We see later how to relax some of these assumptions.

As we shall see, we can expand the model (Equation 16.1) to show how the random effects b_{lqijk} vary across levels, in principle systematically with explanatory variables z_{ijk} (characteristics of the cohort member), z_{jk} (characteristics of the area) and z_k (characteristics of the cohort), and hence how the covariance structure is defined at higher levels.

The model does not explicitly include either the siblings or the children of the cohort members, although implicitly they are included; the older siblings as members of cohorts born before the cohort member and their children as members of cohorts born some time after the cohort member. What is missing, however, are any links that bind the cohort members themselves to their siblings on the one hand and to their children on the other. Rather than elaborate an already rather complicated model, we introduce these extensions within the context of an example. One extension not considered here is a model that includes both institutional (schools or hospitals, for example) and neighbourhood effects. This usually requires a cross-classified model (see Goldstein, 2003, Chapter 11) because cohort members from the same neighbourhood can attend different institutions and cohort members in the same institution can come from different neighbourhoods. A cross-classified model would also be generated by a design that either selected the same neighbourhoods for every cohort studied or at least allowed an overlap in terms of neighbourhoods selected.

Nevertheless, this basic model enables us to estimate:

(1) how the variances in mean responses and growth parameters are partitioned between occasions within cohort members (and areas and cohorts), between cohort members within areas (and cohorts), between areas within cohorts and between cohorts;
(2) to what extent the variances of the random effects (and their covariances) can be accounted for by explanatory variables defined at different levels – for example, by sex at the cohort member level, by a measure of area deprivation and by macro-economic circumstances at the cohort level;
(3) to what extent higher level variances – between areas and between cohorts – remain important after allowing for explanatory variables measured at the cohort member level. In other words might, for example, area differences in educational development be accounted for by the different characteristics of the cohort members who choose or are selected to live within these areas?;
(4) the association between the growth parameters for the responses – between individuals, between areas and between cohorts;
(5) how the variances in the responses, and the correlations between them, vary with age.

16.3 THE DATA

The datasets used throughout the rest of this chapter are from two of the four UK birth cohort studies. These studies follow children born in 1946 (National Survey of Health and Development), 1958 (National Child Development Study or NCDS), 1970 (British Cohort Study or BCS70) and 2000/1 (Millennium Cohort Study or MCS). Details about the first three cohorts can be found in Ferri *et al.* (2003) and the more complicated design used in MCS is described in Plewis (2007).

We use data from the 1958 and 1970 cohorts to illustrate the ideas and we focus on educational attainment. These two studies share a number of characteristics:

- their samples are generated by all the births (ca. 17 000) in a single week, which we regard as simple random samples;
- they both collected educational test data on three occasions during childhood, along with a range of fixed and time-varying explanatory variables;
- they have both now collected data from a sample of children of the cohort members.

Details of the ways in which the 1958 and 1970 samples have changed as the cohorts have aged are given in Plewis *et al.* (2004).

In order meaningfully to interpret estimates from growth curve models we need to assume that the measures of our responses are on the same scale. This is not a problem with biological variables like height and weight or with economic variables like income and price. But when we turn to measures of educational attainment, for example, we are faced with the fact that, although children indubitably become better readers as they get older, an appropriate measure or test at age 5, say, cannot also be used at age 15 (because the questions would be too easy). Moreover, a test that is appropriate for a particular age for one cohort is not necessarily suitable for a later cohort (because the school curriculum can change). In other words, we must work with data from tests that are designed for the age group being studied at that time. This problem of not having a fixed scale occurs in other contexts too. For example, a set of questions about mobility addressed to 90-year-olds would not necessarily apply to those aged 55. And questions designed to tap the political attitudes of adolescents in one cohort would not necessarily have the same salience for a cohort born 10 years later.

Lacking a fixed scale, but reasonably confident that the different tests are nevertheless measuring the same underlying attainment or latent variable both across ages and across cohorts, we can adopt the strategy used by Plewis (1996). We fix the mean score for the attainment for the measured sample at each age to be the mean chronological age. In other words, we assume that, on average, children gain one reading or mathematics 'year' each year and so mean growth is linear. This is a relatively innocuous assumption because we are much more interested in variation around this average than in the average itself. It does, however, mean that if we apply this approach to each of the K cohorts we then eliminate intercohort differences in overall levels and rates of growth of attainment and so we cannot account for the possibility that these levels and rates improve or fluctuate with time. We also assume that, on average, different attainments develop in the same way. This is perhaps unlikely: we might suppose that reading develops more quickly when children are young and then approaches an asymptote as they pass through adolescence, whereas mathematics continues to develop in a linear fashion. Again, this is not a serious problem when analysing each response separately but it could affect the interpretation of multivariate models. Finally, the scale is determined by the children in the sample at each occasion; it ignores the possibility that attrition at occasion t might be related to attainment at occasions $t-k$, with $k \geq 0$. Again, this could affect intercohort comparisons if patterns and correlates of attrition vary across cohorts, and also multivariate models if patterns of attrition are different for different responses.

A more problematic issue than fixing the means at each occasion is determining how the variances of the responses should change with age because, as Plewis (1996) shows, this can affect inferences about correlates of change within cohorts. It is reasonable to assume that, as with height, the variability between children in any attainment increases with age but there are very few clues about just how marked an increase we should expect. For the purposes of this chapter, we will assume that the variance increases in such a way that the coefficient of variation for each attainment is constant across occasions so that the standard deviation increases proportionately with the mean. It is, however, important to stress that this is an assumption and other assumptions are possible: for example, that scores are 'z-transformed' to have zero means and unit standard deviations at each age or that they have standard normal distributions at each age. If we were to use the growth curve approach for older people then we might want to assume that variability in functioning declines with age, whereas if we were to model changes in attitudes with age then the assumption of constant variance might be as good as any other.

The assumption about scaling used here implies that a child who is a year ahead in reading at age 7 is 1 SD ahead of the 'average' child then but if they remain just one year ahead at ages 11 and 16 then they are only 0.64 SD units ahead at age 11 and only 0.44 SD units ahead at age 16 (because the distribution of reading attainment becomes more dispersed with age). Another way of expressing this is to say that children who stay 1 SD above or below the mean have nonlinear growth curves, whereas those who stay one reading year above or below the mean have linear growth curves. The departures from linearity with the scaling used here and with just three measurement occasions are, however, slight.

16.4 MODELLING ONE RESPONSE FROM ONE COHORT

We start with a simple model. In terms of the model of Equation (16.1), we drop the subscripts referring to responses (l), areas (j) and cohorts (k) and we assume that growth is linear (i.e. $Q=1$). We do, however, introduce an explanatory variable at the cohort member level (z), one that is hypothesized to explain between-subject (or level 2) variation in the intercept (or overall mean) and the slope (or growth rate). To facilitate the link to the example used below, we assume that this variable is the sex of the cohort member. We write:

$$b_{0i} = b_{00} + b_{01}z_i + u_{0i}, \quad (16.2a)$$

$$b_{1i} = b_{10} + b_{11}z_i + u_{1i}, \quad (16.2b)$$

where $z_i=0$ for males and 1 for females. Substituting Equations (16.2a) and (16.2b) into the simpler version of Equation (16.1) gives us:

$$y_{ti} = b_{00} + b_{01}z_i + b_{10}a_{ti} + b_{11}a_{ti}z_i + u_{0i} + a_{ti}u_{1i} + e_{ti} \quad (16.3)$$

and the effect of sex on the slope is represented by the cross-level interaction, $a_{ti}z_i$. The last three terms in Equation (16.3) show that the overall variance of the response y, conditional on the explanatory variables, will be a function of age. We return to this point shortly. In addition to the assumptions about e_{ti} discussed above we also assume that $\boldsymbol{u} \sim MVN(0, \Omega)$ and $E(u_{ki}e_{ti})=0$ (where $k=0, 1$).

We illustrate the use of this model with NCDS data on reading collected from children at ages 7, 11 and 16. The large sample is a strength of the NCDS but only having three occasions of measurement is a restriction when fitting growth curve models. We include all children with at least one measurement. Table 16.1 gives some descriptive statistics. It shows how the sample size declines with age and also that at each measurement occasion there was variability in age at testing that we can use in the analysis (because our explanatory variable is age and not occasion). The observed correlations between the test score at occasion 1 and the scores at occasions 2 and 3 are both 0.62, and between occasions 2 and 3 it is 0.80.

Table 16.1 NCDS reading data.

Occasion (age)	Sample size	Age (months) at test: mean (SD)	Reading[a]	
			Mean	SD
1 (7)	14 931	85 (1.7)	7	1
2 (11)	14 133	134 (1.8)	11	1.57
3 (16)	11 987	193 (1.4)	16	2.29

[a] Mean fixed at intended mean chronological age, SD fixed to give a constant coefficient of variation of 0.14.

Table 16.2 Estimates from the model of Equation (16.3).

Effects		Parameter	Estimate	SE
Fixed		Intercept: b_{00}	11	0.016
		Slope: b_{10}	1.0	0.0025
		Sex: b_{01}	0.11	0.022
		Age × sex: b_{11}	−0.033	0.0035
Random	Level 2	Intercept: σ^2_{u0}	1.8	0.023
		Slope: σ^2_{u1}	0.028	0.00058
		Covariance: σ_{u0u1}	0.22	0.003
	Level 1	Intercept: σ^2_e	0.58	0.0077

We centre age around its overall mean (essentially age 11) and then we fit the model (Equation 16.3) using the Iterative Generalized Least Squares (IGLS) algorithm built into the *MLwiN* package (Rasbash *et al.*, 2004). A number of other packages are available although they do not all have such an extensive range of modelling capabilities. The estimates are given in Table 16.2. The estimates of the fixed effects b_{00} and b_{10} in Table 16.2 are defined by the scale we have adopted for reading. The estimates of the effects of sex on mean attainment and growth (b_{01} and b_{11}) tell us that at the mean age (age 11) girls are 0.11 units ahead of boys in reading but that this gap narrows over the age range under study such that girls are about a quarter of a standard deviation ahead at age 7 $[0.11-(4\times-0.033)=0.24]$ but slightly behind at age 16 $[0.11+(5\times-0.033)=-0.06]$.

Turning to the random effects, we see that after allowing for differences between boys and girls there is still considerable variation around the mean growth rate of 1, as evidenced by the estimate of σ^2_{u1}: we would expect about 95 % of the sample's growth rates to be in the range $1-2\sqrt{(0.028)}$ to $1+2\sqrt{(0.028)}$, or 0.67 to 1.34. The estimate of the correlation between the intercept at mean age and the growth rate is 0.996. This is improbably close to 1 and indicates that the specification of the model could be improved.

Because we only have three occasions of measurement, it is not advisable to estimate a polynomial model in age (although, as we have seen, nonlinear functions would be more appropriate with this scaling if there were more than three measurement occasions or if there were a greater spread of ages at each occasion). We can, however, allow the between-child variance for boys to be different from girls by estimating an extra random effect at level 2 – a covariance between the intercept and sex (σ_{u0u2}). We can also loosen the restrictions on the level 1 variance to allow it to change with age, which we do by making age random at level 1: we find that estimating a covariance term between the intercept and age (σ_{e0e1}) is sufficient. These two new terms do not, on their own, have a substantive interpretation. Rather they are devices, explained in more detail in Goldstein (2003, Chapter 3), for dealing with residuals that are potentially heteroscedastic.

The between-child variance is now:

$$\sigma_{u0}^2 + a_{ti}^2\sigma_{u1}^2 + 2a_{ti}\sigma_{u0u1} + 2\sigma_{u0u2} \tag{16.4}$$

and the within-child variance is:

$$\sigma_{e0}^2 + 2a_{ti}\sigma_{e0e1}. \tag{16.5}$$

In addition, the covariance between the responses at any two ages t and t' is:

$$\sigma_{u0}^2 + a_{ti}a_{t'i}\sigma_{u1}^2 + (a_{ti} + a_{t'i})\sigma_{u0u1} + 2\sigma_{u0u2} + \sigma_{e0e1}. \tag{16.6}$$

When we extend the model in this way, the estimates of the fixed effects are essentially unchanged from those in Table 16.2. The random effects are given in Table 16.3. We see that the two extra terms are both large when compared with their standard errors but their introduction has little effect on the other four parameters. However, the correlation between the intercept at age 11 and the slope is now a somewhat more plausible 0.93: the higher a child's score at age 11 on this particular scale, the faster their reading grows afterwards.

Plugging the estimates from Table 16.3 into the expressions for the variances and covariances in Equations (16.4)–(16.6), we get the results shown in Tables 16.4 and 16.5.

The estimates in Tables 16.4 and 16.5 correspond rather closely to the observed variances and correlations reported earlier and suggest that we now have a well-specified model, one that brings out differences in the reading development of boys and girls. Whether differences of this kind are replicated in the cohort measured 12 years later is the question we soon turn to. Before we do, we briefly consider some extensions to this simple model. It is straightforward to include some more fixed characteristics of the child and its family – birthweight, mother's age at birth, for example – and doing so could improve the explanatory power of the model. Suppose, instead, that z_i varies across

Table 16.3 Random effects for extended model.

Level	Parameter	Estimate	SE
2	σ_{u0}^2	1.9	0.026
	σ_{u1}^2	0.027	0.00059
	σ_{u0u1}	0.21	0.0032
	σ_{u0u2}	−0.10	0.011
1	σ_{e0}^2	0.58	0.0079
	σ_{e0e1}	0.019	0.0012

Table 16.4 Estimated variances by age and sex.

Age	Between children		Between age within children	Total	
	Boys	Girls		Boys	Girls
7	0.65	0.44	0.43	1.1	0.87
11	1.9	1.7	0.59	2.5	2.3
16	4.7	4.5	0.78	5.5	5.2

Table 16.5 Estimated correlations by age and sex.

	Boys	Girls
(7, 11)	0.65	0.57
(7, 16)	0.65	0.64
(11, 16)	0.81	0.80

occasions (i.e. z_{ti}). We could then use functions of z_{ti} as explanatory variables at level 2 – for example, its mean and a measure of its spread. Alternatively, we could explicitly model the response as a function of both a_{ti} and z_{ti} at level 1, although a drawback of this is that it can lead to problems of interpretation. Further discussion of this point can be found in Singer and Willett (2003, Chapter 5), who also provide a very detailed introduction to hierarchical repeated-measures models.

16.5 MODELLING ONE RESPONSE FROM MORE THAN ONE COHORT

In principle, we now have a three-level model with subscripts t and i (as before) and now k. Rather than Equations (16.2a) and (16.2b), we have:

$$b_{0ik} = b_{00k} + b_{01k}z_{ik} + u_{0ik} \tag{16.7a}$$

$$b_{1ik} = b_{10k} + b_{11k}z_{ik} + u_{1ik} \tag{16.7b}$$

$$b_{00k} = b_{000} + b_{001}z_k^* + v_{00k} \tag{16.7c}$$

$$b_{01k} = b_{010} + b_{011}z_k^* + v_{01k} \tag{16.7d}$$

$$b_{10k} = b_{100} + b_{110}z_k^* + v_{10k} \tag{16.7e}$$

$$b_{11k} = b_{110} + b_{111}z_k^* + v_{11k} \tag{16.7f}$$

Equations (16.7a) and (16.7b) are essentially the same as Equations (16.2a) and (16.2b) but Equations (16.7c)–(16.7f) introduce a number of new issues. First, Equations (16.7c) and (16.7e) will pick up any between-cohort variation in mean attainment or in the mean slope or both, provided that the scaling method used has not eliminated them. This variation might be systematically related to an explanatory variable z_k^*, which could be a function of time. So z_k^* might just take the values $1, 2 \ldots K$. Second, if the relations with the explanatory variable defined at the child level (z_{ik}) vary with cohort (and with z_k^*), then these effects will be estimated from Equations (16.7d) and (16.7f).

In practice, we rarely have a sufficiently large number of cohort studies to estimate Equations (16.7a) – (16.7f). Some administrative systems – for example, the National Pupil Database in England, which links individual pupils' test scores as they pass through the state school system – could eventually permit a detailed examination of between-cohort variation in educational progress. And long-running panel studies could also provide the data needed. But, for now, we compare just two birth cohorts – those children born in 1958 and 1970. We do this by treating the cohort as a dummy variable z_{2i} ($=0$ for NCDS and 1 for BCS70) and estimating fixed effects in the form of interactions with the cohort. Our model now is:

$$\begin{aligned} y_{ti} = {} & b_{00} + b_{01}z_{1i} + b_{02}z_{2i} + b_{10}a_{ti} + b_{11}a_{ti}z_{1i} + b_{12}a_{ti}z_{2i} + b_{03}z_{1i}z_{2i} \\ & + b_{13}a_{ti}z_{1i}z_{2i} + u_{0i} + a_{ti}u_{1i} + e_{ti}. \end{aligned} \quad (16.8)$$

We are particularly interested in the estimates of b_{03} and b_{13} (the fixed-effect equivalents of Equations (16.7d) and (16.7f)) because these will tell us whether the sex differences for the means and slopes found for the 1958 cohort are replicated in the 1970 cohort.

The descriptive data for the 1970 cohort are given in Table 16.6. Attrition is greater than it is for NCDS and there is a substantial amount of missing data at age 16. The variability in age at measurement is much higher at age 16 for BCS70 than it is for NCDS.

All the estimates from the model of Equation (16.8) are given in Table 16.7. We see that the estimates of both b_{03} and b_{13} are substantially different from zero: the age patterns of the sex differences are different for the two cohorts. Substituting values of the explanatory variables into the estimated version of Equation (16.8), we find that between the first two occasions (ages 7 and 11 for NCDS and ages 5 and 10 for BCS70) girls make 0.14 years less progress in reading than boys in the 1958 cohort but 0.17 years more progress in the 1970 cohort. Between ages 11 and 16, girls born in 1958 make 0.18 years less progress than boys but girls born in 1970 make 0.19 years more progress. Aside from the possibility that the observed turnaround represents a genuine cohort effect in sex differences in educational progress, other plausible explanations are that different tests were used for the two cohorts, and that attrition and missing data affect the two cohorts differently. These kinds of considerations will apply to any cross-cohort analysis. We do

Table 16.6 BCS70 reading data.

Occasion (age)	Sample size	Age (months) at test: mean (SD)	Reading	
			Mean	SD
1 (5)	10 338	62 (1.3)	5	0.71
2 (10)	9998	124 (2.7)	10	1.42
3 (16)	4824	199 (4.6)	16	2.29

Table 16.7 Estimates of fixed effects from the model of Equation (16.8).

Parameter	Estimate	SE
Intercept: b_{00}	11	0.015
Sex: b_{01}	0.11	0.021
Cohort: b_{02}	−2.4	0.022
Age: b_{10}	1.0	0.0025
Age × sex: b_{11}	−0.036	0.0035
Age × cohort: b_{12}	−0.093	0.0042
Sex × cohort: b_{03}	−0.074	0.031
Age × sex × cohort: b_{13}	0.068	0.0058

find that the mean scores are higher at occasion 2 for those measured at occasion 3 than they are for those not measured then, and this is particularly so for BCS70. On the other hand, the gap between those measured and not measured is the same for boys and girls in both cohorts.

16.6 MODELLING MORE THAN ONE RESPONSE FROM ONE COHORT

We now consider a model with two responses – attainment in reading and mathematics – and so $L=2$. A detailed description of the multivariate multilevel model can be found in Snijders and Bosker (1999, Chapter 13). Here, we fit a bivariate model to the NCDS data, noting that the descriptive statistics for mathematics are essentially the same as those for reading given in Table 16.1. We find that girls are slightly behind boys in mathematics at age 7 and that this gap widens to nearly one-third of the standard deviation at age 16. We also find slightly more variability in the growth rates for mathematics than we do for reading. Our main interest is in the random effects from which we can generate correlations of interest, and these correlations are given in Table 16.8. We see that the intercepts for the two attainments are closely related, but the slopes somewhat less so. The within-child correlation between the level 1 residuals for reading and mathematics is 0.17. We find that the evidence for different intercept variances for boys and girls for reading (Table 16.3) is not found for mathematics.

As well as estimating the model-based variances and correlations by age and sex for each variable separately (as in Tables 16.4 and 16.5 for reading), we can also estimate

Table 16.8 Correlations between random effects at level 3 for a bivariate growth model.

Correlation	Estimate
Intercepts (R & M)	0.86
Slopes (R & M)	0.65
Int. (R), slope (M)	0.70
Int. (M), slope (R)	0.75
Int., slope (R)	0.94
Int., slope (M)	0.84

Table 16.9 Correlations between random effects at level 3 for a bivariate growth model: z-transformed scores.

Correlation	Estimate
Intercepts (R & M)	0.86
Slopes (R & M)	0.21
Int. (R), slope (M)	0.20
Int. (M), slope (R)	0.18
Int., slope (R)	0.17
Int., slope (M)	0.08

the correlations between the responses for different combinations of age and sex (see Plewis, 2005).

If we use z-transformed scores for reading and mathematics at each occasion rather than the scales based on increasing variation with age then we get a rather different pattern of correlations as shown in Table 16.9. The differences between Tables 16.8 and 16.9 bring out the importance of the assumptions about scaling. When we use z scores, we get less variation in growth rates for both attainments and therefore the correlations are much lower. To suppose, however, that variabilities in attainments do not change with age is perhaps unrealistic. Nevertheless, further investigations of these relations are warranted, making different assumptions about scaling, including those that vary from response to response.

16.7 MODELLING VARIATION BETWEEN GENERATIONS

We now consider the situation where we have not only repeated measures on the cohort members but also at least some data on the cohort members' children. This is the case for both NCDS and BCS70, with data collected from a sample of cohort members who had become parents by age 33 (NCDS) or 34 (BCS70). We use just the NCDS data on mathematics here, collected for all children aged 4–18 in the selected households. We use data from 1702 households – 1019 with just one child, 509 with two children and 174 with three – together with the longitudinal data from all the cohort members themselves.

The children of the NCDS cohort members are not, of course, a probability sample of children in that age range, partly because they are children of those cohort members still in the sample by age 33 but more particularly because the cohort member's age and the child's date of birth are confounded. A child aged x years must have one parent aged $33 - x$ years at birth so children aged 12 at the time of data collection will have one parent who was aged 21 when they were born. As age of mother at birth and later educational attainment are associated – because better educated parents tend to postpone parenthood – it is important to model the relations between the scores of the cohort members' children and age accurately.

We set up the model in the following way. Level 2 is defined by the cohort member as before. Level 1, however, now consists of two kinds of data: the repeated measures for the cohort member *and* their children's scores if they have any. Thus, there can be up to six measures at level 1 for each level 2 case: up to three repeated measures for the cohort members and cross-sectional measures for up to three of their children. We fit different coefficients to the different parts of the model by using indicator variables, δ_0 to δ_3, as follows:

$$y_{ti} = \delta_0 \sum_{q=0}^{1} b_{qi} a_{ti}^q + \sum_{p=1}^{3} \delta_p b_{0pi}^c + \sum_{p=1}^{3} \sum_{q=1}^{2} \delta_p b_{qp}^c a_{pt}^q + e_{ti} \tag{16.9}$$

$\delta_0 = 1$ for repeated measures, 0 otherwise
$\delta_p = 1$ for child p $(p = 1, 2, 3)$, 0 otherwise.

The first term in Equation (16.9) is just the model used for repeated measures on one response for one cohort, the second term allows the test score for the pth child to vary across cohort members and the third term defines the cross-sectional relation with age (a quadratic here) for child p.

The basic data for the cohort members' children are given in Table 16.10. We observe that there is an inverse relation between birth order and both age and score.

The fixed effects from the model of Equation (16.9) are given in Table 16.11 and we see that, for the children of the cohort, the sex effects – which, unlike the effects for their parents, are measured on the raw score scale shown in Table 16.10 – are small relative to their standard errors.

The correlations between the various random parameters are of particular interest and these are given in Table 16.12. We see that the correlations between the mean attainment of the cohort member (CM) and the attainment of each child are around 0.3. Interestingly, the correlations between the cohort members' growth rates and their children's attainment are also around 0.3, suggesting that parents' progress in mathematics during their school years might have as much influence as their level of attainment on their children's attainments. The model does, however, only provide a description of these links rather than an explanation of why they might occur: they could, for example, be related to continuities in social class across generations. The within-family correlations are higher for the adjacently born children than for children born first and third.

If we had repeated measures for the cohort member's children then we could structure the data so that the cohort members and their children would be at level 2 and all the repeated measures at level 1. Then the second and third terms of Equation (16.9) could be combined and correlations between the growth rates of cohort members and their children could be estimated. Extensions to two or more responses and to more than one cohort follow as before.

Table 16.10 NCDS mathematics data for children of the cohort.

Child number	Sample size	Age (months) at test: mean (SD)	Mathematics	
			Mean	SD
1	1702	114 (36)	41	17
2	683	101 (29)	35	15
3	174	90 (25)	29	15

Table 16.11 Estimates of fixed effects for the model of Equation (16.9).

		Parameter	Estimate	SE
Repeated measures	Intercept	b_{00}	8.3	0.011
	Slope	b_{01}	1.0	0.0027
	Sex	b_{10}	−0.090	0.016
	Age × sex	b_{11}	−0.026	0.0039
Child 1	Intercept	b^c_{010}	45	0.33
	Sex	b^c_{011}	−0.49	0.39
	Age	b^c_{11}	5.3	0.070
	Age^2	b^c_{12}	−0.43	0.021
Child 2	Intercept	b^c_{020}	37	0.50
	Sex	b^c_{021}	−0.55	0.59
	Age	b^c_{12}	5.9	0.14
	Age^2	b^c_{22}	−0.41	0.047
Child 3	Intercept	b^c_{030}	61	3.1
	Sex	b^c_{031}	0.35	1.0
	Age	b^c_{13}	1.7	1.0
	Age^2	b^c_{23}	−0.36	0.085

Table 16.12 Correlations between random effects for the model of Equation (16.9).

Correlation	Estimate
Intercepts (CM & C1)	0.34
Intercepts (CM & C2)	0.32
Intercepts (CM & C3)	0.29
Intercepts (C1 & C2)	0.38
Intercepts (C2 & C3)	0.44
Intercepts (C1 & C3)	0.24
Int., slope (CM)	0.57
Int. (C1), slope (CM)	0.31
Int. (C2), slope (CM)	0.28
Int. (C3), slope (CM)	0.33

16.8 CONCLUSION

This chapter has shown how growth curve models can be used to analyse repeated measures data. The approach has a number of advantages, especially if the primary aim of the analysis is to describe growth or change rather than to explain it. The models use all the available data rather than being restricted to cases with complete data at each occasion and this is important when dealing with studies that span a long period of time or many waves when losses from the sample are almost inevitable. If the data are 'missing at random' – implying that differences between the means for those measured and not measured at occasion t disappear, having conditioned on the known values at occasions

$t-k, k \geq 1$ – then estimates from the growth curve models are efficient. The models are not restricted to the continuous responses used to illustrate the approach here; models with more than one binary (Plewis, 2005) and ordered categorical responses (Plewis *et al.*, 2006) can be specified and estimated.

We have seen how to integrate the analyses of repeated-measures data from more than one cohort, from more than one generation, from all the children in the household and for more than one response. Space considerations precluded coverage of geographical and institutional influences (neighbourhoods and schools) on development.

We have also seen how the absence of fixed scales can create difficulties for social and behavioural scientists as they attempt to model change, in that results of interest can vary according to the scale adopted. Methods of equivalization across age, based on item response theory, can be applied if there is some overlap in the test items used across ages, but convincing applications of the method are hard to find. One way of avoiding the difficulties is to adopt the so-called conditional or regression approach described in, for example, Plewis (1985, 1996) and employed by Goldstein (1979b) for an analysis of some of the NCDS data used here. Rather than model change as a function of time – which reduces to analysing difference scores for just two measurement occasions – change is defined in terms of the expected value of a variable at time t conditional on earlier values of a 'similar' variable. This can be particularly useful when seeking causal inferences from observational data, as a properly specified conditional or regression model can generate a sufficiently close approximation to what would have been observed if it had been possible to randomize the explanatory variable of interest. We have not considered regression modelling in this chapter but note that, with several measurement occasions, it is possible to combine it with the growth curve approach by modelling growth conditional on an initial measurement (see Plewis, 2001).

We should never lose sight of the 'fitness for purpose' criterion when fitting statistical models to longitudinal data. There are a number of statistical approaches and an increasing number of useful texts to guide analysts in their search for the model best suited to the substantive question of interest. Nevertheless, the hierarchical or multilevel structures generated by many longitudinal study designs, and accessible statistical software to fit the models, mean that the methods set out here, and variants of them, have applications well beyond the educational data used as illustrations in this chapter.

REFERENCES

Ferri, E., Bynner, J. and Wadsworth, M. (2003). *Changing Britain, Changing Lives: Three Generations at the Turn of the Century*. London: Institute of Education, University of London.

Goldstein, H. (1979a). *The Design and Analysis of Longitudinal Studies*. London: Academic Press.

Goldstein, H. (1979b). Some models for analysing longitudinal data on educational attainment. *Journal of the Royal Statistical Society A*, 142, 407–442.

Goldstein, H. (2003). *Multilevel Statistical Models* (3rd edn). London: Edward Arnold.

Plewis, I. (1985). *Analysing Change: Measurement and Explanation using Longitudinal Data*. Chichester: John Wiley & Sons, Ltd.

Plewis, I. (1996). Statistical methods for understanding cognitive growth: a review, a synthesis and an application. *British Journal of Mathematical and Statistical Psychology*, 49, 25–42.

Plewis, I. (2001). Explanatory models for relating growth processes. *Multivariate Behavioral Research*, 36, 207–225.

Plewis, I. (2005). Modelling behaviour with multivariate multilevel growth curves. *Methodology*, 1, 71–80.

Plewis, I. (2007). *Millennium Cohort Study First Survey: Technical Report on Sampling* (4th edn). London: Institute of Education, University of London.

Plewis, I., Nathan, G., Calderwood, L. and Hawkes, D. (2004). *National Child Development Study and 1970 British Cohort Study Technical Report: Changes in the NCDS and BCS70 populations and samples over time* (1st edn). London: Centre for Longitudinal Studies, Institute of Education, University of London.

Plewis, I., Tremblay, R. E. and Vitaro, F. (2006). Modelling repeated responses from multiple informants. *Statistical Modelling*, 6, 251–262.

Rasbash, J., Steele, F., Browne, W. and Prosser, B. (2004). *A User's Guide to MLwiN, Version 2.0.* London: Centre for Multilevel Modelling, Institute of Education, University of London.

Singer, J. D. and Willett, J. B. (2003). *Applied Longitudinal Data Analysis.* Oxford: Oxford University Press.

Snijders, T. A. B. and Bosker, R. J. (1999). *Multilevel Analysis: An Introduction to Basic and Advanced Multilevel Modeling*. London: Sage.

CHAPTER 17

Using Longitudinal Surveys to Evaluate Interventions

Andrea Piesse, David Judkins and Graham Kalton
Westat, USA

17.1 INTRODUCTION

Longitudinal surveys are often used in evaluation studies conducted to assess the effects of a program or intervention. There are several reasons why this is the case. One is that the evaluation may need to examine the temporal nature of the effect, whether it is instant or delayed, persistent or temporary. A second reason is to distinguish between confounding variables that causally precede the intervention and mediating variables that are in the causal path between the intervention and the effect. Also, the identification of mediating variables is sometimes an important objective in order to obtain a fuller understanding of the causal process. A third reason is to better control for confounding variables in the evaluation. A longitudinal survey affords the opportunity to collect data on a wider range of potential confounders; also, the analysis of changes in the outcome variables may eliminate the disturbing effects of unmeasured confounders that are constant over time.

There are additional reasons for using a longitudinal survey for the evaluation when the intervention is an ongoing process, such as a long-term television advertising campaign. The evaluation of an ongoing intervention is more complicated than the evaluation of a single-event intervention, such as attendance at a training programme. With an ongoing intervention, the analysis may include the effects of cumulative or incremental exposure, which requires longitudinal data. Also, the intervention may change over time, further complicating the evaluation. Two of the three examples discussed in this chapter are evaluations of ongoing public health communication programmes (primarily television campaigns).

Methodology of Longitudinal Surveys P. Lynn

In preparation for the examples that follow, the next section provides an overview of types of interventions, types of effects, some issues in the design and analysis of evaluation studies and the value of longitudinal data. The subsequent three sections describe a selection of the analyses employed in three evaluation studies that were based on data collected in longitudinal surveys. Section 17.3 describes the design and some analyses of the US Youth Media Campaign Longitudinal Survey (YMCLS), conducted to evaluate a media campaign (the VERB programme) to encourage 9- to 13-year-old Americans to be physically active (Potter *et al.*, 2005). Section 17.4 describes the design and some analyses of the National Survey of Parents and Youth (NSPY), conducted to evaluate the US National Youth Anti-Drug Media Campaign (Orwin *et al.*, 2006). Although the NSPY and the YMCLS were both carried out to evaluate ongoing interventions, the YMCLS collected baseline data before the intervention whereas the NSPY did not. Section 17.5 describes the longitudinal design and analysis methods for evaluating the Gaining Early Awareness and Readiness for Undergraduate Programs (GEAR UP) programme, designed to increase the rate of postsecondary education among low-income and disadvantaged students in the USA. The GEAR UP programme provides various services to students while they are in secondary school, with the final outcomes occurring only after a number of years. Section 17.6 provides some concluding remarks.

17.2 INTERVENTIONS, OUTCOMES AND LONGITUDINAL DATA

Survey-based evaluation studies are of many different forms depending on the nature of the intervention (or treatment), on the nature of the possible outcomes (or effects) to be studied and on the conditions under which the evaluation is conducted. The following discussion briefly reviews some of the key considerations for evaluation studies.

17.2.1 Form of the Intervention

Many interventions are set events, such as attendance at a job training programme. In this case, the intervention is often of limited duration. However, the intervention can span an extended period, as with the GEAR UP programme. Other interventions are ongoing in nature, as with the two media campaigns discussed in this chapter. Interventions can also vary in uniformity of treatment. For example, there may be local variation if the intervention is loosely specified. In such cases, the results of an evaluation might address the effect of funding for the programme or intervention, rather than any specific implementation.

Interventions may be applied at an individual level or a group level, such that all individuals in a group receive the treatment. Although interventions at the individual level are common, the examples discussed here are not of this form. The GEAR UP programme is a group intervention, with the programme being applied to all children in a given cohort in a chosen school. Some interventions are even national in scope, such as VERB and the National Youth Anti-Drug Media Campaign (NYAMC).

There is also a distinction between interventions for which treatment status must be self-reported and those for which it can be ascertained independently. A child in a school that is implementing the GEAR UP programme is in the treatment group by definition. In contrast, since there was no planned geographic variation in exposure for the NYAMC, treatment status in the NSPY was defined in terms of the reports that individuals gave on their perceived exposure level.

17.2.2 Types of Effects

The effects of an intervention may differ in a number of respects that influence the type of evaluation study that is needed. Some of the key differences are described below.

Instant and Delayed Effects
Instant effects may be observable immediately after the intervention, whereas delayed effects occur only after a gestation period. For example, an evaluation of a job training programme may aim to measure an instant effect in terms of immediate employment and a delayed effect in terms of later promotion. It has been postulated that delayed effects, even on cognitive outcomes, may occur with media campaigns (see Section 17.4). Moreover, when the focus is on long-delayed effects, the evaluators may seek to obtain an initial indication of the effectiveness of the intervention by measuring early surrogate markers of the effect (as is also often done in clinical trials); the GEAR UP evaluation has used this approach.

Persistent and Transient Effects
When analysing the effects of an intervention, evaluators may be interested in whether any effects are lasting or transient. For example, does a short-term media campaign continue to affect behaviours over time, or does the effect decay? And, if it decays, how quickly does it decay?

Cumulative and Incremental Effects
With an ongoing intervention, evaluators are likely to be interested in how effects change over time. Do effects strengthen with cumulative exposure? What is the incremental effect of additional exposure? The identification of these effects is, of course, related to whether they are persistent or transient.

Individual and Group Effects
In some studies, only the direct effect of the intervention on the individual is of interest. In others, particularly if the intervention is applied at a group level, indirect effects on the individual (i.e. effects mediated through other individuals or groups) may also be relevant. For example, since the GEAR UP programme is a school-level intervention, its effect on one student may be affected by the fact that the student's peers are also in the programme. In some cases, the intervention may also have an effect on the untreated. For example, a job training scheme may increase the employment chances of participants but, if the number of jobs available is limited, consequently reduce the employment chances of nonparticipants.

17.2.3 Conditions for the Evaluation

Two main features of an evaluation influence the forms of analysis of possible intervention effects.

Randomization
Evaluation studies that randomly assign units (individuals or groups) to treatment groups provide much stronger inferences about the effects of the treatment than those that do not. With a random assignment, the treatment groups are equated – within the bounds of random chance – on all characteristics prior to the intervention. Without a random assignment, systematic differences between the treatment groups must be addressed by matching on key potentially confounding variables in the design, where possible, and/or by analytical methods (see Section 17.2.4). None of the three studies discussed in this chapter employed a random assignment. Despite the great benefits of a randomized study, few evaluation studies have been conducted this way, for what are seen to be practical and/or ethical reasons. However, greater use can be made of randomization with the creative involvement of evaluators at the very early stages of an intervention (Boruch, 1997) and randomized social experiments are receiving renewed interest (Sherman, 2003; Angrist, 2004; Michalopoulos, 2005).

Timing of the Evaluation
A notable difference between the YMCLS and the NSPY is that the former started before the intervention began whereas the latter was introduced only after the intervention was underway. By collecting baseline data before the intervention was in place, the YMCLS was able to identify some pre-existing differences between the control and treatment groups that were likely to be important confounders. When data are collected for two or more waves of a longitudinal study before the initiation of treatment, prior measures of change may provide even greater confounder control. When no true baseline measures are available, the assessment of causal effects is more challenging (see Section 17.4).

17.2.4 Controlling for Confounders in the Analysis

The major threat to the validity of nonrandomized evaluations is that any associations observed between the treatment and an outcome may be due to selection bias. There are a variety of statistical techniques that can be used in an attempt to remove the effects of selection bias, such as regression methods, propensity scoring, instrumental variables and regression discontinuity analysis (Trochim, 1984; Rosenbaum, 2002; Imbens, 2004). Since propensity score weighting was employed in all three of the evaluations discussed here, only that approach (Rosenbaum and Rubin, 1983) will be outlined below.

With propensity scoring, the analyst builds a statistical model for the probability of receiving treatment in terms of potential confounders, and the associations between treatment and outcomes are conditioned on the estimated treatment probabilities (known as propensity scores). An attractive feature of propensity scoring is that there are formal tests (balance tests) for assessing whether confounders have been effectively controlled. Another major advantage of the technique applies when there are many different outcome variables, as in the evaluations discussed here. The propensity scores are developed only once and can then be applied in the analysis of each outcome. In contrast, in the regression framework each of these outcomes would have to be modelled separately. Finally, propensity scoring does not require parametric assumptions about outcome distributions.

Propensity scoring can be viewed as a natural extension of survey weighting. With two groups – a treatment group and a control group – the analyst first develops a logistic (or other) model to predict group membership based on the potential confounders. One approach uses inverses of the predicted propensity scores as weights in the analysis. An alternative approach, which reduces variance while eliminating slightly less bias, is to divide the sample into a small number of strata (say five) based on the propensity scores. The groups are then reweighted so that the weighted count in each stratum is the same in the treatment and control groups. The reweighting thus approximately equates the groups in terms of the confounders. If each group is weighted to conform to the overall estimated population distribution, the estimate of the outcome for each group can be viewed as an estimate of the outcome had the whole population been in that group. Using an ordinal logit model, the propensity scoring methodology can be extended to cover several treatment groups (Joffe and Rosenbaum, 1990).

When the same outcome measure is collected at multiple points in time, the question of whether to use an endpoint or a change score analysis arises. The dependent variable in an endpoint analysis is typically the latest measurement of the outcome, whereas the dependent variable in a change score analysis is the difference between measurements of the outcome at two waves. An endpoint analysis often includes the baseline measure as one of the confounders, and the change score analysis often defines the change score as the difference between the latest outcome measurement and the baseline (or earliest) measurement. When an outcome is measured on an ordinal scale with only a few response categories, like the Likert items in the YMCLS, a change score analysis can be problematic because of floor and ceiling limitations.

An endpoint analysis without control on the baseline outcome measure is more powerful than a change score analysis if the correlation of the measured outcomes at the two time points is less than 0.5. However, there are also bias considerations. If the baseline outcome measure is both correlated with treatment and subject to measurement error, then an endpoint analysis with control on the baseline outcome can yield biased effect estimates (Cook and Campbell, 1979, pp. 160–164), although the bias may be reduced if multiple baseline measurements are used (Cook and Campbell, 1979, pp. 170–175). Analysing change scores without controlling for baseline outcomes can remove this bias under some but not all conditions (Allison, 1990; Bonate, 2000; Senn, 2006).

17.2.5 Value of Longitudinal Surveys

The need for a longitudinal survey in an evaluation depends on the characteristics of the intervention, the potential effects to be examined and whether randomization is employed. If the intervention does not last over an extended period, if the effects of interest are only instantaneous ones and if a random assignment is used there may be no need for a longitudinal survey. However, longitudinal data will generally be required if: the intervention is spread over a long period; there is a desire to examine possible increasing or decreasing effects over time, delayed effects or cumulative or incremental effects of additional exposure in an ongoing intervention; or treatment is not randomly assigned and baseline confounders are deemed important. Unless the evaluators are willing to rely on retrospective recall (with its well-known defects), a longitudinal survey is needed to produce these data.

A longitudinal survey can help to determine the causal ordering of variables and hence, in an evaluation setting, whether the intervention caused the effect or the 'effect' caused the treatment value. It can also help to distinguish between confounder variables that precede the intervention and mediating variables that are on the path between the intervention and the effect (see Section 17.4). Furthermore, a longitudinal survey can help to provide an understanding of *how* an intervention leads to an effect. For example, a media campaign to discourage the use of drugs may operate by raising awareness of their harmful effects.

These benefits of longitudinal surveys are illustrated by the analyses described in the following three sections. However, it should also be acknowledged that longitudinal surveys have their own problems, in particular panel attrition and possible panel conditioning. Like cross-sectional surveys, longitudinal surveys suffer from unit nonresponse at the first round. They also suffer from sample attrition at later rounds. The overall response rates for the original panel of youth in the YMCLS dropped from 42.8 % in the first round, to 38.1 % in the second round and to 31.2 % in the third round. For the NSPY, the overall youth response rates across the four rounds were 64.8 %, 56.7 %, 54.8 %, and 52.2 %. For GEAR UP, the overall youth response rates in the first two rounds were 86.8 % and 79.6 %. Attrition raises the concern that the increasing loss of cases over waves of the panel may threaten the representativeness of the sample at later points. This concern is shared with other longitudinal surveys and was addressed through attrition nonresponse weighting adjustments in the surveys reported here. A special and more challenging aspect of attrition arises in evaluation studies where cases may drop out of the study as a direct result of their reaction to the treatment.

The concern about panel conditioning is that panel members in a longitudinal survey become atypical as a result of participating in the survey (see Cantor, 2008; Sturgis *et al.*, Chapter 7 of this volume). This concern is potentially more serious in intervention evaluation studies. For example, panel members may become sensitized to the existence of the intervention through their participation in the study, resulting in changes in true exposure. Measurement biases in self-reported exposure may also be subject to conditioning. Section 17.3 provides evidence of this type of panel conditioning in the YMCLS. The measurement bias in self-reported exposure can sometimes be affected by the way in which exposure is captured. For example, with an ongoing media campaign the exposure measurement could be anchored to time-specific advertising content.

In passing, it should be noted that trends in outcomes over time are sometimes used to evaluate ongoing interventions like VERB and the National Youth Anti-Drug Media Campaign. Longitudinal surveys can provide trend data, but so can a series of repeated cross-sectional surveys. For the analysis of trends, a series of repeated surveys has the considerable advantage of avoiding problems of attrition and panel conditioning. An interrupted time series design can provide strong evidence on abrupt intervention effects provided that there is a series of measurements both before and after the intervention (Shadish *et al.*, 2002). Without such a time series, there are many other factors that could explain a trend.

17.3 YOUTH MEDIA CAMPAIGN LONGITUDINAL SURVEY

In June 2002, the US Centers for Disease Control and Prevention launched a national media campaign to encourage 9- to 13-year-old Americans to be physically active. The VERB campaign combined paid advertising (e.g. on television and radio) with school and community activities to disseminate its health promotion messages. The Youth Media Campaign Longitudinal Survey (YMCLS) – a telephone survey based on a stratified, list-assisted, random-digit-dialled sample of households – was designed to assess the impact of the campaign. The YMCLS consisted of two panel surveys. The first panel started with a baseline survey conducted in mid-2002, prior to any campaign advertising; interviews were completed with 3114 parent–child dyads. The second, refresher, panel was introduced in 2004 during the second year of VERB; 5177 parent–child interviews were completed. Both panels were then interviewed on an annual basis until 2006.[1]

The introduction of the second panel provided a benchmark against which to assess the effects of attrition and panel conditioning in the first panel. Key analytical variables were compared across the two panels, with the analysis limited to children aged 11–13 in 2004 (since only children in this age range were included in both panels in that year). Significant differences between the estimates were found for only two of the cognitive outcomes relating to physical activity, and they were small. No significant differences were found for behavioural outcomes. However, panel conditioning did have some impact on the exposure measure; children in the original panel were significantly more likely to report awareness of VERB than were children in the second panel (84 % vs. 76 %). This finding is not surprising, since asking about VERB exposure necessarily involved some respondent education.

The prime focus of the YMCLS analyses has been the confounder-controlled associations between VERB exposure and a range of physical activity-related behaviours and cognitions in the original panel (since the second panel lacked baseline measures). The only certain consequence of the apparent conditioning effect on VERB awareness for the association analyses was the loss of power due to the reduction in sample size in the lowest category of the exposure scale (i.e. cases reporting a lack of awareness of VERB). It is unclear whether the conditioning had any other effect on the associations of exposure with outcomes.

VERB exposure was measured on an ordinal scale reflecting perceived frequency of seeing or hearing campaign advertisements. Outcome data on physical activity were collected by asking respondents to recall engagement in specific activities on the day prior to the interview and during the previous week. Several activity outcome measures were then formed using binary indicators of activity/nonactivity and counts of physical activity sessions. Outcome data on cognitions about physical activity were collected by asking respondents to indicate their level of agreement with various statements, using a series of four-point scale items. These items were combined to form Likert scale outcome measures. A large set of outcomes with varying measurement scales was created.

The correlation between an outcome measure at one round and the same measure at the next round was low, generally in the range of 0.15–0.30. Many potential covariates

---

[1] An additional cross-sectional sample was introduced and surveyed in 2006.

were available from the baseline interviews with youth and their parents and from neighbourhood and geographic data. Given these factors, an endpoint analysis with propensity scoring was chosen for confounder control. Propensity score-adjusted weights were prepared by first forming an ordinal logit model for exposure and then, to achieve closer balance on certain key variables, by raking these weights on demographic, geographic and baseline physical activity variables (Judkins *et al.*, 2007). These weights are termed primary weights below. The associations between exposure and the various outcomes were measured by weighted gamma coefficients, with confidence intervals calculated using jack-knife methods (Nadimpalli *et al.*, 2003).

Table 17.1 presents results for four important outcomes. The figures in the table are gamma association statistics or differences between them. Column 2 gives the gamma coefficients of association between the 2004 exposure measure and the 2002 baseline outcome measures based on the primary weights. This column provides evidence of balance since, after the propensity scoring and raking, the gamma coefficients confirm that there is minimal association between these measures. All the gamma coefficients for the endpoint analyses based on the primary weights in column 3 are positive and significant, suggesting that VERB is having a positive impact.

There was, however, a concern with the endpoint analysis that measurement error in the baseline outcome measures could bias the results. To address this concern, a

Table 17.1 Gamma coefficients (and 95 % confidence intervals) for associations between 2004 VERB exposure and outcomes related to physical activity, by year and method of confounder control, YMCLS.

Outcome	Balance test: primary weights 2002	Endpoint analysis: primary weights 2004	Balance test: alternative weights[a] 2002	Endpoint analysis: alternative weights[a] 2004	Double difference analysis: alternative weights[a] 2004–2002
Expectations scale[b]	0.01 (−0.01, 0.02)	0.12* (0.07, 0.17)	0.06* (0.01, 0.11)	0.15* (0.11, 0.20)	0.09* (0.03, 0.16)
Free-time physical activity count for previous week	0.01 (−0.00, 0.02)	0.09* (0.04, 0.13)	0.04 (−0.01, 0.09)	0.09* (0.05, 0.14)	0.05 (−0.01, 0.12)
Overall physical activity/inactivity[c] in previous week	0.00 (−0.00, 0.00)	0.19* (0.06, 0.32)	0.06 (−0.05, 0.17)	0.21* (0.09, 0.33)	0.15 (−0.01, 0.31)
Physical activity/ inactivity on previous day	0.00 (−0.00, 0.00)	0.19* (0.11, 0.26)	0.07 (−0.01, 0.16)	0.18* (0.11, 0.25)	0.11 (−0.01, 0.23)

*Statistically significant difference from zero.
[a]Alternative weights were created without controlling for baseline values of outcomes.
[b]The expectations scale is a Likert scale composed of five items about the perceived outcomes of participating in physical activity.
[c]Overall physical activity is any organized physical activity session or three or more free-time sessions.
Source: Potter *et al.*, 2005.

sensitivity analysis, motivated by change score analysis, was performed. Exposure propensity models were refitted, omitting baseline values of outcomes, to produce alternative propensity score-adjusted weights. The gamma coefficients in column 4 of Table 17.1 show that baseline outcome measures were not as well-balanced through control on socio-economic, demographic and geographic variates alone. However, the fact that the balance did not deteriorate more was most likely because of the large number of other covariates that were available. All the gamma coefficients of association between concurrent exposure and outcomes with the alternative weights remain positive and significant, as shown in column 5. As a final step to compensate for the imbalance on the baseline outcome measures, the gamma coefficients given in column 4 were subtracted from the gamma coefficients given in column 5 to form the double difference estimates of campaign effects in column 6.

As expected, given the low autocorrelation in the outcomes, a comparison of the confidence intervals for the double difference estimates with those for the primary endpoint gamma coefficients shows that the double difference estimates are subject to larger sampling errors. The estimates in column 6, while smaller, are still positive; only one is significant. However, given the consistency in the direction and relative size of the gamma coefficients between the endpoint and sensitivity analyses, it was concluded that there was good evidence of campaign effects.

17.4 NATIONAL SURVEY OF PARENTS AND YOUTH

In 1998, the Office of National Drug Control Policy launched the National Youth Anti-Drug Media Campaign (NYAMC) to reduce and prevent drug use among young Americans. The major components of the campaign were television, radio and other advertising, complemented by public relations efforts that included community outreach and institutional partnerships. The primary tool for the evaluation was the National Survey of Parents and Youth (NSPY), a household panel survey with four annual rounds of data collection that began in late 1999, after the campaign was underway. Interviews with 8117 youth and 5598 parents were obtained at the first round. The main outcomes of interest were marijuana-related attitudes and beliefs, perceived social norms and self-efficacy, intentions to use marijuana and initiation of use. Participants were asked whether they had seen or heard general anti-drug advertisements in several media (television, radio, internet, billboard, etc.). To measure campaign-specific exposure, they also viewed/listened to a sample of recently broadcast television and radio campaign advertisements and were asked whether or not they had seen/heard each advertisement and, if so, approximately how often. These responses were then recoded into an ordinal measure of exposure to campaign advertisements.

Although one component of the campaign evaluation examined trends in outcomes, the primary analyses in the NSPY focused on confounder-controlled associations between outcomes and exposure. To increase the power of the analyses of associations, the data collected in the different rounds of the panel were combined. Thus, for analyses that involved exposures and outcomes at the same round, the propensity models and effect estimates were computed from a pooled dataset that incorporated data collected at all rounds.

Propensity score-adjusted weights were constructed in a manner similar to that described for the YMCLS, but the procedure for admitting variables into the exposure propensity modelling was different. Without pre-campaign baseline data, a priori judgment by a panel of experts had to be made to distinguish between confounder variables that should be included in the propensity scoring and mediators that should not. Admissible confounders included variables such as general demographics, hours of television viewing per week, internet use, academic performance, participation in extra-curricular activities and school attendance. Variables such as beliefs about the consequences of drug use and participation in anti-drug classes or programmes were treated as mediators. If campaign advertisements led youth to change their beliefs, which in turn led to changes in intentions to use marijuana, allowing these belief changes to enter the propensity models would have hindered the ability to detect campaign effects. Similarly, there was some risk that participation in anti-drug programmes might reflect access to campaign advertising, particularly since campaign advertisements were shown extensively on an in-school network.

Findings about campaign efficacy obtained from the exposure–outcome analyses did not always concur with those suggested by the trend estimates. For example, there was a significant increase between 2000 and the first half of 2003 in the mean index score for self-efficacy to refuse marijuana, for 12- to 18-year-old nonusers. However, using data through the first half of 2003, the concurrent confounder-controlled association between exposure and self-efficacy was not significant (Hornik *et al.*, 2003). It was concluded that influences other than the NYAMC campaign were likely to be responsible for the upward trend in the self-efficacy index.

Intervention effects estimated from associations between exposure and outcomes measured at the same time are, however, themselves vulnerable to threats to their validity. One threat is often termed 'reverse causation' or 'ambiguous temporal precedence' (Shadish *et al.*, 2002). If an association is established between exposure and some possible 'outcome' on a contemporaneous basis, there is the possibility that the direction of causation is from the 'outcome' to the exposure – or, even more likely, to the recalled exposure – rather than from exposure to 'outcome'. A second threat to validity is that an absence of a concurrent effect may arise because the exposure has a delayed effect that is not immediately apparent. Each of these threats was a concern in the NSPY evaluation. Both were investigated by taking advantage of the established temporal ordering made possible by the longitudinal data. Analyses of associations between exposure at one round of the panel and outcomes at the following round can be used to examine delayed effects and also to investigate the risk that an observed association might be due to 'outcome' causing exposure. Hence, cross-round analysis was an important part of the campaign evaluation.

The confounder-controlled concurrent associations between parent exposure and parent outcomes in the NSPY suggested positive instant effects of the media campaign on parents. Once follow-up data became available, cross-round analyses were conducted to examine associations between exposures at one round and outcomes at the next, while controlling for previous round confounders, including some outcomes. For these analyses, the propensity models and effect estimates were computed from a dataset that incorporated all pairs of consecutive rounds. Not surprisingly, lagged associations were much weaker than concurrent associations, as shown in Table 17.2. Controlling on prior round outcomes in the cross-round analysis meant that an association could be found

Table 17.2 Gamma coefficients (and 95 % confidence intervals) for concurrent and lagged confounder-controlled associations between exposure and outcomes for parents of 12- to 18-year-olds, 2000–2003 NSPY.

Outcome	Concurrent association	Lagged association
Mean score on parent talking behaviour scale	0.11* (0.06, 0.16)	0.04 (−0.01, 0.10)
Mean score on parent talking cognitions index	0.08* (0.04, 0.11)	0.05* (0.01, 0.08)
Mean score on parent monitoring behaviour scale	0.04 (−0.01, 0.08)	0.00 (−0.03, 0.04)
Mean score on parent monitoring cognitions index	0.04* (0.02, 0.07)	0.00 (−0.04, 0.04)
Mean score on doing fun activities scale	0.14* (0.09, 0.20)	0.06* (0.01, 0.12)

*Statistically significant difference from zero.
Source: Hornik *et al.*, 2003.

only if (a) there existed a delayed effect that was not observed until the next round and/or (b) exposure at one round was correlated with exposure at the next and there was an incremental effect as a result of more exposure. The lagged associations and concurrent associations thus estimate different effects of the campaign. Nonetheless, the existence of lagged associations would suggest that the campaign had an effect and would support the direction of causation from exposure to outcomes in the concurrent associations. The significant lagged associations in Table 17.2 support the case that the campaign had a positive effect on parent talking cognitions and on the frequency with which they engaged in fun activities with their children.

In contrast to the parent analyses, the NSPY youth analyses found very little evidence of instant effects on youth based on the confounder-controlled concurrent associations between exposure and outcomes. However, it has been postulated that it may take time for anti-drug messages to resonate with youth and for effects to be observed. This hypothesis was investigated through cross-round analyses, similar to those performed for parents. While there was no evidence of instant effects of campaign exposure on the four main youth outcomes, there were significant *negative* delayed effects for two of these outcomes. The results are summarized in Table 17.3.

Table 17.3 Gamma coefficients (and 95 % confidence intervals) for concurrent and lagged confounder-controlled associations between exposure and outcomes for 12- to 18-year-olds who have never used marijuana[a], 2000–2002 NSPY.

Outcome	Concurrent association	Lagged association
Percent definitely not intending to try marijuana	−0.03 (−0.14, 0.09)	−0.12* (−0.21, −0.02)
Mean score on Beliefs/Attitudes Index	−0.02 (−0.06, 0.02)	−0.03 (−0.08, 0.02)
Mean score on Social Norms Index	−0.02 (−0.06, 0.02)	−0.5 (−0.11, 0.00)
Mean score on Self-efficacy Index	0.01 (−0.04, 0.07)	−0.08* (−0.15, −0.02)

*Statistically significant difference from zero.
[a]Analyses based on nonusing youth at current round for concurrent associations and on nonusing youth at earlier round for delayed associations.
Source: Hornik *et al.*, 2002.

The negative delayed effects for youth may be the result of bias due to selection maturation (Cook and Campbell, 1979) that was not accounted for by the covariates in the exposure propensity model. Another, unwelcome, conclusion is that the negative effects are real, perhaps because youth do not like being told what to do or because they came away with the message that drug use is expected behaviour. Hornik *et al.* (2002) provide an in-depth discussion of these unexpected youth findings.

17.5 GAINING EARLY AWARENESS AND READINESS FOR UNDERGRADUATE PROGRAMS (GEAR UP)

Introduced in 1998, the Gaining Early Awareness and Readiness for Undergraduate Programs (GEAR UP) programme operates through grants to coalitions of nongovernmental organizations and local schools. Although grantees have wide discretion on which services to provide, their services must be available to all students in a cohort, from 7th grade through high school. (In the USA, this period roughly corresponds to ages 12 through 18 years.) A longitudinal survey was chosen as the evaluation tool because of the long span between the start of the intervention and the final outcomes to be measured after the end of postsecondary education, as well as the need to control on baseline covariates that could not be measured well retrospectively. Twenty GEAR UP and 20 control schools were purposively selected for the study. The control sample was drawn from a matched set of schools geographically close to the GEAR UP schools and similar to them in terms of poverty levels and racial composition. Because some schools withdrew from the study, the original matching was ignored (except during variance estimation) so that data from unmatched schools could be included in the analysis, and a finer level of control could be achieved by including measured student characteristics in a propensity scoring approach. In this section, we highlight three issues arising in this evaluation: treatment measurement, confounder control and surrogate markers.

Since grantees had latitude in how they operated the programme, there was substantial interest in collecting data on the services received by students. However, no satisfactory way of measuring service receipt was devised. One problem was that services such as tutoring were delivered in small packets over a six-year period. Another problem was that some services, such as curriculum reform and staff training, were invisible to students and hence could not be reported by them. A student-level service log system was developed for the grantees, but their reports proved to be of poor quality. Accordingly, treatment was simply defined as a binary indicator of attendance at a school where a grant was active.

The GEAR UP evaluation included a 'baseline' round of data collection, with the interviews carried out as close to the beginning of the 7th grade school year as practical (given the need to collect and sample from student rosters and obtain informed parental consent). However, enough time had elapsed for the grantees to have initiated GEAR UP activities, and there was strong evidence that some of the baseline measures were contaminated by these early activities. The clearest example involved field trips by 7th graders to local colleges, which were reported much more frequently by GEAR UP students than control students even though GEAR UP students generally came from less well-off families with lower levels of education. As a result, the evaluators had to make a collective judgement about the admissibility of potential confounders collected at baseline on a variable-by-variable basis.

The final outcomes of the programme occur in 2007 or later, when the postsecondary education experiences of the sample students become observable. To provide useful feedback to policy makers on a more timely basis, follow-up interviews were conducted near the end of 8th grade to collect data on a wide set of potential indicators of final outcomes. Using data from the US National Education Longitudinal Survey – an earlier survey that followed students and their parents from 8th grade to early adulthood – models were formed to identify 8th grade indicators that are most predictive of postsecondary education enrolment. These indicators, along with variables that were highly correlated with them, were then combined into a College Orientation Index that was analysed as a surrogate marker for the final outcomes.

17.6 CONCLUDING REMARKS

Longitudinal surveys are a valuable tool for evaluating the effects of programmes or interventions, particularly if the evaluation is not a randomized experiment. However, careful consideration must go into the planning stage. Accurate measurements of treatment and outcomes are essential but not sufficient by themselves. Measuring confounders is equally critical. In particular, the estimation of causal effects may be more precise and less prone to bias if some data can be collected before the intervention is put in place; baseline measures of outcomes are natural choices as covariates. Consideration should also be given to capturing variables that predict exposure. If longitudinal surveys are to be used to construct trends, the questionnaire must be kept consistent over time and refresher panels may be required to address possible attrition and conditioning biases.

It would be valuable to estimate possible persistent, transient, cumulative and incremental effects associated with ongoing interventions such as VERB and the NYAMC, in addition to the instant and delayed effects discussed in this chapter. However, even with time-delimited exposure measures, it is extremely difficult to disentangle these effects without substantial sample sizes for persons with different levels of exposure over time.

Some of the issues that make the evaluation of interventions problematic are unique to longitudinal surveys. While all surveys are subject to nonresponse, sample attrition (unit nonresponse after the first interview) further lowers the overall response rate. Nonresponse may render the responding sample unrepresentative of the original survey population and may thereby introduce bias into the estimation of causal effects. Panel conditioning may also affect the estimation of causal effects. Although researchers need to be aware of these issues, longitudinal surveys nevertheless have considerable advantages over cross-sectional surveys for evaluation studies.

REFERENCES

Allison, P. D. (1990). Change scores as dependent variables in regression analysis. *Survey Methodology*, 20, 93–114.

Angrist, J. D. (2004). American education research changes tack. *Oxford Review of Economic Policy*, 20, 198–212.

Bonate, P. L. (2000). *Analysis of Pretest-Posttest Designs*. Boca Raton: Chapman & Hall/CRC.

Boruch, R. F. (1997). *Randomized Experiments for Planning and Evaluation: A Practical Guide*. Thousand Oaks: Sage.

Cantor, D. (2008). A review and summary of studies on panel conditioning. In S. Menard (Ed.), *Handbook of Longitudinal Research: Design, Measurement and Analysis* (pp. 123–138). San Diego: Elsevier.

Cook, T. D. and Campbell, D. T. (1979). *Quasi-Experimentation: Design and Analysis Issues for Field Settings*. Boston: Houghton Mifflin.

Hornik, R., Maklan, D., Cadell, D., Barmada, C. H., Jacobsohn, L., Henderson, V. R., *et al.* (2003). *Evaluation of the National Youth Anti-Drug Media Campaign: 2003 Report of Findings*. Report prepared for the National Institute on Drug Abuse. Retrieved May 9, 2008, from http://www.nida.nih.gov/despr/westat/#reports

Hornik, R., Maklan, D., Cadell, D., Barmada, C. H., Jacobsohn, L., Prado, A., *et al.* (2002). *Evaluation of the National Youth Anti-Drug Media Campaign: Fifth Semi-Annual Report of Findings*. Report prepared for the National Institute on Drug Abuse. Retrieved May 9, 2008, from http://www.nida.nih.gov/despr/westat/#reports

Imbens, G. W. (2004). Nonparametric estimation of average treatment effects under exogeneity: a review. *Review of Economics and Statistics*, 86, 4–29.

Joffe, M. M. and Rosenbaum, P. R. (1990). Propensity scores. *American Journal of Epidemiology*, 150, 327–333.

Judkins, D., Morganstein, D., Zador, P., Piesse, A., Barrett, B. and Mukhopadhyay, P. (2007). Variable selection and raking in propensity scoring. *Statistics in Medicine*, 26, 1022–1033.

Michalopoulos, C. (2005). Precedents and prospects for randomized experiments. In H. S. Bloom (Ed.), *Learning More from Social Experiments* (pp. 1–36). New York: Russell Sage Foundation.

Nadimpalli, V., Judkins, D. and Zador, P. (2003). Tests of monotone dose-response in complex surveys. *Proceedings of the Section on Survey Research Methods of the American Statistical Association* (pp. 2983–2988). Washington, DC: American Statistical Association.

Orwin, R., Cadell, D., Chu, A., Kalton, G., Maklan, D., Morin, C., *et al.* (2006). *Evaluation of the National Youth Anti-Drug Media Campaign: 2004 Report of Findings*. Report prepared for the National Institute on Drug Abuse. Retrieved May 9, 2008, from http://www.nida.nih.gov/despr/westat/#reports

Potter, L. D., Duke, J. C., Nolin, M. J., Judkins, D., Piesse, A. and Huhman, M. (2005). *Evaluation of the CDC VERB Campaign: Findings from the Youth Media Campaign Longitudinal Survey, 2002, 2003, 2004*. Report prepared for the Centers for Disease Control and Prevention. Rockville, MD: Westat.

Rosenbaum, P. R. (2002). *Observational Studies* (2nd edn.). New York: Springer.

Rosenbaum, P. R. and Rubin, D. B. (1983). The central role of the propensity score in observational studies for causal effects. *Biometrika*, 70, 41–55.

Senn, S. (2006). Change from baseline and analysis of covariance revisited. *Statistics in Medicine*, 25, 4334–4344.

Shadish, W. R., Cook, T. D. and Campbell, D. T. (2002). *Experimental and Quasi-Experimental Designs for Generalized Causal Inference*. Boston: Houghton Mifflin.

Sherman, L. W. (2003). Misleading evidence and evidence-led policy: making social science more experimental. *Annals of the American Academy of Political and Social Science*, 589, 6–233.

Trochim, W. (1984). *Research Design for Program Evaluation: The Regression-Discontinuity Approach*. Beverly Hills: Sage.

CHAPTER 18

Robust Likelihood-Based Analysis of Longitudinal Survey Data with Missing Values

Roderick Little and Guangyu Zhang
University of Michigan, USA

18.1 INTRODUCTION

One approach to the analysis of longitudinal survey data with missing values is to multiply-impute the missing values using draws from their predictive distribution given the observed data (Rubin, 1987; Little and Rubin, 2002, chapter 10). An advantage of the method is that predictors that are not relevant or not available for the final analysis model can be used in the imputation model. In this chapter, a robust model-based multiple imputation approach is proposed that extends the work of Little and An (2004). Imputations are based on a model that models the relationship with the response propensity nonparametrically by a penalized spline, and models relationships with other variables parametrically. This approach has a form of double robustness that will be described, and simulation comparisons with other methods suggest that the method works well in a wide range of populations, with little loss of efficiency relative to parametric models when the latter are correct. Extensions to general patterns of missing data and to parameters other than unconditional means are outlined.

Missing values arise in longitudinal surveys for many reasons. A primary source is *attrition*, when subjects drop out prior to the end of the study. In longitudinal surveys, some individuals provide no information because of noncontact or refusal to respond (*unit* nonresponse). Other individuals are contacted and provide some information, but fail to answer some of the questions (*item* nonresponse). Often indices are constructed by summing values of particular items. For example, in economic studies, total net worth is

Methodology of Longitudinal Surveys P. Lynn

a combination of values of individual assets or liabilities, some of which may be missing. If any of the items that form the index are missing, some procedure is needed to deal with the missing data.

The missing-data *pattern* indicates which values in the dataset are observed and which are missing. Specifically, let $Y = (y_{ij})$ denote an $(n \times p)$ rectangular dataset without missing values, with ith row $y_i = (y_{i1}, \ldots y_{ip})$, where y_{ij} is the value of variable Y_j for subject i. In repeated-measures problems variables measured at each time point form blocks of columns; for a single outcome Y_j denotes the value of a variable at the jth time point. With missing values, the pattern of missing data is defined by the *missing-data indicator matrix* $M = (m_{ij})$ with ith row $m_i = (m_{i1}, \ldots m_{ip})$, such that $m_{ij} = 1$ if y_{ij} is missing and $m_{ij} = 0$ if y_{ij} is present. We assume throughout that (y_i, m_i) are independent over i.

The performance of missing-data methods depends strongly on the missing-data mechanism, which concerns the reasons why values are missing, and in particular whether missingness depends on the values of variables in the dataset. For example, subjects in a longitudinal intervention may be more likely to drop out of a study because they feel the treatment was ineffective, which might be related to a poor value of an outcome measure. Rubin (1976) treated M as a random matrix, and characterized the missing-data mechanism by the conditional distribution of M given Y, say $f(M \mid Y, \phi)$, where ϕ denotes unknown parameters. When missingness does not depend on the values of the data Y, missing or observed, that is, $f(M \mid Y, \phi) = f(M \mid \phi)$ for all Y, ϕ, the data are called missing completely at random (MCAR). With the exception of planned missing-data designs, MCAR is a strong assumption, and missingness often does depend on recorded variables. Let Y_{obs} denote the observed values of Y and let Y_{mis} denote the missing values. A less restrictive assumption is that missingness depends only on values of Y_{obs} that are observed, and not on values of Y_{mis} that are missing, that is, $f(M \mid Y, \phi) = f(M \mid Y_{\text{obs}}, \phi)$ for all Y_{mis}, ϕ. The missing-data mechanism is then called missing at random (MAR). Many methods for handling missing data assume that the mechanism is MCAR or MAR, and yield biased estimates when the data are not MAR (NMAR).

18.2 MULTIPLE IMPUTATION FOR REPEATED-MEASURES DATA

Two model-based approaches to repeated-measures data with missing values are (a) to compute maximum likelihood (ML) or restricted ML (REML) estimates, based on the model, with standard errors based on the information matrix, the sandwich estimator or the bootstrap (e.g. Little and Rubin, 2002, chapter 6); or (b) to create a rectangular data matrix by multiply-imputing the missing values, analysing the imputed datasets using a procedure for complete data and then combining the estimates and standard errors using multiple imputation combining rules (Rubin, 1987, 1996; Little and Rubin, 2002, chapter 10). The ML approach is widely available in statistical software packages for the case where the missing-data mechanism is MAR. Methods are available for normal models (Pinheiro and Bates, 2000; SAS, 2004, PROC MIXED) and for generalized linear mixed models (McCulloch and Searle, 2001; SAS, 2004, PROC NLMIXED). Strengths of these methods are that they can handle a wide range of modelling situations, they are principled and have optimal asymptotic properties under the assumed model and, by constructing likelihoods based on the observed data, they avoid the need to impute the missing values. Weaknesses are that the inferences are asymptotic and potentially vulnerable to model misspecification. Also, current widely available software does not handle missing data in covariates or nonignorable models (although the ML approach can be applied to handle such problems).

An alternative to these methods is to multiply-impute the missing values with draws from their predictive distribution under a model. Multiple imputation (MI) allows the imputation uncertainty to be propagated in inferences by using MI combining rules. A principled version of this approach, assuming MAR, is to impute using the posterior predictive distribution $p(Y_{\text{mis}} \mid Y_{\text{obs}})$ based on the model for Y and a prior distribution $p(\theta)$ for the parameters θ of this model. This approach uses the Bayesian paradigm to propagate uncertainty in the parameters. The imputations can be used for MI inferences, or can be viewed as a stepping stone to simulating directly the posterior distribution $p(\theta \mid Y_{\text{obs}})$ of the parameters, as in data augmentation (Tanner and Wong, 1987; Schafer, 1997; Little and Rubin, 2002, chapter 10). When the imputations and complete-data inferences are based on the same model, say $f(Y \mid \theta)$, and the prior distribution $p(\theta)$ has support over the full parameter space of θ, this approach is asymptotically equivalent to the ML or REML approach mentioned above. This fact follows from the well-known large-sample equivalence of Bayes and ML, discussed for example in Little and Rubin (2002, chapter 6). The Bayes/MI approach can be a useful advance over ML for small samples; for example, in simple settings it yields inferences based on the Student t distribution that allow for uncertainty in estimating the variance, which are not a feature of the ML approach.

A useful feature of MI is that there is no requirement that the model for creating the imputations be the same as the analysis model applied to the filled-in data. This idea featured prominently in Rubin's initial proposal of MI, which was in the context of imputation of public-use datasets where the imputer could not necessarily be expected to have the same analysis in mind as the analyst. A number of thorny theoretical issues arise when the imputation and analysis models differ and may not be mutually "congenial" (Rubin, 1987, 1996; Meng, 1994; Fay, 1996; Rao, 1996; Robins and Wang, 2000), but the added flexibility opens up some attractive missing-data solutions for the practitioner. The resulting methods are not in all cases theoretically "pristine", but a well-thought-out imputation model is likely to yield inferences with good properties. Here are some examples:

(1) *Missing data in covariates.* As noted above, the ML approach as currently implemented in standard software does not allow for missing covariates. One possibility for dealing with missing data in the covariates is to multiply-impute them conditional on the value of Y_{obs} and observed values of the covariates, using software such as IVEware (Raghunathan *et al.*, 2001). If there are not many missing covariate values, then a relatively simple imputation method that conditions on key predictors of the missing values should suffice. Then a repeated-measures model such as PROC MIXED can be applied to the imputed datasets and the results combined using MI combining rules. Multiple imputations of Y_{mis} might be generated in the process of imputing the missing covariates, but these can be ignored for the PROC MIXED analysis since it allows missing Y values.

(2) *Auxiliary variables included in the imputation model.* Sometimes there are variables that are related to the missing variables and potentially related to the missing-data mechanism that are not included in the final repeated-measures model. For example, in a clinical trial, the variables measuring side effects of treatments may relate to dropout and the outcome variables but are not included in the final analysis model since they are post-treatment variables, and hence should not be included as covariates. Such variables might be included in the imputation model but not in the final analysis model. Another example is the intention-to-treat analysis, where variables indicating which treatments are actually received can be included in the imputation model, even though they are not included in the primary analysis for comparing the treatments (Little and Yau, 1996).

(3) *Combining nonparametric complete-data analyses with parametric imputations.* Any imputations require assumptions about the predictive distribution of the missing values, and these can be explicitly formulated and assessed within a parametric modelling framework. On the other hand, some favour less parametric approaches to the final analysis, such as nonparametric tests based on ranks, and these are often specified as the main analysis in a clinical trial. Methods of imputation for rank data are not highly developed. A compromise is to base MI on a parametric model, and then allow the final analysis of the multiply-imputed data to be nonparametric, with results from the tests combined using MI combining rules. The advantage of this approach is that the parametric assumptions are confined to the imputations, and violations of these assumptions have much less impact than if the same model was applied for the analysis of the filled-in data. This is particularly true if the fraction of missing information in the missing values is small, since then the impact of model misspecification is relatively small.

Model misspecification remains a concern for the imputation model, particularly when the fraction of missing information is high and the sample size is large. We discuss here MI based on penalized spline of propensity prediction (PSPP), an approach that provides robustness to misspecification of the mean structure in the model. We assume that the missing-data mechanism is MAR, but make some comments about modifications of the approach for NMAR missing data in the concluding section.

18.3 ROBUST MAR INFERENCE WITH A SINGLE MISSING OUTCOME

We first consider the problem of imputing missing values of a single continuous outcome variable Y_p given fully observed covariates $X_1, \ldots X_q$ and prior repeated measures $Y_1, \ldots Y_{p-1}$. MI then requires a regression model for the conditional distribution of Y_p given $X_1, \ldots X_q$ and $Y_1, \ldots Y_{p-1}$. A standard parametric approach is to assume a normal linear model, such as

$$[Y_p | X_1, \ldots X_q, Y_1, \ldots Y_{p-1}, \beta, \sigma^2] \sim N[\mu(X_1, \ldots X_q, Y_1, \ldots Y_{p-1}; \beta), \sigma^2],$$
$$\mu(X_1, \ldots X_q, Y_1, \ldots Y_{p-1}; \beta) = \beta_0 + \sum_{j=1}^{q} \beta_j X_j + \sum_{k=1}^{p-1} \beta_{q+k} Y_k, \qquad (18.1)$$

where $N(\mu, \sigma^2)$ denotes a normal distribution with mean μ and variance σ^2.

We consider an elaboration of this model that provides robustness against misspecification of the mean function μ. Other robust approaches, such as replacing the normal distribution by a longer tailed distribution like the t (Lange *et al.*, 1989) can provide protection against outliers, but arguably misspecification of the mean structure is more important, at least in large samples. A standard accepted way of elaborating the model of Equation (18.1) is to add nonlinear terms and interactions to the model. With a single predictor, a more flexible approach is to model the relationship of that variable with Y_p via a spline function (e.g. Cheng, 1994). This can be extended to handle a small number of covariates, but is subject to the "curse of dimensionality" when the number of covariates is large. The PSPP imputation model addresses this problem by confining the spline to a particularly important direction in covariate space, namely the direction defined by the propensity that Y_p is missing, and allowing a parsimonious form for other covariates.

Specifically, we define the propensity score as the logit of the propensity for Y_p to be observed, namely:

$$Y_p^* = \text{logit}[\Pr(M_p = 0|X_1, \ldots X_q, Y_1, \ldots Y_{p-1})], \tag{18.2}$$

where M_p is the indicator for whether Y_p is missing. We can estimate Y_p^* by a logistic regression of M on $(X_1, \ldots X_q, Y_1, \ldots Y_{p-1})$, yielding an estimated propensity $\hat{Y}_p^*$; this logistic regression can include nonlinear terms and interactions in $(X_1, \ldots X_q,\ Y_1, \ldots Y_{p-1})$, as necessary. We then predict the missing values of Y_p using the following PSPP model:

$$\begin{aligned}
&[Y_p | Y_p^*, Y_2, \ldots Y_{p-1}, \beta, \phi, \sigma^2] \sim N[\mu(X_1, \ldots X_q, Y_1, \ldots Y_{p-1}),\ \sigma^2], \\
&\mu(X_1, \ldots X_q, Y_1, \ldots Y_{p-1}) = s_p(\hat{Y}_p^*; \phi) + g(X_1, \ldots X_q, Y_1, Y_2, \ldots Y_{p-2}, \beta),
\end{aligned} \tag{18.3}$$

where $s_p(\hat{Y}_p^*; \phi)$ is a penalized spline (discussed below) and g is a parametric function indexed by unknown fixed parameters β, with the property that

$$g(\hat{Y}_p^*, 0, \ldots 0, \beta) = 0 \text{ for all } \beta. \tag{18.4}$$

One variable (here Y_{p-1}, or better a covariate that is less highly correlated with Y_p) is left out of the function g in Equation (18.3) to avoid multicollinearity with $\hat{Y}_p^*$. In general g should be chosen to include variables that are predictive of Y_p after conditioning on $\hat{Y}_p^*$. A simple example of g that satisfies Equation (18.4) is the linear additive model

$$g(X_1, \ldots X_q, Y_1, Y_2, \ldots Y_{p-2}, \beta) = \beta_0 + \sum_{j=1}^{q} \beta_j X_j + \sum_{j=1}^{p-2} \beta_{j+q} Y_j, \tag{18.5}$$

but the model also allows interactions with Y_p*, for example:

$$g(X_1, \ldots X_q, Y_1, Y_2, \ldots Y_{p-2}, \beta) = \beta_0 + \sum_{j=1}^{q} \beta_j X_j + \sum_{j=1}^{p-2} \beta_{j+q} Y_j + \beta_{p+q-1} X_1 \hat{Y}_p^*.$$

The spline $s_p(\hat{Y}_p^*; \phi)$ could be based on any method that provides a flexible shape for the regression of Y_p and $\hat{Y}_p^*$. We have used a penalized spline with a large number of fixed knots, based on the mixed model with a power-truncated spline basis (Eilers and Marx, 1996; Ruppert and Carroll, 2000):

$$s_p(\hat{Y}_p^*, \phi) = \phi_0 + \sum_{j=1}^{d} \phi_j \hat{Y}_p^{*j} + \sum_{k=1}^{K} \phi_{d+k} \left(\hat{Y}_p^* - \tau_k\right)_+^d, \tag{18.6}$$

where d is the degree of polynomial, $(x)_+^d = x^d I(x \geq 0)$, $\tau_1 < \ldots < \tau_K$ are selected fixed knots and K is the total number of knots. The parameters $(\phi_0, \ldots \phi_d)^{\mathrm{T}}$ in Equation (18.6) are treated as fixed (assigned as noninformative prior in Bayesian terms), and smoothing is achieved by treating $(\phi_{d+1}, \ldots \phi_{d+K})^{\mathrm{T}}$ as normal with mean zero and variance $\lambda\sigma^2$, where λ is a tuning parameter estimated from the data. Some comments on this model follow:

(1) The key idea of the model is that the relationship between Y_p and Y_p^* is modelled flexibly by a spline. It is important to get that relationship right, since misspecification of this aspect leads to bias. Conditional on Y_p^*, the distribution of the predictors is the same for respondents and nonrespondents, by the basic balancing property of the propensity score (Rosenbaum and Rubin, 1983). This means that misspecification of the form of g does not lead to bias for the mean of Y_p given Y_p^* (although it may lead to bias for other parameters). Thus for imputation, the form of g is less important than the form of the regression on Y_p^*, although it is useful to include in g predictors of Y_p that improve the precision of the imputations. More generally, g should include variables with relationships with Y_p that are of interest in the final analysis of the imputed data.

(2) ML or REML estimates $(\hat{\beta}, \hat{\phi}, \hat{\sigma}^2, \hat{\lambda})$ of the parameters $(\beta,\phi,\sigma^2,\lambda)$ can be obtained using widely available software such as PROC MIXED in SAS (SAS, 2004) and lme( ) in S-plus (Pinheiro and Bates, 2000). Missing values of Y_p can then be drawn from their predictive distribution of Y_p given Y_{obs} and $(\hat{\beta}, \hat{\phi}, \hat{\sigma}^2, \hat{\lambda})$. This MI approach is improper (Rubin, 1987) in that it does not propagate uncertainty in estimating the parameters. A proper MI procedure is to compute ML estimates of each set of imputations from a bootstrapped sample of cases. Alternatively, a fully Bayesian approach is to draw values $(\beta^{(d)}, \phi^{(d)}, \sigma^{2(d)}, \lambda^{(d)})$ of the parameters from their posterior distribution using the Gibbs' sampler, and then draw the missing values $Y_{\mathrm{mis}}^{(d)}$ of Y_p from the predictive distribution of Y_p given Y_{obs} and $(\beta^{(d)}, \phi^{(d)}, \sigma^{2(d)}, \lambda^{(d)})$.

(3) One can think of $\hat{Y}_p^*$ as replacing one of the predictors in the model. Thus the function g should usually exclude one of the predictors to avoid multicollinearity problems.

(4) Uncertainty in the logistic regression coefficients that determine $\hat{Y}_p^*$ can be ignored as a first approximation, or propagated by drawing them from their posterior distribution or computing them on a bootstrapped sample for each set of multiple imputations.

(5) The "double robustness" property is that the estimated mean of Y_p for this model is consistent if either the function μ is correctly specified or the response propensity is correctly specified and the spline correctly models the relationship between Y_p and the response propensity. In the original formulation of the model (Little and An, 2004), this "double robustness" property was shown to hold when the variables in the parametric function g were centred by replacing them by residuals from regressions

on $\hat{Y}_p^*$. Subsequently, Zhang and Little (2008) showed that the property still holds for the simpler version of the model (Equation 18.3) that omits this centring.

(6) Another way to achieve robustness for the mean of Y_p is to model the relationship between Y_p and the predictors parametrically, and to calibrate the predictions from the model by adding a weighted mean of the residuals, with weights that are the inverse of the response propensity. This leads to a 'calibration' estimator of the form

$$\hat{\mu}_p = n^{-1}\left[\sum_{i=1}^{r} w_i(y_{ip} - \hat{y}_{ip})\right] + n^{-1}\left(\sum_{i=1}^{n} \hat{y}_{ip}\right). \tag{18.7}$$

where $\hat{y}_{ip}$ is the prediction for case i. This estimator has properties of semiparametric efficiency and "double-robustness" (Robins *et al.*, 1994; Robins and Rotnitsky, 2001; Scharfstein and Irizarry, 2003), in the sense that the estimate is consistent if just one of the models for prediction and weighting is correctly specified. However, since the calibration of the predictions is to correct effects of model misspecification, we believe that the calibration of the predictions (Equation 18.7) is unnecessary if the prediction model does not make strong parametric assumptions, as in Equation (18.4). This conjecture is supported by the simulation studies in Little and An (2004), where the PSPP method has similar performance to the calibration approach and dominates it for some choices of simulation parameters.

18.4 EXTENSIONS OF PSPP TO MONOTONE AND GENERAL PATTERNS

For longitudinal data subject to attrition, the resulting data have a monotone missing-data pattern for the repeated measures. Let $(X_1, \ldots X_P, Y_1, Y_2)$ denote a $(P + 2)$-dimensional vector of variables with $\underline{X} = (X_1, \ldots X_P)$ fully observed covariates and Y_1, Y_2 with missing values in a monotone pattern. We can divide the dataset into three parts based on the missing-data pattern. The first part contains subjects with both Y_1 and Y_2 present, denoted as P00; the second part contains cases with Y_1 present and Y_2 missing, denoted as P01; the third part contains cases with both Y_1 and Y_2 missing, denoted as P11. We can then define the propensity score for Y_1,

$$Y_1^* = \text{logit}[\Pr(M_1 = 0|X_1, \ldots X_P)], \tag{18.8}$$

and its estimate $\hat{Y}_1^*$ computed on all the cases in P00, P01 and P11, and the propensity score for Y_2 given that Y_1 is observed,

$$Y_2^* = \text{logit}[\Pr(M_2 = 0|M_1 = 0, Y_1, X_1, \ldots X_P)], \tag{18.9}$$

and its estimate $\hat{Y}_2^*$ computed on all the cases in P00 and P01. Then PSPP can be applied as before to impute missing values of Y_1, by the regression of Y_1 on $\underline{X}$ and the spline of $\hat{Y}_1^*$, estimated using cases in P00 and P01. The generalization of PSPP for imputing missing values of Y_2 is less clear. A natural approach is to impute missing values of Y_2 in P01 and P11 by a regression of Y_2 on $\underline{X}$, Y_1 and a spline of $\hat{Y}_2^*$, where missing values of Y_1 in P11 are replaced by estimates from Equation (18.8). This method does not yield a double robustness (DR) estimate of the mean of Y_2, since the estimates of the imputed values of Y_1 in P11 need to be based on a correct prediction model, and the propensity score for Y_2

to be observed, $\hat{Y}_2^*$, only applies to cases in P01 and not to cases in P11, since conditional on Y_1 being missing the probability that Y_2 observed is 0, given the monotone pattern. We propose the following alternative approach that preserves the DR property for the mean of Y_2. Missing values of Y_1 are imputed as in the univariate missing data setting, that is, with predictions from a regression of Y_1 on $\underline{X}$ and the spline of $\hat{Y}_1^*$. The imputed values are denoted as $\hat{Y}_1$. Missing values of Y_2 are imputed in two steps:

- Step 1: Missing values of Y_2 in P01 are imputed using predictions from a regression of Y_2 on $\underline{X}$, Y_1 and splines of $\hat{Y}_1^*$ and $\hat{Y}_2^*$, that is

$$Y_2 \sim s(\hat{Y}_1^*) + s(\hat{Y}_2^*) + g(\underline{X}, Y_1),$$

  estimated using the cases in P00. Let $Y_{2\_imp} = (Y_2, \hat{Y}_2)$ be the vector of Y_2, which contains the observed Y_2 of P00 and the imputed Y_2 of P01, then the second step follows.

- Step 2: Regress Y_{2_imp} on $\underline{X}$, Y_1 and a spline of $\hat{Y}_1^*$, that is

$$Y_{2\_imp} \sim s(\hat{Y}_1^*) + g(\underline{X}, Y_1).$$

  The model is fit to cases in P00 and P01, and missing values of Y_2 in P00 are imputed by substituting the observed covariates, the imputed Y_1 and the propensity score $\hat{Y}_1^*$ into the above formula.

For a general pattern of missing data, the sequential imputation methods of Raghunathan *et al.* (2001) can be extended to provide PSPP imputations that condition on the spline of the propensity that each variable is missing. An (2005) discusses these extensions in detail.

18.5 EXTENSIONS TO INFERENCES OTHER THAN MEANS

The PSPP method provides robustness for estimating the marginal mean of Y, but this robustness property does not necessarily apply for other parameters, such as subgroup means. In particular in longitudinal clinical trials, we are often primarily concerned with comparisons of means between treatment groups, which are contrasts of means in subgroups defined by the treatment variable. Zhang and Little (2008) show in simulations that the PSPP method does not provide estimates of treatment contrasts that are protected against model misspecification, even if the treatment variable is included as a covariate in the g function. Zhang and Little (2008) propose extensions of the PSPP – stratified PSPP and bivariate PSPP – to address this issue.

Specifically, suppose that missing values are confined to Y_p, and interest is in the mean of Y_p given X_1, where X_1 is a categorical variable with C categories. The stratified PSPP method computes a separate spline on the propensity to respond in each of the categories of X_1, as follows:

- Estimate the response propensity:

$$Y_p^* = \text{logit}\left[\Pr(M_p = 0 | X_1, X_2, \ldots X_q, Y_1, \ldots Y_{p-1})\right].$$

- Let $I_c = 1$ if $X_1 = c$; $I_c = 0$ if $X_1 \neq c$, $c = 1, \ldots C$, and form the propensity in each category of $X_1 : Y_p^{*c} = Y_p^* \times I_c$, $c = 1, \ldots, C$. Then model the conditional distribution of Y_p as in Equation (18.3) with mean in category $X_1 = c$:

  $$\mu(X_1 = c, X_2, \ldots X_q, Y_1, \ldots Y_{p-1}) = s_{pc}(\hat{Y}_p^{*c}; \phi) + g(X_1, \ldots X_q, Y_1, Y_2, \ldots Y_{p-1}, \beta),$$

  where $s_{pc}(Y^{*c})$, $c = 1, \ldots C$, is a spline for the regression of Y on Y_p^{*c} and g is as in Equation (18.3).

This method yields a DR property for the conditional mean of Y_p given X_1 (Zhang and Little, 2008). The marginal mean of Y_p is a weighted average of conditional means, which again has the consistency property.

Zhang and Little (2008) also consider estimating the conditional mean of Y_p given a continuous variable X_1, assuming a linear regression function $E(Y_p \mid X_1) = \gamma_0 + \gamma_1 X_1$. To achieve the DR property for estimates of γ_0, γ_1, we need to replace the univariate spline on the propensity by a bivariate spline on the propensity and X_1.

18.6 EXAMPLE

To illustrate these methods we use data from an online weight loss study conducted by Kaiser Permanente, the analysis of which is discussed in Couper *et al.* (2005). Approximately 4000 subjects were randomly assigned to the treatment or the control group after they filled out an initial survey, which contained baseline measurements such as weight, motivation to weight loss, etc. For the treatment group, the weight loss information provided online was tailored to the subjects based on their answers to the initial survey; for the control group, information provided online was the same for all the subjects but they could search for additional weight loss information using hyperlinks in the text. At 3, 6 and 12 months, follow-up invitations for web surveys were sent to the participants, which collected measurements such as current weight. Our goal is to compare short-term and long-term treatment effects; in particular, we compare the reduction of the body mass index (BMI), defined as the difference of the follow-up BMI and the baseline BMI.

There were 2059 subjects in the treatment group and 1956 subjects in the control group at baseline. At 3 months, 623 subjects in the treatment group and 611 subjects in the control group responded to the second survey. At 6 months, 438 subjects in the treatment group and 397 subjects in the control group remained in the study. At 12 months, 277 subjects in the treatment group and 314 subjects in the control group responded to the last survey. Comparisons of the baseline measurements between subjects who remained in the study and those who dropped out at 3 months indicate that subjects who remained in the study have much lower baseline BMI than those who dropped out of the study for the treatment group ($P < 0.001$), but this difference is not seen in the control group ($P = 0.47$); On the other hand, subjects who remained in the study at 3 months have better baseline health, as shown by the incidence of previous disease, than those who dropped out of the study for the control group ($P < 0.01$), but this difference is not seen in the treatment group ($P = 0.56$). Similarly, subjects who remained in the study at 6 months have much lower baseline BMI than those who dropped out of the study for the treatment group ($P < 0.001$), but this difference is not seen in the control group

($P = 0.82$). These differences suggest that interactions between treatment and baseline covariates are included when estimating the propensity scores.

We assume MAR in our analysis, and estimate the propensity scores by a logistic regression, with the inclusive criterion of retaining all variables with P values less than 0.20. The MAR assumption is supported by the fact that we have rich baseline information for characterizing the respondents and nonrespondents, although it would be advisable to conduct a sensitivity analysis to assess the impact of violations from MAR, for example treating dropouts as treatment failures.

We apply the PSPP methods to the data to derive the BMI reduction at 3, 6 and 12 months. For BMI reduction at 3 months, we fit two stratified PSPP models:

(1) Stratified PSPP method with null g function, denoted as $\left[\sum_{c=1}^{2} I_c s_c(\hat{Y}_1^*)\right]$, where $\hat{Y}_1^*$ is the estimated 3-month propensity score.
(2) Stratified PSPP method with baseline covariates in the g function, denoted as $\left[\sum_{c=1}^{2} I_c s_c(\hat{Y}_1^*) + g(\text{baseline covariates})\right]$.

For BMI reduction at 6 months, we apply two stepwise stratified PSPP models:

(1) Stepwise stratified PSPP method with null g function, denoted as $\left\{\text{step1}: \sum_{c=1}^{2}\left[I_c s_c(\hat{Y}_1^*) + I_c s_c(\hat{Y}_2^*)\right],\ \text{step2}: \sum_{c=1}^{2} I_c s_c(\hat{Y}_1^*)\right\}$, where $\hat{Y}_2^*$ is the estimated 6-month propensity score and $\hat{Y}_1^*$ is the estimated 3-month propensity score.
(2) Stepwise stratified PSPP method with the g function, denoted as $\left\{\text{step1}: \sum_{c=1}^{2}[I_c s_c(\hat{Y}_1^*) + I_c s_c(\hat{Y}_2^*)] + g(\text{covariates}),\ \text{step2}: \sum_{c=1}^{2} I_c s_c(\hat{Y}_1^*) + g(\text{covariates})\right\}$.

For BMI reduction at 12 months, we fit two stepwise stratified PSPP models:

(1) Stepwise stratified PSPP method with null g function, denoted as $\left\{\text{step1}: \sum_{c=1}^{2}[I_c s_c(\hat{Y}_1^*) + I_c s_c(\hat{Y}_2^*) + I_c s_c(\hat{Y}_3^*)];\ \text{step2}: \sum_{c=1}^{2}[I_c s_c(\hat{Y}_1^*) + I_c s_c(\hat{Y}_2^*)];\ \text{step3}: \sum_{c=1}^{2} I_c s_c(\hat{Y}_1^*)\right\}$, where $\hat{Y}_3^*$ is the estimated 12-month propensity score, $\hat{Y}_2^*$ is the estimated 6-month propensity score and $\hat{Y}_1^*$ is the estimated 3-month propensity score.
(2) Stepwise stratified PSPP method with the g function, denoted as $\left\{\text{step1}: \sum_{c=1}^{2}[I_c s_c(\hat{Y}_1^*) + I_c s_c(\hat{Y}_2^*) + I_c s_c(\hat{Y}_3^*)] + g;\ \text{step2}: \sum_{c=1}^{2}[I_c s_c(\hat{Y}_1^*) + I_c s_c(\hat{Y}_2^*)] + g;\ \text{step3}: \sum_{c=1}^{2} I_c s_c(\hat{Y}_1^*) + g\right\}$.

For the parametric function g in model (2) at 3, 6 and 12 months we include baseline covariates that predict the outcome, the reduction of BMI.

Table 18.1 BMI reduction within groups compared to complete case (CC) analysis, based on 200 bootstrap samples.

		Treatment		Control	
		Mean SE	95 % CI	Mean SE	95 % CI
3 months	CC	−0.91(0.09)	(−1.09, −0.73)	−0.45(0.10)	(−0.65, −0.25)
	Stratified PSPP, null g function	−1.01 (0.11)	(−1.23, −0.78)	−0.40 (0.10)	(−0.60, −0.20)
	Stratified PSPP, with non-null g function	−1.00 (0.10)	(−1.21, −0.80)	−0.42 (0.09)	(−0.61, −0.23)
	Sequential regression	−0.97 (0.09)	(−1.16, −0.78)	−0.46 (0.10)	(−0.66, −0.27)
	Weighted CC analysis	−1.04 (0.11)	(−1.27, −0.82)	−0.40 (0.09)	(−0.59, −0.21)
6 months	CC	−0.88 (0.15)	(−1.18, −0.58)	−0.63 (0.15)	(−0.93, −0.33)
	Stepwise stratified PSPP, null g function	−1.07 (0.18)	(−1.43, −0.72)	−0.57 (0.18)	(−0.92, −0.21)
	Stepwise stratified PSPP, with non-null g function	−1.17 (0.18)	(−1.53, −0.81)	−0.57 (0.17)	(−0.91, −0.24)
	Sequential regression	−1.01 (0.13)	(−1.27, −0.76)	−0.61 (0.15)	(−0.92, −0.30)
	Weighted CC analysis	−1.11 (0.22)	(−1.55, −0.67)	−0.54 (0.17)	(−0.88, −0.19)
12 months	CC	−1.24 (0.17)	(−1.58, 0.91)	−0.93 (0.23)	(−1.40, −0.47)
	Stepwise stratified PSPP, null g function	−1.55 (0.69)	(−2.93, 0.17)	−1.00 (0.31)	(−1.62, 0.38)
	Stepwise stratified PSPP, with non-null g function	−1.27 (0.25)	(−1.77, −0.77)	−1.04 (0.28)	(−1.60, −0.47)
	Sequential regression	−1.24 (0.17)	(−1.59, −0.89)	−0.91 (0.22)	(−1.34, −0.48)
	Weighted CC analysis	−1.42 (0.37)	(−2.15, −0.69)	−0.65 (0.54)	(−1.73, 0.43)

In addition to the PSPP methods, we fit regression models to impute the missing BMI reductions at 3, 6 and 12 months sequentially. The covariates in the regression models are the same as the covariates in the *g* functions of the PSPP methods, but with an extra categorical variable that indicates whether the subject is in the treatment or the control group. We also include weighted complete case analysis for comparison. We estimate the weights based on the same logistic regression model as in the PSPP methods and apply the weighting procedure to the treatment or the control group separately.

We compare our method with complete case analysis. Empirical standard errors (SE) and the corresponding normal confidence intervals are obtained from 200 bootstrap samples. Results are summarized in Table 18.1. The treatment group has a larger reduction of BMI after 3 months (−0.91 (0.09)) compared to the control group (−0.45 (0.10)) based on the complete case analysis. The 95 % confidence intervals for the treatment group do not overlap with the control group, suggesting a treatment effect on the weight loss. At 6 and 12 months, the difference between the two groups is not statistically significant. The PSPP methods yield the same conclusions, except that treatment effects at 3 and 6 months are stronger after imputation and the BMI reduction for the treatment and control groups increases monotonically. We did not find much difference for the methods with and without the parametric function in the weight loss study, except for the 12-month BMI reduction, where adding parametric function *g* reduces variance of the estimates significantly. Weighted complete case analysis and the sequential regression models yield conclusions similar to the PSPP methods.

The results show that the treatment group has a fast response in terms of BMI reduction. But later on, subjects in the control group catch up and the two groups show similar levels of weight loss. These results suggest that tailoring the information does help subjects to lose weight, especially in the beginning. Later on, if the subjects continue trying to lose weight, then tailored and untailored information have similar effects on weight loss. This seems plausible, since the effect of tailoring might be expected to be minor in individuals who have a longlasting motivation to lose weight.

18.7 DISCUSSION

We have described a multiple imputation approach to the analysis of longitudinal data with missing values, which provides flexibility by allowing the analysis model and the imputation model to differ. A good MI model should make relatively weak assumptions about the predictive distribution of the missing values, particularly when samples are large and missing data are extensive. In this vein, we have discussed MI models that include a spline on the propensity to be missing, which provide a form of double robustness and hence some protection against misspecification of the mean structure. We have considered normal models here, but for non-normal outcomes the same approach to the mean structure can be applied to generalized linear models, with the systematic component of the link function replacing the mean.

Throughout the chapter we have assumed that the missing data are MAR, so the question remains how to handle situations where the mechanism is thought to be NMAR. The model-based MI approach to missing data can be applied to NMAR models by formulating a model for the joint distribution of the data and the missing data mechanism (e.g. Diggle and Kenward, 1994; Little, 1995; Little and Rubin, 2002, chapter 15). However, lack of identifiability of the parameters is a serious issue in fitting these models, as deviations from MAR cannot be detected from the data without making strong structural or distributional assumptions. The impact of NMAR nonresponse can be reduced by the following strategies:

- At the design stage, record values of covariates that are predictive of nonresponse, and condition on these in imputing the missing values.
- Consider following up a subsample of nonrespondents to recover at least the key missing information on these cases. These data can then be used to multiply-impute the information on incomplete cases that are not followed up.
- When there are various mechanisms of missingness, attempt to determine which of the missing values of a variable are likely to be MAR and which are likely not to be MAR, and then use MAR methods to multiply-impute the former. This reduces the scope of the NMAR problem for this variable, as compared with an analysis that fits a NMAR model to all the missing values.

For the missing values that are thought to be NMAR, we recommend a sensitivity analysis under plausible alternative models for nonignorable nonresponse (Rubin, 1977; Little and Wang, 1996; Scharfstein *et al.*, 1999). As exemplified in Rubin (1977), the Bayesian framework is convenient for this modelling since it explicitly recognizes the need for prior information. A simple approach, which may be as effective as more complex alternatives, is to introduce offsets into the means of the predictive distributions used to multiply-impute the missing values that are thought to be NMAR. For example, in the sequential IVEware approach of Raghunathan *et al.* (2001), imputations for a missing variable Y_j at a particular iteration t are based on the predictive distribution of Y_j given the other variables, with missing values of the other variables imputed with draws from previous steps of the sequence. Suppose some set $\{y_{ji}, i = 1, \ldots r\}$ of the missing values of Y_j are thought to be NMAR, and let $\{\mu_{ji}^{(t)} : i = 1, \ldots r\}$ and $\{\sigma_{ji}^{(t)} : i = 1, \ldots r\}$ be the means and standard deviations of the predictive distributions of these values at iteration t under a MAR model. One possible sensitivity analysis consists of replacing these predictive means by $\{\mu_{ji}^{(t)} + \delta_j \sigma_{ji}^{(t)} : i = 1, \ldots r\}$ for one or more values of δ_j that are thought to represent plausible deviations from MAR, for example $\delta_j = 0.5$ if a positive deviation of half a standard deviation is reasonable, $\delta_j = -0.5$ if a negative deviation of that size is thought reasonable or both $\delta_j = 0.5$ and $\delta_j = -0.5$ if the sign of the deviation is unknown. This change is easily accomplished by replacing the MAR imputations $\hat{y}_{ji}^{(t)} : i = 1, \ldots r$ by NMAR imputations $\{\hat{y}_{ji}^{(t)} + \delta_j \sigma_{ji}^{(t)} : i = 1, \ldots r\}$. The size of the perturbations need to reflect the fact that draws condition on other variables. For an application of this idea, see Van Buuren *et al.* (1999). What this method lacks in sophistication may be compensated by the transparency of the underlying assumptions. A Bayesian variant is to assign δ_j a prior distribution and draw values of that parameter at each iteration.

ACKNOWLEDGEMENTS

This research was supported by grant P50 CA101451 from the National Institutes of Health.

REFERENCES

An, H. (2005). *Robust likelihood-based inference for multivariate data with missing values*. PhD Dissertation, Department of Biostatistics, University of Michigan, Ann Arbor, MI.

Breslow, N. E. and Clayton, D. G. (1993). Approximate inference in generalized linear mixed models. *Journal of the American Statistical Association*, 88, 9–25.

Cheng, P. E. (1994). Nonparametric estimation of mean functionals with data missing at random. *Journal of the American Statistical Association*, 89, 81–87.

Couper, M. P., Peytchev, A., Little, R. J. A., Strecher, V. J. and Rothert, K. (2005). Combining information from multiple modes to reduce nonresponse bias. In *Proceedings of the American Statistical Association, Section on Survey Research Methods* (pp. 2910–2917). Washington, DC: American Statistical Association.

Diggle, P. and Kenward, M. G. (1994). Informative drop-out in longitudinal data analysis, *Journal of the Royal Statistical Society, C*, 43, 49–73.

Eilers, P. H. C. and Marx B. D. (1996). Flexible smoothing with B-splines and penalties (with discussion). *Statistical Science*, 11, 89–121.

Fay, R. E. (1996), Alternative paradigms for the analysis of imputed survey data, *Journal of the American Statistical Association*, 91, 490–498.

Lange, K., Little, R. J. A. and Taylor, J. M. G. (1989). Robust statistical inference using the t distribution. *Journal of the American Statistical Association*, 84, 881–896.

Little, R. J. (1995). Modeling the drop-out mechanism in longitudinal studies. *Journal of the American Statistical Association*, 90, 1112–1121.

Little, R. J. A. and An, H. (2004). Robust likelihood-based analysis of multivariate data with missing values. *Statistica Sinica*, 14, 949–968.

Little, R. J. and Rubin, D. B. (1999). Comment on 'Adjusting for non-ignorable drop-out using semiparametric models' by D. O. Scharfstein, A. Rotnitsky and J. M. Robins. *Journal of the American Statistical Association*, 94, 1130–1132.

Little R. J. A. and Rubin, D. B. (2002). *Statistical Analysis with Missing Data*. Wiley, New York.

Little, R. J. and Wang, Y.-X. (1996). Pattern-mixture models for multivariate incomplete data with covariates. *Biometrics*, 52, 98–111.

Little, R. J. A. and Yau, L. (1996). Intent-to-treat analysis in longitudinal studies with drop-outs. *Biometrics*, 52, 1324–1333.

McCulloch, C. E. and Searle, S. R. (2001). *Generalized, Linear, and Mixed Models*. New York: John Wiley & Sons, Inc.

Meng, X.-L. (1994). Multiple imputation inferences with uncongenial sources of input (with discussion). *Statistical Science*, 9, 538–573.

Pinheiro, J. C. and Bates, D. M. (2000). *Mixed-Effects Models in S and S-PLUS*. New York: Springer-Verlag.

Raghunathan, T. E., Lepkowski, J. M., VanHoewyk, J. and Solenberger, P. (2001). A multivariate technique for multiply imputing missing values using a sequence of regression models. *Survey Methodology*, 27, 85–95.

Rao, J. N. K. (1996). On variance estimation with imputed data. *Journal of the American Statistical Association*, 91, 499–506.

Robins, J. M. and Rotnitsky, A. (2001). Comment on 'Inference for semiparametric models: some questions and an answer' by P. Bickel and J. Kwon. *Statistica Sinica*, 11, 920–936.

Robins, J. M., Rotnitsky, A. and Zhao, L. P. (1994). Estimation of regression coefficients when some regressors are not always observed. *Journal of the American Statistical Association*, 89, 846–866.

Robins, J. M. and Wang, N. (2000). Inference for imputation estimators. *Biometrika*, 87, 113–124.

Rosenbaum, P. R. and Rubin, D. B. (1983). The central role of the propensity score in observational studies for causal effects. *Biometrika*, 70, 41–55.

Rubin, D. B. (1976). Inference and missing data. *Biometrika*, 63, 581–592.

Rubin, D. B. (1977). Formalizing subjective notions about the effect of nonrespondents in sample surveys. *Journal of the American Statistical Association*, 72, 538–543.

Rubin, D. B. (1987). *Multiple Imputation for Nonresponse in Surveys*. New York: John Wiley & Sons, Inc.

Rubin, D. B. (1996). Multiple imputation after 18+ years, *Journal of the American Statistical Association*, 91, 473–489.

Ruppert, D. and Carroll R. J. (2000). Spatially adaptive penalties for spline fitting. *Australia and New Zealand Journal of Statistics*, 42, 205–223.

SAS (2004). *SAS OnlineDoc® 9.1.3*. Cary, NC: SAS Institute Inc.

Schafer, J. L. (1997). *Analysis of Incomplete Multivariate Data*. New York: CRC Press.

Scharfstein, D. and Irizarry, R. (2003). Generalized additive selection models for the analysis of studies with potentially nonignorable missing outcome data. *Biometrics*, 59, 601–613.

Scharfstein, D., Rotnitsky, A. and Robins, J. (1999). Adjusting for nonignorable dropout using semiparametric models (with discussion). *Journal of the American Statistical Association*, 94, 1096–1146.

Tanner, M. A. and Wong, W. H. (1987). The calculation of posterior distributions by data augmentation (with discussion), *Journal of the American Statistical Association*, 82, 528–550.

VanBuuren, S., Boshuizen, H. C. and Knook, D. L. (1999). Multiple imputation of missing blood pressure covariates in survival analysis. *Statistics in Medicine*, 18, 681–694.

VanBuuren, S. and Oudshoorn, C .G. M (1999). Flexible multivariate imputation by MICE. Leiden: TNO Preventie en Gezondheid, TNO/VGZ/PG 99.054. For associated software see http://www.multiple-imputation.com.

Yu, Y. and Ruppert, D. (2002). Penalized spline estimation for partially linear single-index models. *Journal of the American Statistical Association*, 97, 1042–1054.

Zhang, G. and Little, R. J. (2008). Extensions of the penalized spline propensity prediction method of imputation. *Biometrics*, in press.

CHAPTER 19

Assessing the Temporal Association of Events Using Longitudinal Complex Survey Data

Norberto Pantoja-Galicia
Harvard University, USA

Mary E. Thompson
University of Waterloo, Canada

Milorad S. Kovacevic
Statistics Canada

19.1 INTRODUCTION

The longitudinal nature of surveys such as the National Population Health Survey (NPHS) and the Survey of Labour and Income Dynamics (SLID), conducted by Statistics Canada, allows the use of event history analysis techniques to study relationships among events. For example, outcomes from the NPHS questionnaire offer the necessary information to explore the relationship between smoking cessation and pregnancy, while SLID provides the possibility of analysing the association between job loss and divorce. Let T_1 and T_2 be the times to the occurrence of events of interest; in the first case these events correspond to becoming pregnant and ceasing to smoke, while in the second instance the events represent the times of losing a job and becoming divorced respectively.

Establishing causation in observational studies is seldom possible. Nevertheless, if one of two events tends to precede the other closely in time, a causal interpretation of an association between these events can be more plausible. The role of longitudinal surveys is crucial, then, since they allow sequences of events for individuals to be observed.

Methodology of Longitudinal Surveys P. Lynn

Thompson and Pantoja-Galicia (2003) discuss, in this context, notions of temporal association and ordering and propose a formal nonparametric test for a partial order relationship.

In this paper the nonparametric test is implemented to account for complexities of the sample design and to allow the observed times to events to be interval censored. The mentioned test requires estimation of the joint density for T_1 and T_2. For this purpose, we extend the ideas of Duchesne and Stafford (2001) and Braun, Duchesne and Stafford (2005), who develop kernel and local likelihood density estimation for interval-censored data, to the bivariate case. Since we deal with survey data that have been collected with a complex design, we also make adaptations to the methodology to account for this complexity. Modifications to the test are made through appropriate weighting and variance estimation techniques, whose only difficulty may lie in the implications for the numerical computations due to the high variability of weights. In our examples, such a difficulty has occurred very seldom.

Different patterns of missing longitudinal data were taken into account either through weighting or through treating the missing responses as censored observations.

The techniques developed in this research may be applicable to other problems in analysis of complex longitudinal survey data. To make this chapter self-contained, we begin with an overview of a notion of temporal order along with a formal nonparametric test for a partial order relationship presented by Thompson and Pantoja-Galicia (2003). Estimation of the densities involved in the test is discussed in Section 19.3, taking account of the interval censoring nature of the data. The complexities of the survey design are also incorporated in Section 19.4. An important element of the nonparametric test is the standard error, which is assessed in Section 19.4.1. Sections 19.5 and 19.6 present the applications. We close in Section 19.7 with some discussion and possible extensions.

19.2 TEMPORAL ORDER

Let E_1 and E_2 be two types of lifetime events. Let T_1 be the time to occurrence of event E_1 and let T_2 be the time to occurrence of event E_2, considering a specified time origin. Knowledge of the exact times of occurrence of each event would provide the appropriate elements to model a temporal relationship through their joint intensities.

19.2.1 Close Precursor

From Thompson and Pantoja-Galicia (2003), a local association of T_1 and T_2 is implied by the following concept: T_1 is a *close precursor* of T_2 if, for some positive numbers δ and $\kappa(t_1)$, we have

$$\frac{F_2[t_1 + \kappa(t_1)|T_1 = t_1]}{F_2(t_1|T_1 = t_1)} < \frac{F_2[t_1 + \kappa(t_1)]}{F_2(t_1)} - \delta \tag{19.1}$$

for all t_1 in a a specified interval (a, b), with $a, b \in \mathrm{R}$. Here $F_i(t) = \Pr(T_i > t)$ for $i = 1, 2$. In other words, T_1 is a close precursor of T_2 if the occurrence of the first event E_1 at T_1 decreases the probability of having to wait longer than $\kappa(t_1)$ to observe the occurrence of the second event E_2, and this happens with some uniformity in (a, b). The decrease is seen relative to the analogous probability if T_1, T_2 are independent.

The interval length $\kappa(t_1)$ may be thought of as the duration of an effect and would come from subject-matter considerations. We have allowed it to depend in general on t_1, anticipating that the effect of T_1 on the hazard of T_2 might not have constant duration.

Although we have given the definition of the close precursor in terms of survivor functions, it can be expressed approximately in terms of hazard functions, as follows: T_1 is a close precursor of T_2 if, for suitably chosen $\kappa(s)$ and $\delta > 0$,

$$h_2(u|T_1 = s) > h_2(u) + \delta,$$

for $u \in (s, s + \kappa(s))$. Then δ is seen to correspond to an "additive" lower bound to a short-term change in the hazard function.

In either formulation, the motivation is that the more closely T_2 tends to follow T_1, the greater the plausibility for a causal connection might be.

Since Equation (19.1) reflects approximately a short-term raising of the hazard function for T_2, it is not difficult to formulate an analogue for point process intensities, giving us an alternative way of modelling events less tied to a time origin. Blossfeld and Mills (2003) use interdependent point processes to model interrelated family events, namely entry into marriage (for individuals in a consensual union) and first pregnancy/childbirth; see also Lawless (2003) for some discussion of intensity models.

19.2.2 Nonparametric Test for Close Precursor

A formal test for a close precursor relationship between T_1 and T_2 as indicated in Section 19.2.1 (T_1 a close precursor of T_2) is given by the following: for suitable $\kappa(t_1)$, let

$$Q = \int \left(\frac{\hat{F}_2[t_1 + \kappa(t_1)|T_1 = t_1]}{\hat{F}_2(t_1|T_1 = t_1)} - \frac{\hat{F}_2[t_1 + \kappa(t_1)]}{\hat{F}_2(t_1)} \right) d\hat{F}_1(t_1). \tag{19.2}$$

Under the null hypothesis of independence of T_1 and T_2, the mean of Q will be close to zero. Thus in order to test the null hypothesis, the value of Q may be compared with twice its estimated standard error, $se(Q)$. Note that the difference within Equation (19.2) approximates to the difference between the hazard function conditional on $T_1 = t_1$ and the unconditional hazard.

To compute Q, we first would obtain an estimate of the joint density of (T_1, T_2) within the observation window. Then, we would obtain numerically the corresponding marginal probability density functions and consequently the respective survivor functions (conditional and unconditional versions).

As a check on the finite sample distribution of Q, we have performed simulations (available on request from the authors) using our density estimation method and models and parameter settings appropriate to the NPHS application in Section 19.5. These show a mean for Q under independence that is very close to zero (and validates the significance test of Section 19.5).

19.3 NONPARAMETRIC DENSITY ESTIMATION

In order to apply the test described in Section 19.2.2, we proceed to estimate the joint density of (T_1, T_2).

Silverman (1986), Scott (1992) and Wand and Jones (1995) cover material on kernel density estimation of univariate and multivariate density functions for independent and identically distributed random variables. In the case of univariate interval-censored data, nonparametric methods are proposed for density estimation by Duchesne and Stafford (2001) and Braun *et al.* (2005), and for estimation of the survivor function by Turnbull (1976), Gentleman and Geyer (1994) and Li, Watkins and Yu (1997). In addition, Betensky and Finkelstein (1999) and Wong and Yu (1999) propose nonparametric maximum likelihood estimators for multivariate interval-censored data. Density estimation is treated in the area of complex surveys by Bellhouse and Stafford (1999), Bellhouse, Goia and Stafford (2003) and Buskirk and Lohr (2005).

Pantoja-Galicia, Thompson and Kovacevic (2005) present an extension to the bivariate case of the procedure proposed by Duchesne and Stafford (2001) and Braun *et al.* (2005) to obtain a simple kernel density estimate for interval-censored data. An overview is included here to serve as an introduction to the local likelihood approach presented in Section 19.3.2.

19.3.1 Kernel Density Estimation

In the presence of complete (noncensored) data $Y_i = (Y_{i,1}, Y_{i,2})$, where $i = 1, \ldots n$, the bivariate kernel density estimator with kernel $K_h(y)$, $y = (y_1, y_2)$ and bandwidth $h = (h_1, h_2)$ is given by

$$\hat{f}_{nc}(y) = n^{-1}\sum_{i=1}^{n} K_h(Y_i - y).$$

In the context of interval-censored data, $X_i = (X_{i,1}, X_{i,2})$ lies within the two-dimensional interval $I_i = (A_{i,1}, B_{i,1}) \times (A_{i,2}, B_{i,2})$, and $A_{i,1}, B_{i,1}, A_{i,2}, B_{i,2} \in R$. Therefore, for $x = (x_1, x_2)$, a generalization of the univariate approach proposed by Braun *et al.* (2005) to the bivariate scenario gives the following estimator:

$$\hat{f}(x) = n^{-1}\sum_{i=1}^{n} E_{\hat{f}}\left[K_h(X_i - x)\middle|X_i \in I_i\right], \tag{19.3}$$

which involves the conditional expectation of the kernel, given that X_i lies within I_i (the information we know about X_i). Here, the conditional expectation is with respect to the density $\hat{f}$.

Then, in terms of iterated conditional expectation, a solution to Equation (19.3) should be

$$\hat{f}_j(x) = n^{-1}\sum_{i=1}^{n} E_{\hat{f}_{j-1}}\left[K_h(X_i - x)\middle|X_i \in I_i\right]. \tag{19.4}$$

The expectation in Equation (19.4) is with respect to the conditional density

$$\hat{f}_{j-1|I_i}(u) = \delta_i(u)\hat{f}_{j-1}(u) \Big/ \int_{I_i} \hat{f}_{j-1}(s)ds$$

over I_i, where $\delta_i(u) = 1$ if $u \in I_i$ and 0 otherwise.

Equation (19.4) implies that the conditional expectation with respect to $\hat{f}_{j-1}$ is employed to obtain $\hat{f}_j$. Note that we need to have an initial estimate $\hat{f}_0$ of the density.

In Equation (19.4) computation of the corresponding conditional expectation for each interval-censored observation X_i is needed. Let us define

$$\mu_{j-1|I}(x) = E_{\hat{f}_{j-1}}\left[K_h(X-x)\middle|X \in I\right]. \tag{19.5}$$

In Pantoja-Galicia *et al.* (2005), an importance sampling scheme proposed by Duchesne and Stafford (2001) is extended to the bivariate case and is employed to estimate Equation (19.5) by using

$$E_f\left[K_h(X-x)\middle|X \in I\right] = E_g\left[K_h(X-x)w(X)\right], \tag{19.6}$$

where g is a suitable distribution over the interval I, and $w(X)=\hat{f}_{j-1|I}(X)/g(X)$ is the importance sampling weight. Therefore, Equation (19.5) may be approximated in the following manner:

$$\hat{\mu}_{j-1|I}(x) = \sum_{b=1}^{B}\left[K_h(X_b^u - x)w_b^u\right]. \tag{19.7}$$

Here we let $g(X)$ be a bivariate uniform density and therefore the X_b^u values are generated over the interval I using a bivariate uniform sampling scheme derived from the orthogonal array-based Latin hypercubes described by Tang (1993), and $w_b^u = w(X_b^u)/\sum_{k=1}^{B} w(X_k^u)$, with $b=1,\ldots B$. Tang's procedure establishes that B has to be a perfect square, i.e. $B \in \{1^2, 2^2, 3^2, \ldots\}$. At the jth step, an estimate of Equation (19.4) may be obtained by

$$\hat{f}_j(x) = n^{-1}\sum_{i=1}^{n} \hat{\mu}_{j-1|I_i}(x). \tag{19.8}$$

19.3.2 Local Likelihood Approach

Kernel density estimation may present increased bias at and near the boundary. Wand and Jones (1995) present a discussion on this issue. One way to overcome this is by using a local likelihood approach, which we present next.

In the presence of univariate noncensored data $Y_1, Y_2, \ldots Y_n \in R$, Hjort and Jones (1996) and Loader (1996) define equivalent local log-likelihood functions for density estimation. These are based on the concept of having an approximating parametric family, $f(y)=f(y,\theta)=f(y,\theta_1,\theta_2,\ldots\theta_p)$, which may be locally estimated at y by maximizing the following local log-likelihood function:

$$\frac{1}{n}\sum_{i=1}^{n} K_h(Y_i - y)\log f(Y_i) - \int K_h(t-y)f(t)\mathrm{d}t. \tag{19.9}$$

Maximization of Equation (19.9) amounts to solving

$$\frac{1}{n}\sum_{i=1}^{n} K_h(Y_i - y)A(y, Y_i, \theta) = \int K_h(t-y)A(y,t,\theta)f(t)\mathrm{d}t,$$

with the following score function of the model of dimension $p \times 1$:

$$A(y,t,\theta) = \left[\frac{\partial}{\partial\theta_1}\log f(t,\theta), \ldots \frac{\partial}{\partial\theta_p}\log f(t,\theta)\right]^T.$$

Braun *et al.* (2005) take this concept to the context of univariate interval-censored data $X_1, X_2, \ldots X_n \in R$, proposing the following local log-likelihood function:

$$\frac{1}{n}\sum_{i=1}^{n} E\left\{K_h(X_i - x)\log\left[f(X_i)\right]\middle|X_i \in I_i\right\} - \int K_h(t-x)f(t)\mathrm{d}t.$$

Loader (1996) supposes that $\log f(t)$ can be approximated by a low-degree polynomial around x. That is:

$$\log f(t) \approx P(t-x) = \sum_{i=0}^{p} \theta_i (t-x)^i.$$

Then, we generalize the local likelihood function to the bivariate scenario as follows:

$$n^{-1}\sum E\left[K_h(X_i - x)P(X_i - x)\middle|X_i \in I_i\right] - \int K_h(t-x)\exp\left[P(t-x)\right]\mathrm{d}t, \qquad (19.10)$$

with the assumption that $\log f(t)$ can be approximated by

$$\log f(t) \approx P(t-x) = \theta_0 + \theta_1(t_1 - x_1) + \theta_2(t_2 - x_2). \qquad (19.11)$$

Then, maximization of Equation (19.10) amounts to solving

$$n^{-1}\sum_{i=1}^{n} E\left[K_h(X_i - x)A(x, X_i, \theta)\middle|X_i \in I_i\right] - \int K_h(t-x)A(x,t,\theta)\mathrm{e}^{P(t-x)}\mathrm{d}t, \qquad (19.12)$$

where $\theta = (\theta_0, \theta_1, \theta_2)$ and the corresponding score function $A(x, t, \theta) = (1, t_1 - x_1, t_2 - x_2)^T$.

Solving this system of log-likelihood equations for the coefficients of Equation (19.11) leads to a local Expectation–Maximization (EM) algorithm as described in Braun *et al.* (2005).

Let us suppose that the logarithm of the density is locally constant, i.e. $\log f(t) = \theta_0$. From Equation (19.12), solving the system of one local likelihood equation with one unknown coefficient θ_0 results in the following estimator for $f(x)$:

$$n^{-1}\sum_{i=1}^{n} E\left[K_h(X_i - x)\middle|X_i \in I_i\right] = \int K_h(t-x)e^{\tilde{\theta}_0}\mathrm{d}t = \mathrm{e}^{\tilde{\theta}_0} = \tilde{f}(x).$$

This corresponds to the kernel density estimate in Equation (19.4), which is estimated by Equation (19.8).

If the polynomial approximation is taken as in Equation (19.11) and the product normal kernel is employed, we have to solve the following system of three local likelihood equations with three unknown coefficients θ_0, θ_1 and θ_2:

$$\begin{aligned} &n^{-1}\sum_{i=1}^{n} E\left[K_h(X_i - x)(1, X_{i,1} - x_1, X_{i,2} - x_2)^T\middle|X_i \in I_i\right] \\ &= \int K_h(t-x)(1, t_1 - x_1, t_2 - x_2)^T \mathrm{e}^{\theta_0 + \theta_1(t_1 - x_1) + \theta_2(t_2 - x_2)}\mathrm{d}t. \end{aligned} \qquad (19.13)$$

If the solutions are $\tilde{\theta}_0, \tilde{\theta}_1$ and $\tilde{\theta}_2$, the local likelihood density estimate of $f(x)$ is given by

$$\tilde{f}(x) = e^{\tilde{\theta}_0}. \tag{19.14}$$

The first part of Equation (19.13) leads to

$$n^{-1}\sum_{i=1}^{n} E\left[K_h(X_i - x)\middle|X_i \in I_i\right] = \int K_h(t - x)e^{\tilde{\theta}_0+\tilde{\theta}_1(t_1-x_1)+\tilde{\theta}_2(t_2-x_2)}dt.$$

Using the product Gaussian kernel, this yields

$$n^{-1}\sum_{i=1}^{n} E\left[K_h(X_i - x)\middle|X_i \in I_i\right] = e^{\tilde{\theta}_0}\int K_{h_1}(t_1 - x_1)e^{\tilde{\theta}_1(t_1-x_1)}dt_1\int K_{h_2}(t_2 - x_2)e^{\tilde{\theta}_2(t_2-x_2)}dt_2$$

or

$$n^{-1}\sum_{i=1}^{n} E\left[K_h(X_i - x)\middle|X_i \in I_i\right] = e^{\tilde{\theta}_0}e^{\frac{1}{2}(h_1\tilde{\theta}_1)^2}e^{\frac{1}{2}(h_2\tilde{\theta}_2)^2}$$

or

$$n^{-1}\sum_{i=1}^{n} E\left[K_h(X_i - x)\middle|X_i \in I_i\right] = e^{\tilde{\theta}_0}m(h_1\tilde{\theta}_1)m(h_2\tilde{\theta}_2), \tag{19.15}$$

where $m(h_k\tilde{\theta}_k)$ is the moment-generating function of $y_k = t_k - x_k$, with $y_k \sim N(0, h_k^2)$, $k = 1, 2$. This implies that

$$e^{\tilde{\theta}_0} = n^{-1}\sum_{i=1}^{n} E\left[K_h(X_i - x)|X_i \in I_i\right]\left[m(h_1\tilde{\theta}_1)m(h_2\tilde{\theta}_2)\right]^{-1}$$

or

$$\tilde{f}(x) = \hat{f}(x)e^{-\frac{1}{2}[(h_1\tilde{\theta}_1)^2+(h_2\tilde{\theta}_2)^2]}, \tag{19.16}$$

where $\hat{f}(x)$ is obtained as in Equation (19.8). To obtain $\tilde{\theta}_1$ and $\tilde{\theta}_2$ we proceed to solve the second and third parts of Equation (19.13), which eventually yields

$$\tilde{\theta}_1 = \frac{\frac{\partial}{\partial x_1}\hat{f}(x_1, x_2)}{\hat{f}(x_1, x_2)},$$

$$\tilde{\theta}_2 = \frac{\frac{\partial}{\partial x_2}\hat{f}(x_1, x_2)}{\hat{f}(x_1, x_2)}.$$

Therefore, in terms of iterated conditional expectation, the explicit expression for linear adjustments to the kernel density estimate is as follows:

$$\tilde{f}_j = \hat{f}_j(x)\exp\left\{-\frac{1}{2}\sum_{k=1}^{2} h_k^2\left[\frac{\partial}{\partial x_k}\hat{f}_j(x_1, x_2)/\hat{f}_j(x)\right]^2\right\}, \tag{19.17}$$

which is parallel to the result of Hjort and Jones (1996) but with

$$\frac{\partial}{\partial x_k}\hat{f}_j(x_1, x_2) = n^{-1}\sum_{i=1}^{n} E_{\hat{f}_{j-1}}\left[\frac{\partial}{\partial x_k}K_b(X - x)|X \in I_i\right]. \tag{19.18}$$

Let us define

$$\mu^k_{j-1|I}(x) = E_{\hat{f}_{j-1}}\left[\frac{\partial}{\partial x_k}K_b(X - x)|X \in I\right]. \tag{19.19}$$

In the same way as in Section 19.3.1, Equation (19.19) may be approximated by the following expression

$$\hat{\mu}^k_{j-1|I}(x) = \sum_{b=1}^{B}\left[\frac{\partial}{\partial x_k}K_b(X^u_b - x)w^u_b\right], \tag{19.20}$$

where w^u_b and X^u_b are obtained as described in Section 19.3.1. Therefore, Equation (19.18) can be approximated by

$$n^{-1}\sum_{i=1}^{n}\hat{\mu}^k_{j-1|I_i}(x). \tag{19.21}$$

19.4 SURVEY WEIGHTS

The estimates in Sections 19.3.1 and 19.3.2 do not consider the complexities of survey design. The purpose of this section is to take into account some of these complexities by incorporating survey weights into these estimates.

Let w^l_i be the longitudinal weight derived for the ith individual in the survey sample. This weight is broadly interpretable as the number of subjects represented by subject i in the population at the time of recruitment. The survey weights are constructed to compensate for nonresponse, selection bias, stratification and poststratification. Thompson (1997, pp. 160–163) gives a further description of survey weights. An important feature of the longitudinal weights is that they always add up to the size of the population from which the longitudinal sample was selected. This implies that the longitudinal weights change over time, reflecting the changes of the sample due to attrition and other types of longitudinal nonresponse. When doing analysis involving data from two or more cycles of a longitudinal survey, one has to decide which weights to use. Using the weights from the first cycle means dealing with the unit nonresponse in later cycles. On the other hand, using the final weights (i.e. weights from the last cycle) means that the dataset is cleared of dropouts. In the analysis presented hereinafter the final longitudinal weights are used.

Let w^*_i be the normalized weight for the ith individual in the survey sample, rescaled so that $\Sigma_{i\in S} w^*_i = 1$, where S corresponds to the longitudinal sample. By replacement of the population totals in Equation (19.8) by weighted totals, a kernel density estimate that accounts for some of the complexities of the survey design is given by Pantoja-Galicia *et al.* (2005) in the following form

$$\hat{f}_j^w(x)\sum_{i\in S}\hat{\mu}_{j-1|I_i}(x)w_i^*. \tag{19.22}$$

For the estimate of Equation (19.21), we propose the following weighted expression

$$\hat{f}_j^{k,w}(x)\sum_{i\in S}\hat{\mu}_{j-1|I_i}^k(x)w_i^*. \tag{19.23}$$

Consequently, the corresponding weighted estimate for local linear adjustments to the kernel density estimate in Equation (19.17) is given by

$$\tilde{f}_j^w = \tilde{f}_j^w(x)\exp\left\{-\frac{1}{2}\sum_{k=1}^{2}h_k^2\left[\hat{f}_j^{k,w}(x)/\tilde{f}_j^w(x)\right]^2\right\}. \tag{19.24}$$

The test statistic Q in Equation (19.2) may be computed from Equation (19.24) as outlined in Section 19.2.2.

19.4.1 Assessing the Standard Error

To test the null hypothesis in Section 19.2.2, it is necessary to obtain an estimate of the standard error of Q in Equation (19.2). There are several different methods available for estimation of the design-based variance of nonlinear statistics such as Equation (19.2). Good reviews of these methods are given in Wolter (1985), Rust and Rao (1996) and Shao and Tu (1996). The method we are using in this paper is the survey bootstrap as outlined in Rao *et al.* (1992). Statistics Canada produces a large number (500 and more) of bootstrap replicates of the survey weights for most of its national longitudinal surveys. These bootstrap weights allow for calculation of correct design-based variance estimators (Rao and Wu, 1988; Yung, 1997).

Let $w_i^{(b)}$ be the normalized bootstrap weight of the bth replicate for individual i, such that $\Sigma_{i\in S}\, w_i^{(b)} = 1$. If we employ B of these bootstrap weight replicates, the required standard error can be evaluated as follows.

For each set b of replicates:

1. Obtain Equation (19.24) using $w_i^{(b)}$ instead of w_i^* for $i \in S$. Let $\tilde{f}_j^{w(b)}$ be the corresponding estimate.
2. Calculate Equation (19.2) using $\tilde{f}_j^{w(b)}$ and call it Q_b^*.
3. Finally, compute $v^*(Q) = \frac{1}{B-1}\sum_{b=1}^{B}\left(Q_b^* - \overline{Q}^*\right)^2$, where $\overline{Q}^* = B^{-1}\sum_{b=1}^{B} Q_b^*$, and obtain $se(Q) = \sqrt{v^*(Q)}$.

19.5 APPLICATION: THE NATIONAL POPULATION HEALTH SURVEY

Our first application involves time to pregnancy and time to smoking cessation using data from the National Population Health Survey (NPHS). The National Population Health Survey (NPHS) is Statistics Canada's national longitudinal survey on health and health behaviour. Responses have been collected starting with the first cycle in 1994–1995. Subsequent cycles have been collected every second year thereafter and this process is planned to continue until completion of about 10 cycles. The sampling design for the initial cycle selected one individual at random from each of about 17 000 households across the country.

The NPHS employed a stratified two-stage design. Hence, homogeneous strata were created in the first stage and independent samples of clusters were drawn from each stratum. During the second stage, dwelling lists were prepared for each cluster and then households were selected from these lists.

Only one member in each sampled household responds to the in-depth health questions and therefore the probability of being selected as a respondent is inversely related to the number of people in the household. For a detailed description of the design and other important features of this survey, Larry and Catlin (1995) and Swain *et al.* (1999) are useful references.

19.5.1 Pregnancy and Smoking Cessation

The longitudinal nature of the NPHS makes it possible to observe changes in the responses of the surveyed people across cycles. We pay special interest to women's responses to questions related to pregnancy and smoking cessation. Therefore, at every cycle we are able to collect information such as: date of birth and gender of every participant, date of interview, whether a female respondent is pregnant and the smoking status for each respondent (daily or occasional smoker and nonsmoker). Also, each former daily smoker states at what age she stopped smoking cigarettes daily.

Considering the time origin to be the date of the interview at cycle n, let T_1 denote the time until a pregnancy begins and let T_2 denote the time until a smoking cessation occurs that lasts until the next interview at cycle $n + 1$. The way we determine these times, which are interval censored for each subject, is revealed in Section 19.5.3. The unit of measurement for time is years.

19.5.2 Subsample

We need responses at two successive cycles, n and $n + 1$, to determine the intervals for the beginning of the pregnancy and the smoking cessation.

Our analysis is based on the following subsample of longitudinal respondents. For cycles 1, 2 and 3, we select individuals who at the moment of the interview in the nth cycle are between the ages of 15 and 49, are regular daily smokers and are not pregnant. The second inclusion criterion is that, among these subjects, we select those at the next cycle, $n + 1$, who report being pregnant or having given birth since the last interview and report having abandoned cigarettes.

19.5.3 Interval-Censored Times

A household file enabled us to obtain the date of birth for every household member living with the longitudinal respondent. From this, inferring the approximate dates of pregnancy for the longitudinal individual is possible by simply subtracting nine months from the date of birth of the appropriate child. More specifically, if we consider 280 days (or 40 weeks) to be the average duration of a pregnancy, plus or minus 14 days (or 2 weeks) (Taylor *et al.*, 2003, p. 67), the date of becoming pregnant falls within an interval that is 28 days in length. More specifically, let: D_n be the date of the interview at cycle n (time origin); B be the date of birth of the appropriate child; P be the inferred date of pregnancy ($B-280$ days); $P_L=P-14$ days; and $P_R=P+14$ days. Therefore, $T_1 \in I_1=(P_L-D_n, P_R-D_n)$.

Note that the time origin is not at the same date for every respondent (see Sections 19.5.1 and 19.5.2). For some respondents it is D_1, for others it is D_2 and for some others it is D_3.

At every cycle, subjects who became former daily smokers were asked the age at which they quit smoking. The approximate duration between interviews is two years. If smoking cessation is reported, T_2 is thus observed to be between two endpoints T_{20} and T_{21}. T_{20} is either 0, which corresponds to the date of the interview in cycle n (D_n), or the duration of time between the date of the interview in cycle n and the corresponding birthday. On the other side, T_{21} is either the time to a birthday or 2 (obtained using the date of the interview in cycle $n+1$).

By way of an example, and recalling the time units to be in years, let us consider the case of a longitudinal individual who at cycle n is 24 years old and who has her interview in cycle $n+1$ exactly 2 years after the interview in cycle n. Let B_{25} denote the duration of time between the date of her interview at cycle n and the date of her 25th birthday (and equivalently for B_{26}). Depending on the reported age at which she quit smoking, T_2 belongs to the interval I_2, which may be: $(0, B_{25})$ if the reported age of quitting is 24; (B_{25}, B_{26}) if the age of smoking cessation is reported to be 25; or $(B_{26}, 2)$ if the subject responded 26 as her age of abandoning cigarettes.

19.5.4 Results

Recall that to compute the test statistic Q presented in Equation (19.2) we first obtain an estimate, within the observation window, of the joint density of (T_1, T_2) using Equation (19.24). Figure 19.1 depicts the contour plot of such estimated joint density of (T_1, T_2) and shows the expected ordering of T_1 and T_2. Subsequently, we obtain numerically the corresponding marginal probability density functions and consequently the respective survivor functions (conditional and unconditional versions).

The sample satisfying the conditions described in Section 19.5.2 consists of 57 individuals. These 57 respondents represent about 68 000 members of our total target population. For every subject the corresponding intervals I_1 and I_2 have been determined according to the description in Section 19.5.3. To obtain Equation (19.24) and Figure 19.1, the following settings were also employed. We let $g(X)$ in Equation (19.6) be a bivariate uniform density and considered $B=7^2$ to be an appropriate choice for the value B required in Equation (19.7). Therefore at every iteration j that estimates $\hat{f}_j$, we generated 7^2 bivariate uniform random values within each rectangle I_i. This value of B was also employed for the results in Section 19.6.4.

The two modes in Figure 19.1 appear because the dataset being used contains two groups of respondents. Group 1 consists of those who had children and quit smoking but did not relapse between cycle n and $n+1$. Group 2 represents those quitters who have become pregnant and are still pregnant at cycle $n+1$. A feature of the NPHS data means that some of the intervals for these two groups are fixed. The visual representation given by the estimated joint density is a useful tool for seeing the effect of this additional structure.

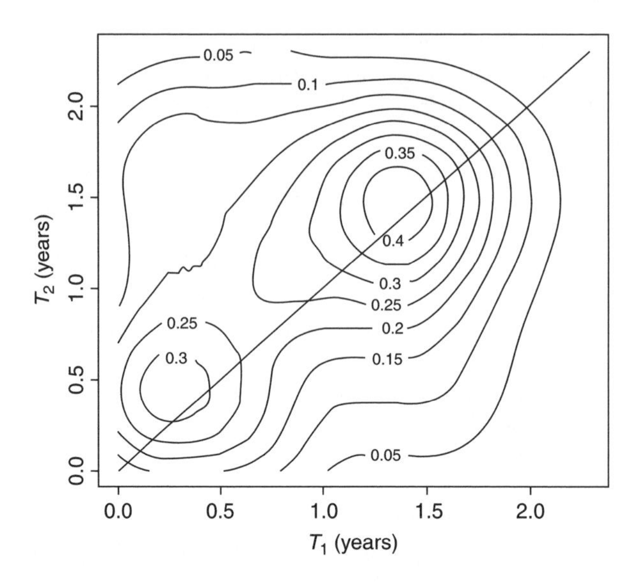

Figure 19.1 Line $T_2 = T_1$ and contour plot of the estimated joint density of T_1 and T_2: time to pregnancy and smoking cessation (NPHS).

Regarding the test for a close precursor as discussed in Section 19.2.2, we let $\kappa(t_1)$ have a constant value of about 2.5 months, based upon a prior assumption that smoking cessation might occur with increased probability within the first trimester of pregnancy. For each t_1, we have $Q = 0.029$ with $se(Q) = 0.004$. After comparing Q with twice its estimated standard error, we reject the null hypothesis, which is effectively that the mean of Q is zero. Therefore, we can argue that there is evidence that T_1 is a close precursor of T_2, i.e. the occurrence of the pregnancy at time T_1 decreases the probability of having to wait longer than 2.5 months for the smoking cessation to occur.

The strong association caused by the two modes in Figure 19.1 results in significance in both directions for our test. If we focus on respondents from group 2 we find that there is evidence to say that T_1 is a close precursor of T_2, but not that T_2 is a close precursor of T_1.

Even though the line 1 : 1 has been included in Figure 19.1, readers should be cautioned that the estimated marginal distributions of T_1 and T_2 do not have the same mean, a fact that the formal test takes into account.

19.6 APPLICATION: THE SURVEY OF LABOUR AND INCOME DYNAMICS

The second application considers time to job loss and time to divorce (or separation) using data from the Survey of Labour and Income Dynamics (SLID). We present an example applying the methodology of Section 19.3 to data of the second panel of the SLID.

The SLID is a longitudinal survey of the labour activity and economic status of Canadians. A longitudinal panel of all persons belonging to the sample of households derived from the Canadian Labour Force Survey is formed and kept for six years. Longitudinal respondents aged 16 years or older are contacted twice a year to answer questions about their labour activities (a January interview) and about their income (a May interview or access with permission to the respondent's tax return).

The longitudinal sample of the SLID is a stratified multistage sample, with strata defined within provinces and generally with two primary sampling units (PSUs) selected from each stratum with probability proportional to size. A sample of households is chosen within each PSU. The first panel was selected in 1993 and originally had about 40 000 persons. The second panel of approximately the same size was selected in 1996, the third in 1999, etc. Here we present analysis of data from the second panel.

Further details regarding design as well as other important issues of the SLID can be found at Statistics Canada (1997, 2005).

19.6.1 Job Loss and Separation or Divorce

According to reports from Marienthal, given in a classic book in unemployment research (Jahoda *et al.*, 1933, p. 86), improvements in the relationship between husband and wife as a result of unemployment are definitely exceptional. As it is a subject of interest for social scientists, the topic of job loss and divorce has been examined more recently by Yeung and Hofferth (1998), Huang (2003) and Charles and Stephens (2004).

Considering the time origin to be the date of the first interview of the respondent (day 0), which takes place at the beginning of the life of the panel (some time in January of 1996), let T_1 denote the time to the termination of the job of the subject and let T_2 be the time to either separation or divorce (whichever comes first, as the result of termination of the marriage or common-law relationship of the individual). The unit of measurement for time is years.

A vector of all the dates of changes of marital status for each respondent, along with associated type of change, can be obtained from a panel of the SLID. In the same manner, a vector of job history with dates of changes in employment status can be retrieved. With this information, an appropriate dataset can be generated for our own subsample, which is described in the next section.

19.6.2 Subsample

We consider subjects from the second panel of the SLID with the following characteristics: employed and married (or in a common-law union) at day 0; with only one marriage (or common-law relationship) and one job during the life of the panel; and with termination of the job and dissolution of the marital relationship occurring during the life of the panel. We further restrict the analysis to cases where the job ended due to involuntary reasons.

19.6.3 Interval-Censored Times

In survey responding, reporting the occurrence of events more recently than they actually happened is known as *forward telescoping*. Such dating error, which might induce measurement error, can also work in the opposite direction (*backward telescoping*) and both have been documented by survey methodologists and cognitive psychologists, starting with Neter and Waksberg (1964). An extensive literature review on these and other types of reporting errors due to inaccurate dating is presented by Tourangeau, Rips and Rasinski (2000), who on page 11 mention that "people may make incorrect inferences about timing based on the accessibility (or other properties) of the memory, incorrectly guess a date within an uncertain range, and round vague temporal information to prototypical values (such as 30 days)". Huttenlocher, Hedges and Bradburn (1990) also show that respondents round estimates to times that are stand-ins for calendar units, that is, 7 days or 30 days.

This background justifies trusting the reported dates of the events to be within a certain period of time instead of a specific day. By way of illustration, let T_J and T_D be the reported times to occurrence of job termination and divorce (or separation) respectively. Therefore, we may say that $T_1 \in I_1 = (T_J - \delta_1, T_J + \delta_2)$ and $T_2 \in I_2 = (T_D - \delta_3, T_D + \delta_4)$, where δ_k is a particular period of time, for $k = 1, \ldots 4$. For instance, if the trusted period of time is 30 days then $\delta_k = 15$ days for all k.

19.6.4 Results

Figure 19.2 shows an estimate of the joint density of (T_1, T_2) obtained according to Equation (19.24). In compliance with the conditions established in Section 19.6.2, 53 individuals were selected. These represent about 40 000 people of the total target population. In accordance with Section 19.6.3, the intervals I_1 and I_2 were determined for each subject.

We choose to let $\kappa(t_1)$ have a constant value of 6.5 months, as this is a censoring half-interval plus one-twelfth of the length of a 6-year panel, thus representing a short term in comparison with the panel length. With $\kappa(t_1) = 6.5$ for every t_1, the test statistic for the

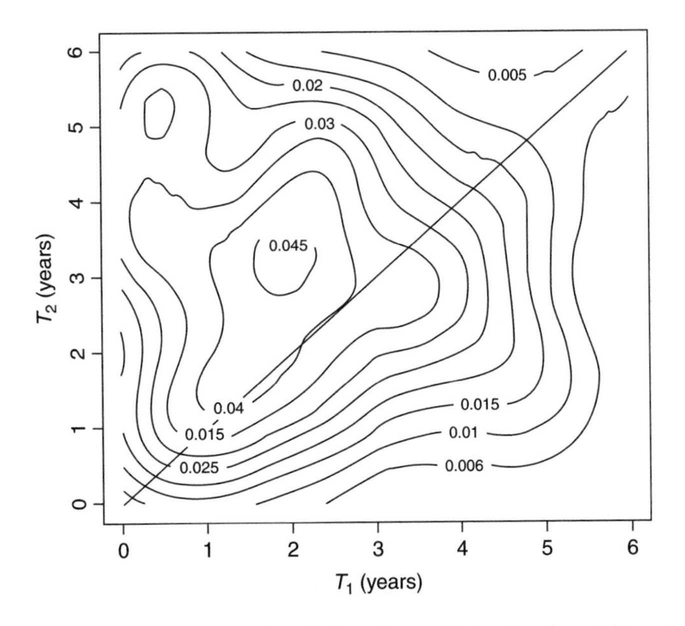

Figure 19.2 Line $T_2 = T_1$ and contour plot of the estimated joint density of T_1 and T_2: time to job loss and separation or divorce (SLID).

close precursor Q from Equation (19.2) results in a value of 0.0304 with a standard error of 0.0008. The comparison of Q with twice its estimated standard error gives evidence to reject the null hypothesis from Section 19.2.2. Therefore, there is evidence that T_1 is a close precursor of T_2, i.e. losing a job at time T_1 decreases the probability of having to wait longer than 6.5 months to observe a separation or divorce.

19.7 DISCUSSION

The tool presented in this paper is proposed as an exploratory technique that would precede other analytical approaches such as survival analysis. A few remarks regarding this method are presented as follows.

As mentioned in Section 19.2.1, the quantity $\kappa(t_1)$, which may be thought of as the duration of an effect, would come from subject-matter considerations. Although Equation (19.1) indicates that it depends on t_1 (anticipating that the effect of T_1 on the hazard of T_2 might not have constant duration), in neither of our applications was there a clear reason not to take κ to be constant.

The estimates from Sections 19.5.4 and 19.6.4 were obtained for the data with the (T_1, T_2) interval censored. We have not used the complete T_1 data for the estimation of F_1. Thus the null hypothesis we are testing is that T_1 and T_2 are independent, given that both are within the observation window. This is a weaker null hypothesis than global independence of T_1 and T_2 and is more appropriate in our context because the data are extracted for joint density estimation conditional on being in the observation window.

It can be noted that Q may be estimated from the usual empirical joint distribution of T_1 and T_2 and its marginals, but it does involve integration of estimated conditional cumulative distribution functions (CDFs) with respect to a marginal measure. Our intuition is that in the interval-censored case the measure and the conditional CDFs are better (because more smoothly) represented if the joint CDF is obtained by integration from an estimated joint density. The estimated joint density also has an easily interpreted visual representation. We can also observe whether the plot is consistent with the inference, and whether it is consistent with the idea of triggering.

In Section 19.2.2 it has been pointed out that the difference within Equation (19.2) approximates to the difference between the hazard function conditional on $T_1 = t_1$ and the unconditional hazard (i.e. a local additive change). However, if it is considered that a scale increase in hazard function is more appropriate, an expected ratio statistic could be used.

In another approach, estimating the CDF of $T_2 - T_1$ and comparing it with what is expected under independence could be used to test various kinds of temporal order. (When interval censoring is not present, this approach does not require estimating the joint distribution of T_1 and T_2, but only the CDFs of T_1, T_2 and $T_2 - T_1$.)

The results presented in Section 19.5.4 do not suggest that pregnancy makes people quit smoking more quickly. They indicate that according to the data there exists evidence to argue that the occurrence of the pregnancy at time T_1 decreases the probability of having to wait longer than 2.5 months to observe the occurrence of the smoking cessation. Correspondingly, an analogous situation applies for the results from Section 19.6.4.

We have focused here on the problem of testing for a close precursor nonparametrically. However, for some related topics using semiparametric approaches some useful references are Jewell *et al.* (2005), Cook *et al.* (2008) and Pantoja-Galicia (2007).

The problem presented here might alternatively be viewed in terms of estimation, rather than hypothesis testing. This alternative would involve writing $Q(\kappa)$ instead of just Q. In this manner, an estimate of the function $Q(\kappa)$ for a range of values of κ, with error bands, would provide more information than a test of the null hypothesis that $Q(\kappa^*)=0$ for some fixed κ^*.

It can be noted that the selection criteria from Section 19.5.2 make it possible to select a respondent more than once over the four cycles. In fact, however, we did not have respondents with more than one contribution to the data.

The window width h used for the nonparametric density estimation that resulted in Figures 19.1 and 19.2 was obtained using a subjective analysis. A formal procedure for an optimal choice of the smoothing parameter h needs further research in the context discussed here. A possible approach would generalize to the bivariate setting (together with considering the interval-censored nature of the data) the method proposed by Breunig (2001), who, in the univariate context of nonparametric density estimation from clustered sample survey data, examined an optimal bandwidth selection using a higher order kernel. Other possibilities for the choice of this window size are to generalize to the bivariate case the methodology of likelihood cross-validation presented by Braun *et al.* (2005), or to use a least-squares cross-validation-based criterion.

The outcome in Section 19.6.4 was obtained using data from the second panel of the SLID. Even though Pantoja-Galicia *et al.* (2005) use data from the first panel of this survey and a different estimation scheme, it is interesting to note that the patterns of the results are qualitatively the same.

In this paper, we do not address the application of the proposed approaches to causal inference. Nevertheless, the topic studied here has significant potential for future research on assessing causal hypotheses from large longitudinal observational studies, such as national longitudinal surveys. This paper offers an impetus for future research on methods that will allow detection and interpretation of the relationships beyond the usual association.

ACKNOWLEDGEMENTS

We are grateful to Statistics Canada, the Mathematics of Information Technology and Complex Systems (MITACS), the National Program on Complex Data Structures (NPCDS) and the National Council on Science and Technology (CONACYT) for financial support. We would like to thank James Stafford for helpful discussion and for providing his paper with John Braun and Thierry Duchesne at the manuscript stage, as well as for allowing us to look at some of their computing programs. We appreciate the comments and suggestions from the reviewers and editor.

REFERENCES

Bellhouse, D., Goia, C. and Stafford, J. (2003). Graphical displays of complex survey data through kernel smoothing. In R. Chambers and C. J. Skinner (Eds), *Analysis of Survey Data* (pp. 133–150). New York: John Wiley & Sons, Inc.

Bellhouse, D. R. and Stafford, J. E. (1999). Density estimation from complex surveys. *Statistica Sinica*, 9, 407–424.

Betensky, R. and Finkelstein, D. (1999). A non-parametric maximum likelihood estimator for bivariate interval censored data. *Statistics in Medicine*, 18, 3089–3100.

Blossfeld, H. and Mills, M. (2003). A causal approach to interrelated family events. *Proceedings of Statistics Canada Symposium 2002: Modelling Survey Data for Social and Economic Research*. Ottawa: Statistics Canada.

Braun, J., Duchesne, T. and Stafford, J. (2005). Local likelihood density estimation for interval censored data. *Canadian Journal of Statistics*, 33, 39–60.

Breunig, R. V. (2001). Density estimation for clustered data. *Econometric Reviews*, 20, 353–367.

Buskirk, T. and Lohr, S. (2005). Asymptotic properties of kernel density estimation with complex survey data. *Journal of Statistical Planning and Inference*, 128, 165–190.

Charles, K. K. and Stephens, M. J. (2004). Job displacement, disability and divorce. *Journal of Labour Economics*, 22, 489–522.

Cook, R. J., Zeng, L. and Lee, K. (2008). A multistate model for bivariate interval-censored failure time data. *Biometrics*, 64, 1100–1109.

Duchesne, T. and Stafford, J. (2001). A kernel density estimate for interval censored data. *Technical Report No. 0106*, University of Toronto.

Gentleman, R. and Geyer, C. (1994). Maximum likelihood for interval censored data: consistency and computation. *Biometrika*, 81, 618–623.

Hjort, N. L. and Jones, M. C. (1996). Locally parametric nonparametric density estimation. *Annals of Statistics*, 24, 1619–1647.

Huang, J. (2003). Unemployment and family behavior in Taiwan. *Journal of Family and Economic Issues*, 24, 27–48.

Huttenlocher, J., Hedges, L. and Bradburn, N. M. (1990). Reports of elapsed time: Bounding and rounding processes in estimation. *Journal of Experimental Psychology: Learning, Memory and Cognition*, 16, 196–213.

Jahoda, M., Lazarsfeld, P. F. and Zeisel, H. (1933). *Marienthal. The Sociography of an Unemployed Community*. Chicago: Aldine Atherton Inc.

Jewell, N., van der Laan, M. and Lei, X. (2005). Bivariate current status data with univariate monitoring times. *Biometrika*, 92, 847–862.

Larry, J.-L. T. and Catlin, G. (1995). Sampling design of the National Population Health Survey. In *Health Reports* (vol. 7, pp. 29–38). Ottawa: Statistics Canada.

Lawless, J. (2003). Event history analysis and longitudinal surveys. In R. Chambers and C. J. Skinner (Eds), *Analysis of Survey Data* (pp. 221–243). New York: John Wiley & Sons, Inc.

Li, L., Watkins, T. and Yu, Q. (1997). An EM algorithm for smoothing the self-consistent estimator of survival functions with interval-censored data. *Scandinavian Journal of Statistics*, 24, 531–542.

Loader, C. R. (1996). Local likelihood density estimation. *Annals of Statistics*, 24, 1602–1618.

Neter, J. and Waksberg, J. (1964). A study of response errors in expenditures data from household interviews. *Journal of the American Statistical Association*, 59, 17–55.

Pantoja-Galicia, N. (2007). Interval censoring and longitudinal survey data. *PhD Thesis*, Department of Statistics and Actuarial Science, University of Waterloo.

Pantoja-Galicia, N., Thompson, M. E. and Kovacevic, M. (2005). A bivariate density estimation approach using data from the survey of labour and income dynamics: an example of testing for temporal order of job loss and divorce. *Proceedings of the Survey Methods Section, Statistical Society of Canada* Annual Meeting, June 2005, at http://www.ssc.ca/survey/documents/SSC2005_N_Pantoja-Galicia.pdf

Rao, J. and Wu, C. (1988). Resampling inference with complex survey data. *Journal of the American Statistical Association*, 83, 231–241.

Rao, J., Wu, C. and Yue, K. (1992). Some recent work on resampling methods for complex survey data. *Survey Methodology*, 18, 209–217.

Rust, K. and Rao, J. (1996). Variance estimation for complex surveys using replication techniques. *Statistical Methods in Medical Research*, 5, 283–310.

Scott, D. W. (1992). *Multivariate Density Estimation: Theory, Practice, and Visualization*. New York: John Wiley & Sons, Inc.

Shao, J. and Tu, D. (1996). *The Jackknife and Bootstrap*. New York: Springer-Verlag.

Silverman, B. W. (1986). *Density Estimation for Statistics and Data Analysis*. London: Chapman & Hall.

Statistics Canada (1997). *Survey of Labour and Income Dynamics Microdata User's Guide*. www.statcan.ca/english/freepub/75M0001GIE/free.htm.

Statistics Canada (2005). *Survey of Labour and Income Dynamics (SLID) – A Survey Overview*. www.statcan.ca/english/freepub/75F0011XIE/free.htm.

Swain, L., Catlin, G. and Beaudet, M. P. (1999). *The National Population Health Survey–its longitudinal nature*. In *Health Reports* (vol. 10, pp. 69–82). Ottawa: Statistics Canada.

Tang, B. (1993). Orthogonal array-based latin hypercubes. *Journal of the American Statistical Association*, 88, 1392–1397.

Taylor, R. B., David, A. K. and Fields, S. A. (2003). *Fundamentals of Family Medicine: The Family Medicine Clerkship Textbook* (3rd edn). New York: Springer.

Thompson, M. E. (1997). *Theory of Sample Surveys*. London: Chapman & Hall.

Thompson, M. E. and Pantoja-Galicia, N. (2003). Interval censoring of smoking cessation in the National Population Health Survey. *Proceedings of Statistics Canada Symposium 2002: Modelling Survey Data for Social and Economic Research*. Ottawa: Statistics Canada.

Tourangeau, R., Rips, L. J. and Rasinski, K. (2000). *The Psychology of Survey Response*. New York: Cambridge University Press.

Turnbull, B. W. (1976). The empirical distribution function with arbitrarily grouped, censored and truncated data. *Journal of the Royal Statistical Society Series B*, 38, 290–295.

Wand, M. and Jones, M. (1995). *Kernel Smoothing*. London: Chapman & Hall.

Wolter, K. (1985). *Introduction to Variance Estimation*. New York: Springer-Verlag.

Wong, G. and Yu, Q. (1999). Generalized mle of a joint distribution function with multivariate interval-censored data. *Journal of Multivariate Analysis*, 69, 155–166.

Yeung, W. J. and Hofferth, S. L. (1998). Family adaptations to income and job loss in the U.S. *Journal of Family and Economic Issues*, 19, 255–283.

Yung, W. (1997). Variance estimation for public use files under confidentiality constraints. In *Proceedings of the American Statistical Association, Section on Survey Research Methods* (pp. 434–439). Washington, DC: American Statistical Association.

CHAPTER 20

Using Marginal Mean Models for Data from Longitudinal Surveys with a Complex Design: Some Advances in Methods

Georgia Roberts

Statistics Canada

Qunshu Ren

Statistics Canada and Carleton University

J. N. K. Rao

Carleton University

20.1 INTRODUCTION

A common feature of most longitudinal studies is that there are repeated observations of the same variables from the same individuals at different points in time. Data from these studies are used for a variety of purposes, including gross flows estimation, event history modelling, conditional modelling of the response at a given time point as a function of past responses and present and past covariables, and modelling of marginal means of responses as functions of covariates. In this chapter, we will focus on marginal logistic modelling of binary response data from a longitudinal survey with a complex sampling design.

Because the data from longitudinal studies arise from repeated observation of the same individuals, the within-subject correlation among repeated measures must be accounted for during analysis; otherwise, inferences on model parameters could be erroneous because of underestimation of standard errors of parameter estimators and P values of tests. One analytical technique that accounts for within-subject dependence uses marginal mean modelling. Marginal models for binary response longitudinal

Methodology of Longitudinal Surveys P. Lynn

data, for example, are a natural extension of logistic regression models for cross-sectional data. The term 'marginal' indicates that the model for the mean response of a subject at a given time point depends only on the associated covariates and not on any subject-specific random effects or previous responses. In the marginal model approach, inferences on the parameters of the marginal model, taking account of within-subject dependence among the repeated responses, are of primary interest. Assumptions about the full joint distribution of repeated observations are not needed and only the model for the marginal mean response needs to be correctly specified.

The case of marginal modelling with a simple random sample, where individuals are considered to be independent and to have equal chances of being selected, has been studied extensively in the literature, especially for applications in biomedical and health sciences (see Diggle *et al.*, 2002; Fitzmaurice *et al.*, 2004). Liang and Zeger (1986) used generalized estimating equations (GEE) to estimate the model parameters, assuming a 'working' correlation structure for the repeated measurements. They also obtained standard errors of parameter estimators, based on 'sandwich' variance estimators that are 'robust' in the sense of validity even if the working correlation is not equal to the true correlation structure. For the binary response case, Lipsitz, Laird and Harrington (1991) used odds ratios to model the working covariance structure in GEE, instead of working correlations used by Liang and Zeger (1986) and others. An odds ratio is a natural measure for capturing association between binary repeated outcomes from the same individual. Moreover, for binary responses the odds ratio approach is less seriously constrained than the correlation approach (Liang *et al.*, 1992).

When data are obtained from a longitudinal survey with a complex sampling design that involves clustering of subjects, there may be cross-sectional correlations among subjects in addition to within-individual dependencies. Because of this additional complexity, new methods are needed to replace the methods used for longitudinal data where subjects are assumed to be independent. Methods that account for within-individual dependence but ignore clustering of subjects and other design features could also lead to erroneous inferences from underestimation of standard errors of parameter estimators and P values of tests. In Section 20.2, we adapt the GEE method for binary responses and marginal logistic regression models to the case of longitudinal survey data and estimate the model parameters as the solution to survey-weighted GEE. We use the odds ratio approach to model the working covariance structure. For the case of complex survey data, where the nonindependence among individuals must be accounted for, variance estimation techniques must be modified. Rao (1998) explained how to obtain an appropriate robust sandwich-type variance estimator for the case of complex survey data, but in practice this can be difficult to carry out with Taylor linearization if you wish to account for nonresponse adjustment and poststratification. As an alternative, a design-based bootstrap method becomes a good choice because it can account for clustering and other survey design features as well as nonresponse adjustment and poststratification in a straightforward manner. At Statistics Canada, design information for variance estimation is released only in the form of bootstrap survey weights for many of its longitudinal surveys based on stratified multistage cluster sampling designs, where the first-stage clusters (or primary sampling units) are selected with probability proportional to size measures attached to the clusters. Bootstrap design weights are first obtained by the method of Rao and

Wu (1988) and then adjusted for unit nonresponse and poststratification in the same manner as the full sample design weights to arrive at the bootstrap survey weights. Rust and Rao (1996) provide an overview of the bootstrap and other resampling methods in the context of stratified multistage cluster sampling. In Section 20.3 we extend the estimating function (EF)–bootstrap method of Hu and Kalbfleisch (2000) to the case of complex longitudinal survey data and obtain EF–bootstrap standard errors that account for within-subject dependence of repeated measurements as well as clustering of subjects and other survey design features. Advantages of the EF–bootstrap over the direct bootstrap are discussed in Section 20.3. Rao and Tausi (2004) similarly developed an EF–jackknife method for inference from cross-sectional survey data.

The GEE approach assumes that the marginal means are correctly specified. It is therefore important to check the adequacy of the mean specification, using goodness-of-fit tests and other diagnostics, before making inferences on model parameters. Several global goodness-of-fit tests have been proposed in the literature for the binary response case and a logistic model, assuming simple random sampling of subjects. In particular, for the case of no repeated measurements on a sample individual, Hosmer and Lemeshow (1980) proposed a Pearson chi-squared statistic after partitioning the subjects into groups based on the values of the estimated response probabilities under the assumed mean specification (null hypothesis). Horton *et al.* (1999) studied the case of longitudinal data under simple random sampling and proposed a score test after partitioning all the response probabilities under the null hypothesis into groups, assuming working independence for the repeated measurements. As noted by Horton *et al.* (1999), since global (or omnibus) tests may not have high power to detect specific alternatives one should not regard a nonsignificant goodness-of-fit test as clear evidence that the assumed marginal mean specification gives a good fit. Data analysts should look for lack of fit in some other ways as well, guided by subject-matter knowledge. The above goodness-of-fit tests are generally not applicable to complex survey data because of cross-sectional dependencies among subjects caused by clustering. For the case of no repeated measurements, Graubard *et al.* (1997) used the survey weights and the associated estimated response probabilities to partition the subjects into groups, and then developed goodness-of-fit Wald tests based on the weighted sum of responses and the weighted sum of estimated response probabilities in the groups, taking account of the survey design. In Section 20.4, we extend the Horton *et al.* (1999) score test to the case of complex longitudinal survey data and show that it is equivalent to an extension of the Graubard *et al.* (1997) Wald test to the longitudinal case. We also develop suitable corrections to a weighted version of the Hosmer and Lemeshow goodness-of-fit statistic that account for the survey design, following Rao and Scott (1981) and Roberts *et al.* (1987). We use design-based bootstrap methods in developing the above goodness-of-fit tests.

Finally, in Section 20.5, we illustrate the proposed method for inference on model parameters and global goodness-of-fit of the marginal mean model, using longitudinal data from Statistics Canada's National Population Health Survey (NPHS).

We should emphasize that our major contribution in this chapter is to apply design-based bootstrap methods for making design-based inferences under marginal mean models and for testing the goodness-of-fit of marginal mean specification when the data are longitudinal and obtained from a complex survey design.

20.2 SURVEY-WEIGHTED GEE AND ODDS RATIO APPROACH

Suppose that a sample, s, of size n is selected by a complex survey design from a population U of size N, and that the same sampled units are observed for T occasions, assuming no wave nonresponse. Let the data have the form $[(y_{it}, x_{it}), i \in s, t = 1, \ldots T]$, where y_{it} is the response of the ith individual on occasion t and x_{it} is a $p \times 1$ vector of fixed covariates associated with y_{it}. In the case of a binary response variable (i.e. $y_{it} = 0$ or 1), the marginal logistic regression model is a natural choice for describing the relationship between y_{it} and x_{it}. For the marginal logistic regression model, the marginal mean of y_{it}, given the covariates x_{it}, is given by $E(y_{it}) = P(y_{it} = 1) = p_{it}$, where

$$\text{logit}(p_{it}) = \log\left[p_{it}/(1 - p_{it})\right] = \beta_0 + x'_{it}\beta_1. \tag{20.1}$$

Assuming independence between sample individuals (simple random sampling with negligible sampling fraction n/N), an estimator of the vector of model parameters $\beta = (\beta_0, \beta_1')'$ is obtained as the solution of the generalized estimating equations (GEE):

$$\hat{u}(\beta) = \sum_{i \in s} D_i' V_i^{-1}\left[y_i - p_i(\beta)\right] = 0, \tag{20.2}$$

where $y_i = (y_{i1}, y_{i2}, \ldots y_{iT})'$, $p_i(\beta) = (p_{i1}, p_{i2}, \ldots p_{iT})'$, $D_i = \partial p_i(\beta)/\partial\beta$, V_i is the 'working' covariance matrix of y_i and $\hat{u}(\beta)$ is a $(p + 1) \times 1$ vector (Liang and Zeger, 1986). It should be kept in mind that, while V_i may differ from the true covariance matrix of y_i, we assume that the mean of y_i is correctly specified, i.e. $E(y_i) = p_i(\beta)$. Note that V_i may be written as $V_i = A_i^{\frac{1}{2}} R_i A_i^{\frac{1}{2}}$, where A_i is a diagonal $T \times T$ matrix with diagonal elements $\text{Var}(y_{it}) = p_{it}(1 - p_{it})$ and R_i is a $T \times T$ working correlation matrix. Under a working independence assumption, R_i is the $T \times T$ identity matrix and Equation (20.2) then reduces to

$$\hat{u}_0(\beta) = \sum_{i \in s}\sum_{t=1}^{T}(y_{it} - p_{it}) = 0, \quad \hat{u}_1(\beta) = \sum_{i \in s}\sum_{t=1}^{T} x_{it}(y_{it} - p_{it}) = 0, \tag{20.3}$$

corresponding to β_0 and β_1, respectively.

In the case of a complex survey design, denote the survey weights by $\{w_i, i \in s\}$. The survey weights $\{w_i, i \in s\}$ may represent design weights $\{d_i, i \in s\}$ adjusted for unit non response and poststratification. We assume no wave nonresponse here. Rao (1998) proposed the following survey-weighted estimating equations (SEEI) for estimating β:

$$\hat{u}_{1w}(\beta) = \sum_{i \in s} w_i D_i' V_i^{-1}\left[y_i - p_i(\beta)\right] = 0. \tag{20.4}$$

Denote the solution of Equation (20.4) as $\hat{\beta}_w$ for a specified V_i. Under working independence, the SEEI are the weighted version of Equation (20.3) obtained by changing $\Sigma_{i \in s}$ to $\Sigma_{i \in s} w_i$. Note that $\hat{\beta}_w$ is a survey-weighted estimator of the census parameter β_N, which is the solution of census estimating equations $u_N(\beta) = \sum_{i \in U} D_i' V_i^{-1}[y_i - p_i(\beta)] = 0$. The census parameter β_N would be a consistent estimator of β if the population, U, of individuals is regarded as a self-weighting sample from a superpopulation obeying the marginal model. The survey-weighted estimator $\hat{\beta}_w$ is consistent for β_N (and hence for β) if $\hat{u}_{1w}(\beta)$ is design-unbiased or consistent for $u_N(\beta)$. We assume that $\sqrt{n}(\beta_N - \beta)$ is small relative to $\sqrt{n}(\hat{\beta}_w - \beta_N)$ so that $\sqrt{n}(\hat{\beta}_w - \beta) \approx \sqrt{n}(\hat{\beta}_w - \beta_N)$, and thus inference on β_N may be regarded as inference on β as well (Skinner *et al.*, 1989, p. 14). Note that the census parameter under working independence is different from the census parameter under a working covariance assumption, but the two parameters should be close if the mean of the superpopulation marginal model is correctly specified.

In the case of a marginal model with binary responses, Lipsitz *et al.* (1991) used the odds ratio as a measure of association between pairs of binary responses of a subject. The major reason for this choice is that the odds ratio is not constrained by the marginal means of the two binary variables, unlike the correlation. Also, we can use a working model for the odds ratios to specify V_i. If we let $y_{ist} = y_{is}y_{it}$ for all $s = 1, \ldots T-1$, $t = s+1, \ldots T$, and $p_{ist} = E(y_{ist}) = \Pr(y_{is} = 1,\ y_{it} = 1)$, then for given $s \neq t$ the odds ratio γ_{ist} is defined as:

$$\gamma_{ist} = \frac{P(y_{is} = 1, y_{it} = 1)P(y_{is} = 0, y_{it} = 0)}{P(y_{is} = 1, y_{it} = 0)P(y_{is} = 0, y_{it} = 1)} = \frac{p_{ist}(1 - p_{is} - p_{it} + p_{ist})}{(p_{is} - p_{ist})(p_{it} - p_{ist})}. \tag{20.5}$$

It follows from Equation (20.5) that p_{ist} can be expressed as $p_{ist} = g(\gamma_{ist}, p_{it}, p_{st})$.

Suppose that the odds ratio γ_{ist} is modelled as a function of suitable covariates (e.g. log odds ratio is a linear function of covariates), and that α is the vector of parameters in that model, i.e. $\gamma_{ist} = \gamma_{ist}(\alpha)$. Then the elements of the working covariance matrix V_i can be written as

$$\begin{aligned} \mathrm{Var}(y_{it}) &= V_{itt} = p_{it}(\beta)\left[1 - p_{it}(\beta)\right] \\ \mathrm{Cov}(y_{is}, y_{it}) &= V_{ist}(\beta, \alpha) = p_{ist}(\beta, \alpha) - p_{is}(\beta)p_{it}(\beta). \end{aligned} \tag{20.6}$$

Note that γ_{ist} is a function of α. Since α and β are both unknown, we need a second set of survey-weighted estimating equations (SEEII). If we let $u_i = (y_{i12}, \ldots y_{i(T-1)T})'$ and $E(u_i) = \theta_i(\beta,\alpha) = [p_{i12}(\beta,\alpha), p_{i13}(\beta,\alpha), \ldots p_{i(T-1)T}(\beta,\alpha)]'$, then SEEII are given by:

$$\hat{u}_{2w}(\beta, \alpha) = \sum_{i \in s} w_i C'_i F_i^{-1}\left[u_i - \theta_i(\beta, \alpha)\right] = 0, \tag{20.7}$$

where $C_i = \partial\theta_i'/\partial\alpha$ and $F_i = \mathrm{diag}\ [p_{ist}(1 - p_{ist})]$. A Newton–Raphson-type iterative method may be used to solve Equations (20.4) and (20.7) simultaneously using some initial values $\hat{\alpha}_{(0)}, \hat{\beta}_{(0)}$. Iterations are given by

$$\hat{\beta}_{(m+1)} = \hat{\beta}_{(m)} - \left(\sum_{i \in s} w_i D'_{i(m)} V_{i(m)}^{-1} D_{i(m)}\right)^{-1} \left\{\sum_{i \in s} w_i D'_{i(m)} V_{i(m)}^{-1} \left[y_i - p_i(\hat{\beta}_{(m)})\right]\right\} \tag{20.8}$$

and

$$\hat{\alpha}_{(m+1)} = \hat{\alpha}_{(m)} - \left(\sum_{i \in s} w_i C'_{i(m)} F_{i(m)}^{-1} C_{i(m)}\right)^{-1} \left\{\sum_{i \in s} w_i C'_{i(m)} F_{i(m)}^{-1} \left[u_i - \theta_i(\hat{\alpha}_{(m)}, \hat{\beta}_{(m+1)})\right]\right\}, \tag{20.9}$$

where the subscripts (m) and $(m+1)$ indicate that quantities are evaluated at $\beta=\hat{\beta}_{(m)}$ and $\alpha=\hat{\alpha}_{(m)}$ in Equation (20.8) and at $\beta=\hat{\beta}_{(m+1)}$ and $\alpha=\hat{\alpha}_{(m)}$ in Equation (20.9). At convergence of the iterations, we obtain $\hat{\beta}_w$ and $\hat{\alpha}_w$. The estimator $\hat{\beta}_w$ is a consistent estimator of β even under mis-specification of the means of the u_i. We assume that $\hat{\alpha}_w$ converges in probability to some α^* that agrees with α only when the working model $\gamma_{ist} = \gamma_{ist}(\alpha)$ is correctly specified. Note that $\hat{\beta}_w$ can be obtained from Equation (20.4) alone if γ_{ist} is estimated empirically: for example, assuming that the odds ratio is a constant within an age group (see Section 20.5).

20.3 VARIANCE ESTIMATION: ONE-STEP EF–BOOTSTRAP

In order to make inferences from an estimated marginal model, we need standard errors of estimated model parameters that take account of clustering of sample subjects as well as within-individual dependencies. We use the design-based bootstrap method of Rao and Wu (1988) and Rao, Wu and Yue (1992) for stratified multistage cluster sampling. Using this method, we first obtain B sets of bootstrap design weights $\{d_i^{(b)}, i \in s\}$ for $b = 1, \ldots B$ from the full-sample design weights $\{d_i, i \in s\}$ by resampling the sample primary clusters and then adjusting the weights $d_i^{(b)}$ for unit nonresponse and poststratification in the same manner as the adjustments to the full-sample design weights d_i, to get B sets of bootstrap survey weights $\{w_i^{(b)}, i \in s\}$ for $b = 1, \ldots B$. Typically, at least $B = 500$ sets of bootstrap survey weights are generated for complex longitudinal surveys at Statistics Canada.

The direct bootstrap method for variance estimation involves obtaining point estimates of model parameters with the full-sample survey weights, w_i, and then, in an identical fashion, with each set of bootstrap survey weights $w_i^{(b)}$. This method, consisting of many repetitive operations, can be computationally intensive and time consuming. Furthermore, Binder, Kovacevic and Roberts (2004) found that, when using this approach for logistic regression, it was possible to have several sets of bootstrap weights for which the parameter estimation algorithm would not converge due to ill-conditioned matrices that were not invertible. To overcome these problems, Binder *et al.* (2004) and Rao and Tausi (2004) proposed estimating function (EF) bootstrap approaches, motivated by the work of Hu and Kalbfleish (2000) for the nonsurvey case. Here, we extend the one-step EF–bootstrap approach of Rao and Tausi (2004) to the marginal logistic regression model.

Let $\{w_i^{(b)}, i \in s\}$, for $b = 1, \ldots B$ be B sets of bootstrap weights for the sample s. Let

$$\hat{u}_{1w}^{(b)}(\hat{\beta}) = \sum_{i \in s} w_i^{(b)} D_i' V_i^{-1} [y_i - p_i(\hat{\beta})] \tag{20.10}$$

and

$$\hat{u}_{2w}^{(b)}(\hat{\alpha}, \hat{\beta}) = \sum_{i \in s} w_i^{(b)} C_i' F_i^{-1} [u_i - \theta_i(\hat{\beta}, \hat{\alpha})] \tag{20.11}$$

be the bootstrap estimating functions corresponding to Equations (20.4) and (20.7) and evaluated at $\beta = \hat{\beta}$ and $\alpha = \hat{\alpha}$, where $\hat{\beta}$ and $\hat{\alpha}$ are obtained from Equations (20.8) and (20.9). Now compute one-step Newton–Raphson solutions to the following EF equations using $\hat{\beta}$ and $\hat{\alpha}$ as starting values:

$$\hat{u}_{1w}(\beta) = \hat{u}_{1w}^{(b)}(\hat{\beta}), \tag{20.12}$$

$$\hat{u}_{2w}(\alpha, \beta) = \hat{u}_{2w}^{(b)}(\hat{\alpha}, \hat{\beta}). \tag{20.13}$$

This is equivalent to Taylor linearization of the left-hand sides of Equations (20.12) and (20.13). The one-step bootstrap estimators from the bootstrap sample b are then given by:

$$\tilde{\beta}^{(b)} = \hat{\beta} - \left(\sum_{i \in s} w_i \hat{D}'_i \hat{V}_i^{-1} \hat{D}_i \right)^{-1} \hat{u}_{1w}^{(b)}(\hat{\beta}) \tag{20.14}$$

and

$$\tilde{\alpha}^{(b)} = \hat{\alpha} - \left(\sum_{i \in s} w_i \hat{C}'_i \hat{F}_i^{-1} \hat{C}_i \right)^{-1} \hat{u}_{2w}^{(b)}(\hat{\alpha}, \hat{\beta}), \tag{20.15}$$

where the matrices $\hat{D}_i, \hat{V}_i, \hat{C}_i$ and $\hat{F}_i$ in Equations (20.14) and (20.15) are obtained by evaluating D_i, V_i, C_i and F_i at $\hat{\beta}$ and $\hat{\alpha}$. Note that for all $b = 1, \ldots B$, the inverse matrices in Equations (20.14) and (20.15) are based on the full sample and hence remain the same, so that further inversion for each bootstrap sample is not needed. The EF–bootstrap estimator of $\text{cov}(\hat{\beta})$ is given by

$$v_{BOOT}^{EF}(\hat{\beta}) = \frac{1}{B} \sum_{b=1}^{B} (\tilde{\beta}^{(b)} - \hat{\beta})(\tilde{\beta}^{(b)} - \hat{\beta})'. \tag{20.16}$$

Diagonal elements of Equation (20.16) give the variance estimators, while the covariance estimators are obtained from the off-diagonal elements. Standard errors of estimated model parameters obtained from Equation (20.16) account for clustering and other design features as well as within-subject dependencies due to repeated measurements. The estimator of Equation (20.16) is algebraically identical to the bootstrap estimator proposed by Binder *et al.* (2004), which is obtained by equating $\hat{u}_{1w}(\beta)$ and $\hat{u}_{2w}(\alpha, \beta)$ to $-\hat{u}_{1w}^{(b)}(\hat{\beta})$ and $-\hat{u}_{2w}^{(b)}(\hat{\alpha}, \hat{\beta})$, respectively, and then using a one-step Newton–Raphson iteration.

It may be noted that the above EF–bootstrap method of variance estimation as well as the direct bootstrap method are generally applicable for stratified multistage cluster samples, provided that the overall first-stage sampling fraction is small. The methods are applicable to both cross-sectional and longitudinal data obtained from complex multistage cluster sampling and they can handle binary responses as well as nominal or ordinal or continuous responses under suitable marginal models.

20.4 GOODNESS-OF-FIT TESTS

Inferences on regression parameter β, outlined in Sections 20.2 and 20.3, assume that the mean of y_{it} is correctly specified as $E(y_{it} \mid x_{it}) = p_{it}(\beta)$, with $logit\ [p_{it}(\beta)] = \beta_0 + x'_{it}\beta_1$.

It is therefore important to check the adequacy of the mean specification using goodness-of-fit tests, before making inferences on β.

All of the goodness-of-fit tests that we consider require our observations to be grouped. We describe how this may be done in Section 20.4.1. In Section 20.4.2 we extend the Horton *et al.* (1999) score test to the case of complex longitudinal survey data and show that it is equivalent to an extension of the Graubard *et al.* (1997) Wald test for cross-sectional survey data. Finally, in Section 20.4.3 we develop suitable corrections to a weighted version of the Hosmer and Lemeshow (1980) goodness-of-fit statistic that account for the survey design.

20.4.1 Construction of Groups

In the case of cross-sectional data (y_i, x_i; $i \in s$), Graubard *et al.* (1997) obtained weighted decile groups $G_1, G_2, \ldots G_{10}$: n_1 subjects with the smallest estimated probabilities $\hat{\pi}_i = p_i(\hat{\beta})$ under the null hypothesis $H_0: p_i = p_i(\beta)$ are in the first group G_1, the next n_2 in the second group G_2, and so forth, until the group G_{10} with n_{10} observations having the largest estimated probabilities is formed. The n_l observations in G_l are chosen such that $\Sigma_{i \in G_l} w_i / \Sigma_{i \in s}\, w_i \approx 1/10$; $l = 1, 2, \ldots 10$. In the special case of equal weights, $w_i = w$, this grouping method reduces to the Hosmer and Lemeshow (1980) method of grouping the subjects.

We now extend the above weighted decile grouping method to the case of complex longitudinal survey data, following Horton *et al.* (1999). We make use of all nT estimated probabilities $\hat{\pi}_{it} = p_{it}(\hat{\beta})$, under the null hypothesis $H_0: p_{it} = p_{it}(\beta)$ with $logit[p_{it}(\beta)] = \beta_0 + x'_{it}\beta$ and working independence, to form the weighted decile groups $G_1, G_2, \ldots G_{10}$. The weights w_i associated with $\hat{\pi}_{it}$ are used in forming the groups. Note that a subject's group membership can change for different time points t because $\hat{\pi}_{it}$ for a subject i can change with time t. In the special case of equal weights $w_i = w$, this grouping method reduces to the Horton *et al.* (1999) method of grouping for the longitudinal case.

20.4.2 Quasi-Score Test

Following Horton *et al.* (1999), we formulate an alternative model, based on the groups $G_1, G_2, \ldots G_{10}$, to test the fit of the null model logit $[p_{it}(\beta)] = \beta_0 + x'_{it}\beta_1$. Let $I_{itl} = 1$ if $\hat{\pi}_{it}$ is in group G_l and $I_{itl} = 0$ otherwise for $l = 1, \ldots 10$. The alternative model is then given by

$$logit(p_{it}) = \beta_0 + x'_{it}\beta_1 + \gamma_1 I_{it1} + \ldots + \gamma_9 I_{it9}, \qquad (20.17)$$

i.e. in different groups the slope is the same but the intercepts may differ.

We treat the indicator variables as fixed covariates even though they are based on the random $\hat{\pi}_{it}$. In the case of independence among sample individuals, Moore and Spruill (1975) provided asymptotic justification for treating the partition $\{G_1, G_2, \ldots G_{10}\}$ as though based on the true p_{it}. Under the set-up of Equation (20.17), our null hypothesis is equivalent to testing of $H_0^*: \gamma_1 = \gamma_2 = \ldots. = \gamma_9 = 0$.

Following Rao *et al.* (1998), we now develop a quasi-score test of H_0^*, taking account of the weights and other aspects of the design. We assume working independence and obtain survey-weighted estimating equations under Equation (20.17) as

$$\hat{u}_w(\beta,\gamma) = \begin{bmatrix} \hat{u}_{1w}(\beta,\gamma) \\ \hat{u}_{2w}(\beta,\gamma) \end{bmatrix} = \begin{Bmatrix} \sum_{i\in s}\sum_{t=1}^{T} w_i x_{it}[y_{it} - p_{it}(\beta,\gamma)] \\ \sum_{i\in s}\sum_{t=1}^{T} w_i I_{it}[y_{it} - p_{it}(\beta,\gamma)] \end{Bmatrix} \tag{20.18}$$

where $I_{it} = (I_{it1}, \ldots I_{it9})'$ and $\gamma = (\gamma_1, \ldots \gamma_9)'$. Under H_0^*, we solve $\hat{u}_{1w}(\beta,0) = 0$ to get the estimator $\tilde{\beta}_w$ of β. Note that $\tilde{\beta}_w$ is identical to $\hat{\beta}_w$ obtained from Equation (20.3) under working independence. We substitute $\tilde{\beta}_w$ into the second component $\hat{u}_{2w}(\beta,\gamma)$ of Equation (20.18) to get $\hat{u}_{2w}(\tilde{\beta}_w,0)$ under H_0^*. We have $\hat{u}_{2w}(\tilde{\beta}_w,0) = o_w - e_w$, where $o_w = \sum_{i\in s}\sum_{t=1}^{T} w_i I_{it} y_{it}$ and $e_w = \sum_{i\in s}\sum_{t=1}^{T} w_i I_{it}\hat{\pi}_{it}$, with $\hat{\pi}_{it} = p_{it}(\tilde{\beta}_w,0) = p_{it}(\hat{\beta}_w)$ and $p_{it}(\beta)$ being the null model. Note that o_w and e_w are the weighted observed and expected counts, respectively, under H_0^*. A quasi-score statistic for testing H_0^* is now given by

$$X_{QS}^2 = \hat{u}_{2w}(\tilde{\beta}_w,0)'\{\mathrm{var}[\hat{u}_{2w}(\tilde{\beta}_w,0)]\}^{-1}\hat{u}_{2w}(\tilde{\beta}_w,0), \tag{20.19}$$

where $\mathrm{var}[\hat{u}_{2w}(\tilde{\beta}_w,0)]$ is a consistent estimator of variance of $\hat{u}_{2w}(\tilde{\beta}_w,0)$ under H_0^*. In our illustration with the NPHS data (Section 20.5) we have followed a direct bootstrap variance approach to obtaining a consistent variance estimator, using Rao *et al.* (1992) bootstrap weights to calculate bootstrap estimates $\hat{u}_{2w}^{(b)} = \hat{u}_{2w}(\hat{\beta}_w^{(b)},0), b = 1,\ldots B$, where B is the number of bootstrap replicates, $w_i^{(b)}, b = 1,\ldots B$ are the bootstrap weights and $\hat{\beta}_w^{(b)}$ is obtained from

$$\hat{u}_{2w}^{(b)}(\beta,0) = \sum_{i\in s}\sum_{t=1}^{T} w_i^{(b)}[y_{it} - p_{it}(\beta)] = 0. \tag{20.20}$$

To simplify the computations we propose a one-step Newton–Raphson iteration to obtain $\hat{\beta}_w^{(b)}$, with $\hat{\beta}_w$ as starting value. The direct bootstrap variance estimator of $\hat{u}_{2w} = \hat{u}_{2w}(\tilde{\beta}_w, 0)$ is then given by

$$v_{BOOT}(\hat{u}_{2w}) = \frac{1}{B}\sum_{b=1}^{B}\left(\hat{u}_{2w}^{(b)} - \hat{u}_{2w}\right)\left(\hat{u}_{2w}^{(b)} - \hat{u}_{2w}\right)'. \tag{20.21}$$

Since the NPHS data file provides the weights $\{w_i, i\in s\}$ and the bootstrap weights $\{w_i^{(b)}, i\in s\}, b = 1,\ldots B$ along with the response variables and covariates, it is straightforward to implement $v_{BOOT}(\hat{u}_{2w})$ from the NPHS data file. For the NPHS illustration (Section 20.5), we did not encounter ill-conditioned matrices in computing $\hat{\beta}_w^{(b)}$. An alternative to Equation (20.21) is obtained by substituting $\hat{u}_{2w}^{(\cdot)} = \frac{1}{B}\sum_{b=1}^{B}\hat{u}_{2w}^{(b)}$ for $\hat{u}_{2w}$ but the result should be close to Equation (20.21). The EF method of variance estimation of Section 20.4 was designed for variance estimation of model parameter estimates $\hat{\beta}_w$, and further theoretical work is needed to study its applicability to quasi-score tests.

We can write X_{QS}^2 as a Wald statistic based on $o_w - e_w$:

$$X_{QS}^2 = (o_w - e_w)'[v(o_w - e_w)]^{-1}(o_w - e_w), \tag{20.22}$$

where $v(o_w - e_w)$ is a consistent variance estimator of $o_w - e_w$. Graubard *et al.* (1997) proposed a Wald statistic similar to Equation (20.21) for the case of cross-sectional data. Under H_0^*, the statistic X_{QS}^2 is asymptotically distributed as a χ^2 variable with nine degrees of freedom. Hence, the P value may be obtained as $\Pr\left[\chi_9^2 \geq X_{QS}^2(obs)\right]$, where $X_{QS}^2(obs)$ is the observed value of X_{QS}^2. The P value could provide evidence against H_0^*.

20.4.3 Adjusted Hosmer–Lemeshow Test

We now consider a survey-weighted version of the Hosmer–Lemeshow (HL) chi-squared statistic for testing the null hypothesis H_0 in the context of longitudinal data, and then adjust the statistic to account for the survey design effect following the method of Rao and Scott (1981). For $l = 1, \ldots 10$, let

$$\hat{p}_{wl} = \left(\sum_{i\in s}\sum_{t=1}^{T} w_i I_{itl} y_{it}\right)\left(\sum_{i\in s}\sum_{t=1}^{T} w_i I_{itl}\right)^{-1},$$
$$\hat{\pi}_{wl} = \left(\sum_{i\in s}\sum_{t=1}^{T} w_i I_{itl} \hat{\pi}_{it}\right)\left(\sum_{i\in s}\sum_{t=1}^{T} w_i I_{itl}\right)^{-1}, \text{ and}$$
$$\hat{W}_l = \left(\sum_{i\in s}\sum_{t=1}^{T} w_i I_{itl}\right)\left(T\sum_{i\in s} w_i\right)^{-1}.$$

Then the survey-weighted, longitudinal data version of the HL statistic is given by

$$X_{HL}^2 = nT\sum_{l=1}^{10} \hat{W}_l \frac{(\hat{p}_{wl} - \hat{\pi}_{wl})^2}{\hat{\pi}_{wl}(1 - \hat{\pi}_{wl})}. \tag{20.23}$$

If the weights w_i are equal and $T = 1$, then X_{HL}^2 reduces to the HL statistic for the case of one time point and simple random sampling. In the latter case, Hosmer and Lemeshow (1980) approximated the null distribution by a χ^2 with $10 - 2 = 8$ degrees of freedom, χ_8^2. This result is not applicable in the survey context and we make corrections to Equation (20.23) using the Rao–Scott (1981) method.

Rao and Scott (1981) proposed a first-order correction and a more accurate second-order correction to a chi-squared statistic for categorical data. Roberts, Rao and Kumar (1987) applied the method to logistic regression of estimated cell proportions, and their results are now applied to our grouped proportions $(\hat{p}_{wl}, \hat{\pi}_{wl}), l = 1, \ldots 10$. Let $\hat{p}_w = (\hat{p}_{w1}, \ldots \hat{p}_{w,10})'$, $\hat{\pi}_w = (\hat{\pi}_{w1}, \ldots \hat{\pi}_{w,10})'$ and $\hat{V}_r$ be the estimated covariance matrix of $\hat{p}_w - \hat{\pi}_w = \hat{r}_w$. We used the Rao–Wu bootstrap method to get an estimator $\hat{V}_r$ using the one-step Newton–Raphson iteration. Let $\hat{p}_w^{(b)}$ be the bth bootstrap version of $\hat{p}_w$ and let $\hat{\pi}_w^{(b)}$ be the corresponding bootstrap version of $\hat{\pi}_w$ using $\hat{\beta}_w^{(b)}$ obtained from solving $\sum_{i\in s} w_i^{(b)} \sum_{t=1}^{T} [y_{it} - p_{it}(\beta)] = 0; \sum_{i\in s} w_i^{(b)} \sum_{t=1}^{T} x_{it}[y_{it} - p_{it}(\beta)] = 0$. Then,

$$v_{BOOT}(\hat{r}_w) = \frac{1}{B}\sum_{b=1}^{B}\left(\hat{r}_w^{(b)} - \hat{r}_w\right)\left(\hat{r}_w^{(b)} - \hat{r}_w\right)' \tag{20.24}$$

is a bootstrap estimator of $V_r = \text{cov}(\hat{r}_w)$, where $\hat{r}_w^{(b)} = \hat{p}_w^{(b)} - \hat{\pi}_w^{(b)}$.

The Rao–Scott (RS) first-order correction uses the factor

$$\hat{\delta}. = \frac{1}{9}\left\{(nT)\sum_{l=1}^{10} \hat{V}_{r,ll} \hat{W}_l [\hat{\pi}_{wl}(1 - \hat{\pi}_{wl})]^{-1}\right\}, \tag{20.25}$$

which represents a generalized design effect, where $\hat{V}_{r,ll}$ is the lth diagonal element of $\hat{V}_r$. The first-order corrected HL statistic is then given by

$$X^2_{RS}(1) = X^2_{HL}/\hat{\delta}., \tag{20.26}$$

which is treated as χ^2_8 under H_0. Since the choice of degrees of freedom is not clearcut, we may follow what is done with the Horton *et al.* score statistic and treat Equation (20.26) as χ^2_9 instead of χ^2_8. We need new theory on the asymptotic null distribution of Equation (20.26).

X^2_{RS} (1) takes account of the survey design, but not as accurately as the second-order corrected statistic, which requires knowledge of the off-diagonal elements $\hat{V}_{r,lm}$ of $\hat{V}_r, l \neq m$. We need the factor

$$\hat{a}^2 = \hat{\delta}.^{-2}\left[\frac{1}{9}\sum\nolimits_{l=1}^{9}(\hat{\delta}_l - \hat{\delta}.)^2\right], \tag{20.27}$$

where

$$\sum\nolimits_{l=1}^{9}\hat{\delta}_l^2 = \sum\nolimits_{l=1}^{10}\sum\nolimits_{m=1}^{10}\hat{V}^2_{r,lm}\frac{(nT\hat{W}_l)(nT\hat{W}_m)}{\hat{\pi}_{wl}\hat{\pi}_{wm}(1-\hat{\pi}_{wl})(1-\hat{\pi}_{wm})} \tag{20.28}$$

The second-order corrected HL statistic is given by

$$X^2_{RS}(2) = X^2_{RS}(1)/(1+\hat{a}^2), \tag{20.29}$$

which is treated as χ^2 with degrees of freedom $8/(1+\hat{a}^2)$ or $9/(1+\hat{a}^2)$ under H_0. Note that $X^2_{RS}(2) \approx X^2_{RS}(1)$ if $\hat{a}^2 \approx 0$.

20.5 ILLUSTRATION USING NPHS DATA

We illustrate the marginal logistic regression model and the EF–bootstrap with data from Statistics Canada's National Population Health Survey (NPHS). The NPHS began in 1994/1995, and collects information every two years from the same sample of individuals. A stratified multistage design was used to select households within clusters, and then one household member 12 years or older was chosen to be the longitudinal respondent. The longitudinal sample consists of 17 276 individuals. Currently, seven cycles of data are available but we just use data from three.

We motivated our example by the research of Shields and Shooshtari (2001), who used NPHS data and cross-sectional logistic regression in order to study the validity of self-perceived health as a measure of a person's true overall health status. The focus of their research was to examine whether the factors that explained a choice of excellent or very good self-perceived health rather than good self-perceived health were the same as the factors that explained a choice of fair or poor self-perceived health rather than good self-perceived health. The factors that they considered were various socio-economic, lifestyle, physical and psycho-social health variables.

For our example, we formulated a marginal logistic regression model for $T = 2$ occasions. We took the same sample of 5380 females as used by Shields and Shoostari (2001), who were 25+ years of age at the time of sample selection, were respondents in all of the first three cycles of the survey and did not have proxy responses to the health component of the questionnaire. For occasion t our binary response variable y_{it} is 1 if self-perceived heath of the ith individual at time t is excellent or very good and is 0 if self-perceived health at time t is good, fair or poor. The associated vector of covariates x_{it} consists of 41 dichotomous variables similar to those used by Shields and Shooshtari. Some of the covariates describe the status of the individual at the previous survey cycle, while other covariates describe changes in status between the previous and current survey cycles. For our example, occasion $t = 1$ is 1996/1997 (so that data from both 1994/95 and 1996/97 are used to generate x_{i1}) and occasion $t = 2$ is 1998/1999 (so that data from both 1996/1997 and 1998/1999 are used to generate x_{i2}). Survey weights $\{w_i, i \in s\}$ appropriate for respondents to the first three cycles of NPHS were chosen, along with B = 500 sets of bootstrap weights $\{w_i^{(b)}, i \in s\}, b = 1, \ldots B$.

20.5.1 Parameter Estimates and Standard Errors

We used the following five approaches for estimating the model parameters β of the marginal logistic model:

(1) SEE-Ind: SEE with a working independence assumption;
(2) SEEII-OR$_{\text{constant}}$: SEE with a constant odds ratio model, $\log(\gamma_i) = \alpha_0$, for the SEE working covariance structure:
(3) SEEII-OR$_{\text{f(age)}}$: SEE with a working odds ratio modelled as a function of an individual's age group by $\log(\gamma_i) = \alpha_0 + \alpha_1 {}^* a_i + \alpha_2 {}^* a_i^2$, where $a_i = 1$ for age 25–34, $a_i = 2$ for age 35–44, ... $a_i = 5$ for age 65–74 and $a_i = 6$ for age > 75.
(4) SEE- OR$_{\text{constant}}$-E: empirical constant odds ratio;
(5) SEE- OR$_{\text{f(age)}}$-E: empirical constant odds ratio within each of the six age groups.

In approaches (2) and (3), a second set of estimating equations, $\hat{u}_{2w}(\beta,\alpha) = 0$, was used to estimate the unknown parameter vector α associated with the working odds ratios simultaneously with β, as described in Section 20.2. For approach (2), $\alpha = \alpha_0$, while for approach (3), $\alpha = (\alpha_0,\alpha_1,\alpha_2)'$. Another option is to use empirical odds ratios to estimate α directly from the data, and to use the estimate to approximate V_i so that $\hat{u}_{2w}(\beta, \alpha)$ is not needed. Approaches (4) and (5) use this option. For all five approaches the one-step EF-bootstrap approach was used to obtain variance estimates.

Table 20.1 illustrates the coefficient estimates and associated standard errors under the five different approaches for the following four binary covariates included in the

Table 20.1 Coefficient estimates and associated standard error estimates for four binary covariates.

	Functionally restricted		Heavy smoker		L/M income		Low self-esteem	
Method	Estimate	SE	Estimate	SE	Estimate	SE	Estimate	SE
SEE-Ind	− 1.13	0.14	− 0.32	0.12	− 0.35	0.08	− 0.49	0.11
SEEII-OR$_{\text{constant}}$	− 0.99	0.13	− 0.34	0.11	− 0.29	0.08	− 0.55	0.10
SEEII-OR$_{\text{f(age)}}$	0.99	0.13	0.31	0.11	0.29	0.08	0.55	0.10
SEE-OR$_{\text{constant}}$-E	− 0.94	0.12	− 0.35	0.11	− 0.27	0.08	− 0.57	0.10
SEE-OR$_{\text{f(age)}}$-E	− 0.95	0.13	− 0.35	0.11	− 0.27	0.08	− 0.56	0.10

model, one each from the groups of variables representing physical health, health behaviour, socio-economic factors and psycho-social factors:

- *Functionally restricted (Yes/No).* A respondent was considered functionally restricted at a particular point in time if she reported that, because of health reasons expected to last at least six months, she had an activity limitation or required help in doing everyday activities. Otherwise, she was considered not to be functionally restricted.
- *Heavy smoker (Yes/No).* A respondent was considered to be a heavy smoker at a particular point in time if she reported that she usually smoked 20 or more cigarettes per day. The reference category for the three binary smoking covariates included in the model was 'Never smoked daily'.
- *L/M income (Yes/No).* A respondent was considered to have middle (M) or lower (L) income at a particular point in time if his household income from all sources in the 12 months prior to that time was reported to fall within a specified range determined by household size. Otherwise, he was considered to have upper-middle or high income.
- *Low self-esteem (Yes/No).* A respondent was considered to have low self-esteem at a particular point in time if he scored 17 or less on his responses to six items, each with five possible answers with scores ranging from 0 to 4. Otherwise he was considered not to have low self-esteem.

We decided not to include the results for all of the model variables since the patterns for the ones omitted were similar to those observed for these four.

The first thing to be observed from Table 20.1 is that the parameter estimates do vary slightly over the five estimation approaches, but that the estimates under the working independence assumption stand a bit apart from the others. This could indicate that the superpopulation marginal model is not correctly specified, so that the census parameter being estimated under working independence differs from the census parameters being estimated under the working covariance assumptions for the other approaches. This is not an unreasonable conjecture since we spent little time on diagnosing whether or not we had a good model. Table 20.1 also shows that, for our example, the standard errors are quite similar under the four methods of modelling odds ratios. Also, the gain in efficiency (in terms of SE) over SEE-Ind is quite small in this example. A possible explanation for this small efficiency gain is that the covariates are nearly cluster-specific in the sense of not changing from $t = 1$ to $t = 2$. In the case of a cluster-specific covariate, Lee, Scott and Soo (1993) and Fitzmaurice (1995) have shown that the efficiency gain over the working independence method is small. On the other hand, they have also shown that working independence can lead to a considerable loss of efficiency in estimating parameters associated with time-varying (or within-cluster) covariates.

20.5.2 Goodness-of-Fit Tests

We tested the goodness of fit of our logistic regression model with the 41 dichotomous covariates. Using the quasi-score test (Equation 20.19) based on the bootstrap variance estimator (Equation 20.21), we obtained $X^2_{QS} = 6.57$, with $P(\chi^2_9 \geq X^2_{QS}) = 0.682$, which suggests that there is no evidence against the assumed logistic regression model. Note that the P value is calculated under the framework of the alternative model (Equation 20.17) based on weighted decile groups.

We now turn to the HL chi-squared statistic and its Rao–Scott adjustments for design effect. For the survey-weighted HL statistic (Equation 20.23) we obtained $X_{HL}^2 = 10.11$, but its null distribution is not asymptotically χ_8^2 or χ_9^2. Hence, we use the Rao–Scott adjustments. For the first-order version (Equation 20.26) we obtained $\hat{\delta}. = 1.49$ and $X_{RS}^2(1) = X_{HL}^2/\hat{\delta}. = 6.78$, with $P(\chi_8^2 \geq 6.78) = 0.56$ and $P(\chi_9^2 \geq 6.78) = 0.66$, respectively, suggesting that there is no evidence against the assumed logistic regression model. Note that X_{HL}^2 is substantially larger than $X_{RS}^2(1)$ because it ignores the cross-sectional dependencies. Turning to the more accurate second-order version (Equation 20.29), we obtained $\hat{a}^2 = 0.16$, $X_{RS}^2(2) = 6.78/1.16 = 5.86$ and degrees of freedom $8/(1+\hat{a}^2) = 6.9$ or $9/(1+\hat{a}^2) = 7.8$, with corresponding P values of 0.54 and 0.64, respectively. The above P values again suggest no evidence against the assumed logistic regression model when the alternative is different intercepts but the same slope.

As stated in Section 20.1, one should not regard a nonsignificant goodness-of-fit test as clear evidence that the assumed marginal mean specification gives a good fit. Tests such as the ones used here may not have high power. In fact, further analyses of these data (not shown here) indicated that logistic regression models with occasion-specific coefficients for some of the covariates gave a better fit. Also, the fact that coefficient estimates obtained under different assumptions about the within-person covariance structure tended to differ also indicated mis-specification of the marginal mean.

20.6 SUMMARY

Longitudinal studies have become increasingly important in recent years for researching many scientific issues. One common feature of these studies is that measurements of the same individuals for the same variables are taken repeatedly over time. Since researchers realized that observations from the same individual would be more alike than observations from different individuals, analytical methods were modified to account for this within-person correlation. In the case of a binary response, for example, logistic regression models for cross-sectional data were extended to marginal logistic models for longitudinal data. For most of these methods, it was assumed that data from different individuals were independent.

When data are obtained from a longitudinal survey with a complex design that involves clustering of subjects, there may be cross-sectional correlations among subjects in addition to within-individual dependencies. This means that methods that account for both types of dependency are required in order to avoid erroneous inferences. In this paper, we extend marginal logistic regression modelling techniques for binary responses to longitudinal survey data. We show how the survey weights may be used in order to obtain consistent estimates of the logistic model coefficients. We then describe a design-based bootstrap method for obtaining robust variance estimates of the estimated coefficients. This method makes use of the survey bootstrap weights obtained by the Rao–Wu (1988) approach and can account for the between- and within-individual correlation as well as nonresponse and calibration weight adjustments in a straightforward manner.

Because this method can be computationally intensive and can lead to problems with ill-conditioned matrices, especially for small sample sizes, we propose the one-step EF–bootstrap as an alternative.

Checking the adequacy of an assumed model is a crucial step in any analytical process. For a marginal model, since it is assumed that the marginal means are correct, it is important to check the adequacy of the mean specification before making inferences on model parameters. Goodness-of-fit tests can form a part of this diagnostic process. In this paper, we extend some logistic-model goodness-of-fit tests to the case of longitudinal survey data and point out some areas where more theory needs to be developed with respect to the EF–bootstrap. It should be noted that, as stated by Horton *et al.* (1999), these tests may not have high power to detect specific alternatives and thus other model diagnostics should be used as well. This is a recommendation that holds true whether or not the data originate from a survey with a complex design.

Finally, we took longitudinal data from Statistics Canada's National Population Health Survey to illustrate the methods proposed and developed in this paper. Currently, most of the methods are not accessible for routine use in statistical software packages but will likely be implemented as more researchers analyse longitudinal survey data.

REFERENCES

Binder, D. A., Kovacevic, M. and Roberts, G. (2004). Design-based methods for survey data: alternative uses of estimating functions. In *Proceedings of the Survey Research Methods Section, American Statistical Association* (pp. 3301–3312). Washington, DC: American Statistical Association.

Diggle, P. J., Heagerty, P., Liang, K.-Y. and Zeger, S. L. (2002). *Analysis of Longitudinal Data*, (2nd edn.). New York: Oxford University Press.

Fitzmaurice, G. M. (1995). A caveat concerning independence estimating equations with multivariate survey data. *Biometrics*, 51, 309–317.

Fitzmaurice, G. M., Laird, N. M. and Ware, J. H. (2004). *Applied Longitudinal Analysis*. Hoboken, NJ: John Wiley & Sons, Inc.

Graubard, B. I., Korn, E. L. and Midthune, D. (1997). Testing goodness-of-fit for logistic regression with survey data. In *Proceedings of the Section on Survey Research Methods, American Statistical Association* (pp. 170–174). Washington, DC: American Statistical Association.

Horton, N. J., Bebchuk, J. D., Jones, C. L., Lipsitz, S. R., Catalano, P. J., Zahner, G. E. P. and Fitzmaurice, G. M. (1999). Goodness-of-fit for GEE: An example with mental health service utilization. *Statistics in Medicine*, 18, 213–222.

Hosmer, D. W. and Lemeshow, S. (1980). A goodness-of-fit test for the multiple logistic regression. *Communications in Statistics*, A10, 1043–1069.

Hu, F. and Kalbfleisch, J. D. (2000). The estimating function bootstrap. *Canadian Journal of Statistics*, 28, 449–499.

Lee, A. J., Scott., A. J. and Soo, S. O. (1993). Comparing Liang-Zeger estimates with maximum likelihood in bivariate logistic regression. *Journal of Statistical Computation and Simulation*, 44, 133–148.

Liang, K.-Y. and Zeger, S. L. (1986). Longitudinal data analysis using generalized linear models. *Biometrika*, 73, 13–22.

Liang, K.-Y., Zeger, S. and Qaqish, B. (1992). Multivariate regression analysis for categorical data. *Journal of the Royal Statistical Society, B*, 54, 3–40.

Lipsitz, S. R., Laird, N. M. and Harrington, D. P. (1991). Generalized estimating equations for correlated binary data: using odds ratios as a measure of association, *Biometrika*, 78, 153–160.

Moore, D. S. and Spruill, M. C. (1975). Unified large-sample theory of general chi-squared statistics for tests of fit. *Annals of Statistics*, 3, 599–616.

Rao, J. N. K. (1998). Marginal models for repeated observation: Inference with survey data. In *Proceedings of the Survey Research Methods Section, American Statistical Association* (pp. 76–82). Washington, DC: American Statistical Association.

Rao, J. N. K. and Scott, A. J. (1981). The analysis of categorical data from complex surveys: chi-squared tests for goodness for fit and independence in two-way tables. *Journal of the American Statistical Association*, 76, 221–230.

Rao, J. N. K., Scott, A. J. and Skinner, C. J. (1998). Quasi-score tests with survey data. *Statistica Sinica*, 8, 1059–1070.

Rao, J. N. K. and Tausi, M. (2004). Estimating function jackknife variance estimators under stratified multistage sampling. *Communications in Statistics* 33, 2087–2100.

Rao, J. N. K. and Wu, C. F. J. (1988). Resampling inference with complex survey data. *Journal of the American Statistical Association*, 83, 231–241.

Rao, J. N. K., Wu, C. F. J. and Yue, K. (1992). Some recent work on resampling methods. *Survey Methodology*, 18, 209–217.

Rust, K. and Rao, J. N. K. (1996). Variance estimation for complex surveys using replication techniques. *Statistical Methods in Medical Research*, 5, 283–310.

Roberts, G., Rao, J. N. K. and Kumar, S. (1987). Logistic regression analysis of sample survey data. *Biometrika*, 74, 1–12.

Shields, M and Shooshtari, S. (2001) Determinants of self-perceived health. *Health Reports*, 13, 35–52.

Skinner, C. J., Holt, D. and Smith, T. M. F. (1989). *Analysis of Complex Surveys*. New York: John Wiley & Sons, Inc.

CHAPTER 21

A Latent Class Approach for Estimating Gross Flows in the Presence of Correlated Classification Errors

Francesca Bassi and Ugo Trivellato
University of Padova, Italy

21.1 INTRODUCTION

With panel data, analysts can estimate labour force gross flows, that is, transitions in time between different states. Whereas net flows measure variations in time in the stock of employed and/or unemployed people, gross flows provide information on the dynamics of the labour market. Panel data may be obtained by means of various survey strategies, among which are genuine panel surveys and cross-sectional surveys with retrospective questions, or some combination of the two (e.g. Duncan & Kalton, 1987).

Measurement (or classification) errors in the observed state can introduce substantial bias in the estimation of gross flows, thus leading to erroneous conclusions about labour market dynamics (see Bound *et al.*, 2001, pp. 3792–3802, for a comprehensive survey).

A large body of literature on classification errors and their impact on gross flows estimation is based on the assumption that errors are uncorrelated over time: the so-called Independent Classification Errors (ICE) assumption. ICE implies that classification errors referring to two different occasions are independent of each other conditionally on the true states, and that errors only depend on the present true state. According to that assumption, classification errors produce spurious transitions and consequently induce overestimation of changes (Kuha & Skinner, 1997). Many researchers have proposed methods that adjust estimates of gross flows for measurement error based on the ICE assumption. Frequently, this assumption is coupled with the use of external information on misclassification rates, typically provided by reinterview data (Abowd & Zellner, 1985; Poterba & Summers, 1986; Chua & Fuller, 1987).

Methodology of Longitudinal Surveys P. Lynn

However, for many labour force surveys no such reinterview data are available. Also, the ICE assumption is not realistic in many contexts, because of the survey design and data collection strategies (Skinner & Torelli, 1993; Singh & Rao, 1995). This is especially relevant when panel data are collected by retrospective interrogation, because of the effects of memory inaccuracy (Sudman & Bradburn, 1973; Bernard *et al.*, 1984; Tourangeau *et al.*, 2000). In such circumstances we face correlated classification errors, i.e., errors that violate the ICE assumption: observed states might depend also on true states at other times or on true transitions, or there might be direct effects between observed states. The main implication of correlated classification errors is that, except for some extreme cases, observed gross flows show lower mobility than true ones (van de Pol and Langeheine, 1997).[1]

In this chapter, we use a model-based approach to adjusting observed gross flows for correlated classification errors. It combines a structural submodel for unobserved true transition rates and a measurement submodel relating true states to observed states. A convenient framework for formulating our model is provided by latent class analysis.

We apply our approach to data on young people's observed gross flows among the usual three labour force states – Employed (*E*), Unemployed (*U*) and Out of the labour force (*O*) – taken from the French Labour Force Survey (FLFS), March 1990–March 1992. The model is shown to correct flows in the expected direction: estimated true transition rates exhibit higher mobility than observed ones. In addition, the measurement part of the model has significant coefficient estimates, and the estimated response probabilities show a clear, sensible pattern.

Our approach provides a means of accounting for correlated classification errors across panel data that is less dependent on multiple indicators than previous formulations of latent class Markov models (van de Pol & Langeheine, 1997; Bassi *et al.*, 2000).

The chapter proceeds as follows. Section 21.2 presents our approach. Section 21.3 describes the FLFS and sample data used, and explores evidence on the patterns of gross flows and classification errors. Section 21.4 illustrates our approach by applying it to quarterly gross flows in the French labour market. Section 21.5 concludes.

21.2 CORRELATED CLASSIFICATION ERRORS AND LATENT CLASS MODELLING

Latent class analysis has been applied in a number of studies on panel data to separate true changes from observed ones affected by unreliable measurements. Recent contributions include Bassi *et al.* (1998), Bassi *et al.* (2000) and Biemer and Bushery (2000). The true labour force state is treated as a latent variable, and the observed one as its indicator. The model consists of two parts:

(1) structural, which describes true dynamics among latent variables;
(2) measurement, which links each latent variable to its indicator(s).

[1] This attenuation effect does not occur at some boundaries of the parameter space: when the true transition probability is 0 or when the observed transition probability is 1.

Some restrictions incorporating *a priori* information and/or assumptions are imposed on the parameters of the structural and the measurement part, respectively.

As a starting point, let us consider the simplest formulation of latent class Markov (LCM) models (Wiggins, 1973), which assumes that true unobservable transitions follow a first-order Markov chain. As in all standard latent class model specifications, local independence among the indicators is assumed, i.e. indicators are independent conditionally on latent variables.[2]

Let $X(t)$ denote the true labour force state at time t for a generic sample individual, $Y(t)$ the corresponding observed state, $v_{l_1}(1) = P[X(1) = l_1]$ the initial state of the latent Markov chain and $\mu_{k_t l_{t+1}}(t) = P[X(t+1) = l_{t+1} \mid X(t) = k_t]$ the true transition probability between state k_t and state l_{t+1} from time t to $t+1$, with $t = 1, \ldots T-1$, where T represents the total number of consecutive, equally spaced time points over which an individual is observed. Besides, let $a_{l_t j_t}(t) = P[Y(t) = j_t \mid X(t) = l_t]$ be the probability of observing state j_t at time t, given that the true state is l_t.

It follows that $P[Y(1), \ldots Y(T)]$ is the proportion of units observed in a generic cell of the T-way contingency table. For a generic sample individual, an LCM model is defined as:

$$P[Y(1) = j_1, \ldots, Y(T) = j_T] = \sum_{l_1=1}^{s} \cdots \sum_{l_T=1}^{s} v_{l_1}(1) a_{l_1 j_1}(1) \prod_{t=2}^{T} a_{l_t j_t}(t) \mu_{l_{t-1} l_t}(t-1), \quad (21.1)$$

where j_t and l_t vary over the possible labour force states – in our case, E, U and O, with $s = 3$.

In order to proceed in formulating our model, it is crucial to consider the equivalence between latent class and log-linear models with some unobservable variables (Haberman, 1979). An LCM model may also be specified in the log-linear context through the so-called "modified LISREL approach" proposed by Hagenaars (1990, 1998), which extends Goodman's modified path analysis (Goodman, 1973) to include latent variables. Each conditional probability in Equation (21.1) may be expressed as a function of the log-linear parameters. For example, $a_{l_1 j_1}(1)$ can be written as:

$$a_{l_1 j_1}(1) = \exp\left(\beta_{j_1}^{Y(1)} + \beta_{l_1 j_1}^{X(1)Y(1)}\right) \Big/ \sum_{j_1=1}^{3} \exp\left(\beta_{j_1}^{Y(1)} + \beta_{l_1 j_1}^{X(1)Y(1)}\right), \quad (21.2)$$

where $\beta_{j_1}^{Y(1)}$ and $\beta_{l_1 j_1}^{X(1)Y(1)}$ denote the first- and second-order effects in log-linear parameterization, respectively.

From Equation (21.2), it is apparent that any restriction on conditional probabilities may be equivalently imposed on the log-linear parameters. The specification of a model as the product of conditional probabilities has the advantage of more direct interpretation, whereas log-linear parameterization is more flexible and allows more parsimonious models to be specified. The breakdown of conditional probabilities does imply estimation of the entire set of interaction parameters, while the modified LISREL approach allows us to specify models in which some higher order interactions among variables may be excluded or conveniently constrained.

---

[2] In the LCM with one indicator per latent variable, the assumption of local independence coincides with the ICE condition.

We will exploit the above equivalence: the higher flexibility of log-linear parameterization will allow us to model the correlation of classification errors over time parsimoniously.

Our approach has the advantage that it does not require external information, either on misclassification probabilities from reinterview data or in the form of auxiliary variables affecting transition and/or classification probabilities. Clearly, this is achieved at the cost of having to impose sufficient identifiability conditions on the model. Thus, it rests on a sound formulation of an overidentified model: chiefly, on a sensible and parsimonious specification of the measurement process. For that purpose, it is important to capitalize on the evidence provided by the data and on the suggestions about patterns of measurement error, particularly in retrospective surveys, offered by a fairly large body of literature (Bound *et al.*, 2001, pp. 3743–3748).

As the model involves many unobservables, the issue of identification deserves attention. From the literature on latent Markov chains, it is well known that further information is needed in order to ensure (global) identification, in the form of restrictions on the parameters and/or availability of multiple indicators of the true state (see the seminal paper of Lazarsfeld & Henry, 1968). In the case of fairly complex models with classification errors, it is advisable to check at least local identification. A sufficient condition for local identifiability of a latent class model is that the information matrix be full rank (Goodman, 1974). In practice, one has to work with the estimated information matrix: for a model to be identifiable, all its eigenvalues must be positive.

There are also some potential disadvantages to our approach. The number of unknown parameters may be high, and estimation computation intensive. This indicates that the size of the model should be kept reasonably moderate, first of all in terms of dimensions of the state and time space (i.e. of the number of labour force states and the number of time units, respectively). Furthermore, since the approach involves a model jointly for unobservable true states and classification errors, the results are sensitive to possible model mis-specification. Thus, it is important to check empirical model validity. Three lines of attack are useful:

(1) One may assess the plausibility of the pattern of classifications errors (Biemer & Bushery, 2000, p. 151) and of the structural model, with reference to *a priori* expectations and evidence from previous studies.
(2) One may look at diagnostics of the components of the model: parameter estimates have the expected signs, they are statistically significant, etc.
(3) One may compare the model of interest with alternative, nested or non-nested models. For nested models, the restricted one can be tested with the conditional test, which is asymptotically distributed as χ^2 under weaker conditions than the usual L^2 and X^2 statistics.[3] In order to compare alternative non-nested models, one can resort to indices based on the information criterion, such as AIC or BIC, which weight the goodness of fit of a model against its parsimony (Hagenaars, 1990).

[3] The typical pattern of labour force transitions results in an unbalanced and sparse contingency table. In such a case it is problematic to test the validity of the overall model. When sparse data have to be fitted to a log-linear or latent class model, the usual log-likelihood ratio L^2 and the Pearson X^2 statistics (possibly with correction for the degrees of freedom) cannot be used as an indication of goodness of fit, since their asymptotic χ^2 distribution is no longer guaranteed. Besides, when the amount of sparseness is pretty high – which is the case in our example (we have a sample of 5247 individuals and 3^{10} cells) – bootstrapping goodness-of-fit measures is of no help (Langeheine *et al.*, 1996, pp. 493–497).

21.3 THE DATA AND PRELIMINARY ANALYSIS

The FLFS, *Enquête Emploi*, is conducted yearly by INSEE, the French national statistical agency. The reference population is all members of French households aged 15 years or more in the calendar year in which the survey is carried out. It has a rotating panel design (Elliot *et al.*, chapter 2 of this volume), with one-third of the sample being replaced each year.

Information on labour force participation is collected by means of two sets of questions: retrospective interrogation on a reference period composed of the 12 months preceding the interview month, and a question on actual state during the interview month. Respondents report their monthly labour condition as one of the following categories: self-employed, fixed-term employed, permanently employed, unemployed, on a vocational training programme, student, doing military service, other (retired, housewife, etc.).

We use the information collected in the surveys of March 1991 and March 1992 on a subgroup of the cohort sampled in 1990, made up of young people. The sample consists of 5247 individuals, who were aged between 19 and 29 years in March 1992. At each wave, respondents were asked to report their monthly labour force history from the corresponding month of the previous year to the current month. Thus, there is only one retrospectively observed labour force state at each month, except for March 1991 when two distinct pieces of information are available: one is the concurrent information collected with the March 1991 survey and the other is the retrospective information collected with the March 1992 survey.

Our dataset does not include information on auxiliary variables. Thus, we will specify a model for a homogeneous population. As our reference population consists of a fairly restricted age group of 19–29 years, the assumption is tenable.

In order to keep the model at a manageable size, we carry out our analyses according to a simplified formulation of state and time space. We consider a state space restricted to three labour force states – Employed (E), Unemployed (U) and Out of the labour force (O) – and analyse quarterly transitions from March 1990 to March 1992.

Table 21.1 presents the observed quarterly transition rates for our sample, as well as the yearly transitions March 1991–March 1992. Let us focus on quarterly transition rates, disregard for the moment the distinction of transitions by type and just look at the transitions of the same type – those denoted by WW – over the two years. Observed transition rates exhibit a moderate but neat seasonal pattern, presumably related to the school calendar. From June to September, for example, we observe a proportion of people who enter the labour market (OE and OU rates) greater than the average; instead, a peak of exit rates from employment, chiefly towards inactivity (EO), is documented from September to December.

As regards the pattern of survey responses and classification errors, interesting evidence comes from Tables 21.1 and 21.2. Table 21.2 exploits the double information on the labour force state in March 1991, and presents a cross-classification of people by concurrent and retrospective survey responses. It provides some crude evidence on response error: 7.8 % of respondents declare a different state in the two surveys. This evidence clearly points to the role of memory effects in respondents' accuracy. Inspection of the per cent of row distributions also indicates that response errors vary according to the true state.

Table 21.1 Observed transition rates (%) by type: quarterly transitions from March 1990 to March 1992 and yearly transitions March 1991–March 1992.

Transitions[a]		*EE*	*EU*	*EO*	*UE*	*UU*	*UO*	*OE*	*OU*	*OO*
March90–June90	WW	93.63	2.03	1.34	19.94	77.46	2.60	1.37	0.32	98.40
June90–Sept.90	WW	94.08	4.32	1.60	18.99	79.43	1.58	3.87	3.79	93.34
Sept.90–Dec.90	WW	93.93	4.39	1.98	24.00	72.47	3.53	1.91	0.80	97.30
Dec.90–March91	WW	94.77	4.25	0.98	24.53	72.40	3.07	0.98	0.66	98.36
	BW	91.50	4.86	3.64	31.60	56.84	11.56	4.40	2.10	93.50
March91–June91	WW	96.03	3.02	0.95	23.21	74.32	2.47	1.28	0.68	98.04
	BW	91.48	4.63	3.89	35.01	54.20	10.79	4.84	2.14	93.02
June91–Sept.91	WW	94.29	3.94	1.77	20.93	78.29	0.78	4.71	2.95	92.34
Sept.91–Dec.91	WW	93.73	4.48	1.79	23.63	74.89	1.48	3.22	1.65	95.13
Dec.91–March92	WW	93.90	4.80	1.30	21.67	76.74	1.59	1.70	0.59	97.71
Yearly										
March91–March92	WW	88.50	8.16	3.34	41.23	55.31	3.46	9.56	3.96	86.48
	BW	86.25	8.43	5.32	45.32	45.80	8.87	12.14	5.02	82.84

[a] WW = within-wave; BW = between waves.

Table 21.2 Compatibility between observations of March 1991 and March 1992, referring to labour market state in March 1991 (absolute figures; % overall in italics; % within row in brackets).

March 1991	March 1992								
		E			*U*			*O*	
E		2026			89			103	
	37.3		*(91.3)*	*1.7*		*(4.0)*	*1.9*		*(4.7)*
U		65			287			53	
	1.2		*(16.0)*	*5.3*		*(70.9)*	*1.0*		*(13.1)*
O		69			41			2694	
	1.3		*(2.4)*	*0.7*		*(1.5)*	*49.6*		*(96.1)*

Returning to Table 21.1, let us focus on the December 1990–March 1991 and March–June 1991 transitions. Two types of quarterly flows are observed: within-wave (WW), when information about labour force states is collected in the same survey; and between-waves (BW), when information is collected in two different surveys. Interestingly enough, WW transition rates describe a much more stable labour market than BW ones.

The interpretation of this evidence, with respect to the pattern of classification errors, is not as straightforward as it would have been if BW transitions had resulted from a combination of retrospective and concurrent information, respectively, collected in two subsequent survey waves, and WW transitions had resulted from retrospective information collected within the same survey wave and extending backwards roughly for the same period time (see Martini, 1988). Given the design of the FLFS, the picture is less clear. Here:

(1) WW transition rates result: (a1) for December 1990–March 1991 from the survey wave of March 1991, i.e. from a combination of concurrent (for March) and three-month retrospective information (for December); (a2) for March–June 1991 from

retrospective information quite a way back in time (12 and 9 months, respectively), collected from the survey wave of March 1992.

(2) BW transition rates are estimated on the basis of information collected at the two survey waves of March 1991 (for the initial month) and March 1992 (for the final month), that is: (b1) for December–March, from a combination of 3-month retrospective and 12-month retrospective information; (b2) for March–June, from a combination of concurrent and 9-month retrospective information.

Nevertheless, the overall evidence provides a clear indication that the bias towards stability documented by WW transition rates is caused by survey responses affected by highly correlated classification errors.

Yearly transitions neatly confirm this evidence, obviously in a context of comparatively higher mobility (see Table 21.2, last two rows). WW transitions, observed within the March 1992 survey, document much less change in the labour market than BW transitions, where information is collected in two different surveys one year apart.

In panel surveys with retrospective questions, memory decay is the main source of response error. Respondents tend to forget past events and/or to place them wrongly along the time axis (Sudman & Bradburn, 1973). Abundant evidence from the cognitive psychology and survey methodology literature indicates that the quality of recall declines as the length of the recall period increases, though this relationship is far from being general and stable.[4] The literature on measurement errors in reporting labour market histories documents that short spells are often forgotten and that events (i.e. changes of state) are anticipated or postponed on the time axis towards the boundaries of the reference period (Martini, 1988). More specifically, respondents have a tendency to shift reported changes in labour force towards the interview time – the so-called "forward telescoping effect" – and/or to mechanically report the same condition throughout the whole reference period – the so-called "conditioning effect" (Eisenhower & Mathiowetz, 1991). The overall result of these effects is to induce correlated classification errors, the magnitude of which increases as the recall period extends.

21.4 A MODEL FOR CORRELATED CLASSIFICATION ERRORS IN RETROSPECTIVE SURVEYS

Our purpose is to specify and estimate a model that allows for a seasonal component in true transitions and, based on the empirical findings above and suggestions from the literature on response errors in surveys, parsimoniously describes the pattern of correlated errors.

Figure 21.1 presents the path diagram of a simplified version of our model, which is useful to highlight some of its basic features. Indicators $Y(5)$, $Y(6)$, $Y(7)$, $Y(8)$ and $Y(9)$ represent observed states in the five quarters covered by the March 1992 survey

[4] The recall period interacts with other factors, such as salience of the event, type of response task, etc. (for reviews, see Jabine *et al.*, 1984; Biemer & Trewin 1997; Tourangeau *et al.*, 2000; Bound *et al.*, 2001, pp. 3743–3748).

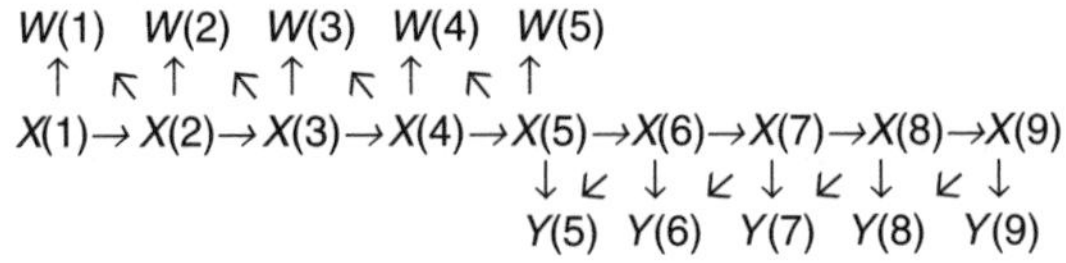

$$
\begin{array}{ccccccccccccccccc}
W(1) & & W(2) & & W(3) & & W(4) & & W(5) & & & & & & & & \\
\uparrow & \nwarrow & \uparrow & \nwarrow & \uparrow & \nwarrow & \uparrow & \nwarrow & \uparrow & & & & & & & & \\
X(1) & \rightarrow & X(2) & \rightarrow & X(3) & \rightarrow & X(4) & \rightarrow & X(5) & \rightarrow & X(6) & \rightarrow & X(7) & \rightarrow & X(8) & \rightarrow & X(9) \\
& & & & & & & & \downarrow & \swarrow & \downarrow & \swarrow & \downarrow & \swarrow & \downarrow & \swarrow & \downarrow \\
& & & & & & & & Y(5) & & Y(6) & & Y(7) & & Y(8) & & Y(9)
\end{array}
$$

Figure 21.1 Basic LCM model of gross flows for 9 time-points

(March, June, September and December 1991 and March 1992, respectively); $W(1)$, $W(2)$, $W(3)$, $W(4)$ and $W(5)$ refer to the sequence of states observed in the preceding survey (March, June, September and December 1990 and March 1991, respectively). As usual, $X(t)$ denotes the true labour force state at time t. It is apparent that, for eight quarters out of nine, there is just one indicator for each latent state: only for $X(5)$ are there two indicators, $Y(5)$ and $W(5)$.

Figure 21.1 distinguishes the first-order Markov process among true states $X(t)$ and the measurement submodel linking indicators $Y(t)$ and $W(t)$ to latent states. Direct and oblique arrows are meant to indicate that responses for time t depend on the true states at both times t and $t + 1$. Clearly, this specification overcomes the traditional ICE assumption.

The relationships described by Figure 21.1 may be formulated as in Equation (21.3), which breaks down the observed proportion in the generic cell of the 10-way contingency table into the following product of conditional probabilities:

$$
\begin{aligned}
& P[W(1)=i_1, W(2)=i_2, W(3)=i_3, W(4)=i_4, W(5)=i_5, Y(5)=j_5, Y(6)=j_6, Y(7)=j_7, Y(8)=j_8 Y(9)=j_9] \\
& = \sum_{l_1,\ldots l_9=1}^{3} v_{l_1}(1) \prod_{t=1}^{8} \mu_{l_t l_{t+1}}(t) a_{l_9 j_9}(9) a_{l_8 l_9 j_8}(8) a_{l_7 l_6 j_7}(7) a_{l_6 l_7 j_6}(6) a_{l_5 l_6 j_5}(5) z_{l_5 i_5}(5) z_{l_4 l_5 i_4}(4) z_{l_3 l_4 i_3}(3) z_{l_2 l_3 i_2}(2) z_{l_1 l_2 i_1}(1),
\end{aligned}
\tag{21.3}
$$

where: $z_{l_5 i_5}(5) = P[W(5) = i_5 \mid X(5) = l_5]$ is the probability of observing state i_5 for March 1992, given that the true state was l_5; $z_{l_1 l_2 i_1}(1) = P[W(1) = i_1 \mid X(1) = l_1, X(2) = l_2]$ is the probability of observing state i_1 for March 1990, given that there was a transition from state l_1 to state l_2 from March 1990 to June 1990; and similarly for the other conditional response probabilities – those denoted by $z_{l_t l_{t+1} i_t}(t) = P[W(t) = i_t \mid X(t) = l_t, X(t + 1) = l_{t+1}]$, $t = 1, \ldots 4$ refer to states observed during the March 1991 interview and those denoted by $a_{l_t l_{t+1} j_t}(t) = P[Y(t) = j_t \mid X(t) = l_t, X(t + 1) = l_{t+1}]$, $t = 5, \ldots 8$ refer to states observed during the March 1992 interview.

Equation (21.3) may be formulated also in terms of a system of multinomial logit equations. In such parameterization more parsimonious models may be specified (e.g. by imposing that the hierarchical log-linear model does not contain third-order interaction parameters, which are implied by Equation (21.3)).

It is important to stress that we have no supplementary information on the measurement process, except for the double observation on labour force state in March 1991. Thus, identification is not ensured unless plausible restrictions are imposed on the parameters. In particular, some assumptions are needed to identify the measurement error process. The final specification of our model results from combining various pieces of information: knowledge of the design and measurement characteristics of the FLFS; theoretical considerations and empirical evidence about the pattern of reporting errors in retrospective surveys, as well as on the true dynamic process, reviewed in Section 21.3; and results from specification searches aimed at obtaining a sensible and parsimonious formulation.

The set of assumptions of the final model, together with some further comments on the reasons motivating them, is as follows:[5]

(1) Quarterly flows among true labour force states follow a first-order nonhomogeneous Markov chain, with transition probabilities among the same calendar quarters in different years set to be equal. This specification amounts to assuming a dominant constant seasonal component for labour market dynamics.
(2) As regards the measurement part of the model, it is worth emphasizing again that only for March 1991 are two distinct observations of the labour force state available – concurrent information $W(5)$ and retrospective information $Y(5)$. Thus, we cannot explicitly model dependencies between indicators.[6] One way of accounting for correlated classification errors in our data is to let the indicators depend also on true states one quarter ahead. That is, we assume that a response given for time t depends not only on the true state at t, but also on the true state at $t + 1$ (this is the meaning of the oblique arrows in Figure 21.1). In other words, a sort of forward telescoping effect is postulated, which is plausible in retrospective surveys.
(3) The literature on memory effects on survey response behaviour suggests that classification error correlation occurs also because response errors depend on the length of recall period. Thus, we impose the following additional assumptions on response probabilities:

— For second-order interaction parameters $\beta_{l_t i_t}^{X(t)W(t)}$, $t = 1, \ldots 5$, and $\beta_{l_t j_t}^{X(t)Y(t)}$, $t = 5, \ldots 9$, describing the association between each latent state and its indicator, we assume a flexible functional form, so that the probability of erroneously reporting the labour force state increases with the distance between reference and survey months. The specification we move from is:

$$\beta_{l_t i_t}^{X(t)W(t)} = \omega_{li}^{XW} + \delta\omega_l^X f(\Delta t)$$

$$\beta_{l_t j_t}^{X(t)Y(t)} = \omega_{lj}^{XY} + \delta\omega_l^X f(\Delta t), \tag{21.4}$$

where ω_{li}^{XW} and ω_{lj}^{XY} are parameters measuring the association between each latent state X and its indicator W or Y; δ is an indicator function having a value of 1 if the true state is correctly reported, 0 otherwise; ω_l^X are proportionality factors depending on the true state – they account for the fact that the three states E, U and O may be perceived differently by respondents; and $f(.)$ is a monotone decreasing function of time distance Δt between reference and survey months.

[5] For the mathematical specification of the model, not reported here for the sake of space, see Bassi and Trivellato (2005), Appendix A.

[6] With essentially one indicator per latent state, a model postulating direct effects between indicators would be trivially underidentified.

— Two additional comments are in order. First, parameters ω_{li}^{XW}, ω_{lj}^{XY} and ω_{l}^{X} are set constant over time, as well as the association between $W(t)$ and $X(t+1)$ for $t = 1, \ldots 4$ and between $Y(t)$ and $X(t+1)$ for $t = 5, \ldots 8$. These restrictions are consistent with the notion that the measurement properties of properly designed survey instruments are stable over time – a result that is well established in the literature. Second, in order to fully specify $f(.)$ we perform a specification search within a set of possible functions in t (linear, squared, exponential). We end up by choosing $f(\Delta t) = exp(t)$ as the best-fit specification, on the basis of L^2.[7]

(4) Lastly, we set up the following further restrictions on parameters of the measurement submodel, motivated both by model parsimony and by evidence on the error generating mechanism.

— The probability of making mistakes is assumed to be constant for the same month across the various years. This implies that $\omega_{li}^{XW} = \omega_{lj}^{XY}$ and $\beta_{l_t i_t}^{X(t)W(t)} = \beta_{l_{t+4} j_{t+4}}^{X(t+4)Y(t+4)}$. The rationale for this assumption again has to do with the concept of time stability of the measurement properties of the survey instruments, but in a slightly different sense. It is based on the fact that the survey waves were carried out in the same month of consecutive years – March 1991 and March 1992 – and on empirical evidence that response errors mainly depend on the period of time elapsing between survey time and the event to be recalled, rather than on the calendar month in which the event took place.

— In the hierarchical log-linear model formulation, all third-order interaction parameters are excluded.

The model has been estimated by maximum likelihood.[8] It is locally identified: estimated information matrix eigenvalues are all positive. In order to evaluate the appropriateness of the model assumptions, we compared it with three alternative specifications – models A, B, and C, respectively:

- Model A is specified as our model for the measurement part, while latent transitions are assumed to follow a first-order stationary Markov chain. Patently, it is nested in our model. When tested by means of the conditional test, it is rejected ($\Delta L^2 = 330.83$, with 18 degrees of freedom). Thus, there is evidence that latent transitions are affected by a significant seasonal component.
- Model B assumes a seasonal first-order Markov chain at the latent level and independent classification errors. That model is not nested in ours. Model comparison based on the BIC index is clearly in favour of our model (−50814.64 for our model vs. −502652.92 for model B), and points to correlated classification errors.

[7] This choice is in accordance with findings from experimental psychology literature and social research, which indicate that, within short or medium recall periods, the process of memory decay is approximately exponential (Sudman & Bradburn, 1973).

[8] Software lEM (Vermunt, 1996) was used. We estimated our model starting from different sets of starting values to check for local maxima.

- Model C specifies that the reported state at time *t* depends on the true state at time *t* and the true state at the time of the interview, while it maintains the assumption of a seasonal first-order Markov chain for the true process. Looking at the value of the BIC index (−501056.61), this model too fits the data worse than our model.

This (albeit small) set of model comparisons confirms the empirical validity of our model. Estimated quarterly transition rates from our model are listed in Table 21.3 and estimated conditional response probabilities in Table 21.4.

When estimated transition rates are compared with the corresponding observed rates in Table 21.1, it emerges that observed transitions are corrected according to expectations. As implied by serially correlated measurement errors in retrospective surveys, true mobility in the French youth labour market is higher than observed mobility.

As regards the measurement part of our model, it is worth noting that all parameters are statistically significant.[9] Specifically, parameters ω_{li}^{XW}, ω_{lj}^{XY} and ω_{l}^{X} are significantly different from zero. This evidence neatly corroborates the hypothesis of correlated classification errors.

Table 21.4 reveals various interesting results. First, no appreciable errors are made in reporting the concurrent condition (Table 21.4, last row). Second, classification errors are unnoticeable also for the quarter immediately preceding the survey (Table 21.4, penultimate row). A comment is in order regarding the interpretation of that result. The opportunity to detect classification errors depends on the information available about the measurement process. As repeatedly noted, in our case study such information is quite limited. Thus, the evidence of no appreciable response errors does not entail that concurrent and one-quarter retrospective responses are intrinsically error-free. Rather, it means that, conditional on available information, potential classification errors in concurrent and one-quarter retrospective responses do not emerge.

As for the other reference months – for which retrospective reporting extends backwards for more than one quarter, as implied by Equation (21.4) the estimated conditional response probabilities show that the longer the recall period, the greater the probability of answering incorrectly. For labour force states involving recall periods of up to one year (Table 21.4, first row), the size of classification errors is substantial – especially for the case of response probabilities conditional to true unemployment.

Table 21.3 LCM model: Estimated quarterly transition rates (%), March 1990–March 1992.[a]

	ee	*eu*	*eo*	*ue*	*uu*	*uo*	*oe*	*ou*	*oo*
March–June	92.76	3.97	3.27	32.60	58.15	9.25	3.39	1.39	95.22
June–Sept.	92.83	5.37	1.80	28.83	70.07	1.09	4.63	2.95	92.42
Sept.–Dec.	93.56	4.53	1.91	24.32	73.10	2.57	2.58	1.21	96.23
Dec.–March	94.31	4.55	1.15	22.98	74.76	2.27	1.32	0.63	98.05

[a] True states: *e* = employment; *u* = unemployment; *o* = out of the labour force.

[9] Parameter estimates of the measurement submodel are in Bassi and Trivellato (2005), Appendix B.

Table 21.4 LCM model: Estimated response probabilities (%) conditional on the true state in the same month (summing over the categories of the true state at $t + 1$) March 1990–March 1992

Reference months	Conditional response probabilities[a]								
	$E\|e$	$U\|e$	$O\|e$	$E\|u$	$U\|u$	$O\|u$	$E\|o$	$U\|o$	$O\|o$
March 1990 and 1991[b]	95.60	2.06	2.34	17.09	72.42	10.49	2.13	0.87	97.00
June 1990 and 1991[b]	97.89	1.34	0.77	7.56	90.48	1.96	1.51	0.77	97.72
Sept.1990 and 1991[b]	99.90	0.06	0.14	0.34	99.48	0.18	0.05	0.02	99.93
Dec. 1990 and 1991[b]	100.00	0.00	0.00	0.00	100.00	0.00	0.00	0.00	100.00
March 1991 and 1992[c]	100.00	0.00	0.00	0.00	100.00	0.00	0.00	0.00	100.00

[a] True states: e = employment; u = unemployment; o = out of the labour force.
[b] Retrospective information collected in March 1991 and March 1992, respectively.
[c] Concurrent information collected in March 1991 and March 1992, respectively.

We obtain further insights on some features of the latent Markov model by jointly considering the parameter estimates of the measurement and structural parts of the model. Since reported states for December and March are not affected by discernible measurement errors, it comes as no surprise that the December–March estimated transitions basically coincide with observed ones, apart from the averaging induced by the assumption that they do not vary over the two years. In addition, the March–June estimated transitions largely reflect the heavy weight assigned to the information in March 1991 in the BW observed transitions – concurrent, which is thus taken as practically error-free. For the other transitions, the general pattern outlined above is clear: there are more true dynamics than it appears from the observed rates; the more the reference month extends backward in time with respect to the survey month, the higher the correction towards mobility; and transitions exhibit a definite seasonal pattern.

It is worth noting that these results are at variance with those obtained by Magnac and Visser (1999), basically for the same sample data, using a quite different model: a time-homogeneous Markov process for true transitions and a weak version of the ICE assumption (see Bassi and Trivellato, 2005, for a comparative assessment).

21.5 CONCLUDING REMARKS

This chapter presents a latent class approach to correct gross flows from correlated classification errors. The emphasis is on the capacity to account for correlated classification errors across panel data without heavily depending on multiple indicators (admittedly, at the price of carefully specifying a parsimonious model with sufficient identifiability conditions).

A case study serves to illustrate our modelling strategy. A model is formulated and estimated within a latent class analysis set-up, to adjust observed quarterly labour force gross flows of French young people, March 1990–March 1992. The data were taken from the FLFS, which collects information on labour force conditions mainly by means of retrospective interrogation. Based on evidence from the same French dataset and on arguments from survey methodology literature, the model accounts for correlated classification errors over time and allows for a seasonal component in the latent Markov process.

Our model corrects observed flows in the expected direction: estimated true transition rates show higher mobility than observed ones. Besides, estimated conditional response probabilities show a neat, sensible pattern: there is no discernible classification error for concurrent information, whereas the size of classification errors is substantial in retrospectively reported labour force states for reference months far from the interview month.

ACKNOWLEDGEMENTS

Research for this paper was supported by grants No. 2001134473 and 2003139334 from the Italian Ministry for Education, University and Research. We are grateful to Michael Visser and Thierry Magnac for providing us with anonymous micro-data from the French Labour Survey. We thank Enrico Rettore, Patrick Sturgis, participants at the MOLS 2006 Conference and two anonymous referees for useful comments and suggestions. The usual disclaimer applies.

REFERENCES

Abowd, J. M. and Zellner, A. (1985). Estimating gross labor force flows. *Journal of Business & Economic Statistics*, 3, 254–283.

Bassi, F., Croon, M., Hagenaars, J. A. and Vermunt, J. K. (2000). Estimating true changes when categorical panel data are affected by correlated and uncorrelated classification errors. An application to unemployment data. *Sociological Methods and Research*, 29, 230–268.

Bassi, F., Torelli, N. and Trivellato, U. (1998). Data and modelling strategies in estimating labour force gross flows affected by classification errors. *Survey Methodology*, 24, 109–122.

Bassi, F. and Trivellato, U. (2005). The latent class approach to estimating gross flows affected by correlated classification errors, with application to data from the French Labour Force Survey. Research Project, *Working Paper* No. 70, Padova, Dipartimento di Scienze Statistiche, Università di Padova [http: valutazione2003.stat.unipd.it/index_ENG.php].

Bernard, H. R., Killworth, P., Krinenfeld, D. and Sailer, L. (1984). The problem of informant accuracy: the validity of retrospective data. *Annual Review of Anthropology*, 13, 495–517.

Biemer, P. P. and Bushery, J. M. (2000). On the validity of latent class analysis for estimating classification error in labor force data. *Survey Methodology*, 2, 139–152.

Biemer, P. and Trewin, D. (1997). A review of measurement error effects on the analysis of survey data. In L. Lyberg *et al.* (Eds), *Survey Measurement and Process Quality* (pp. 603–631). New York: John Wiley & Sons, Inc.

Bound, J., Brown, C. and Mathiowetz, N. A. (2001). Measurement error in survey data. In J. J. Heckman and E. Leamer (Eds), *Handbook of Econometrics* Vol. 5, (pp. 3705–3843). Amsterdam: Elsevier.

Chua, T. C. and Fuller, W. A. (1987). A model for multinomial response error applied to labor flows. *Journal of the American Statistical Association*, 82, 46–51.

Duncan, G. J. and Kalton, G. (1987). Issues of design and analysis of surveys across time. *International Statistical Review*, 55, 97–117.

Eisenhower, D. and Mathiowetz, N. A. (1991). Recall errors: sources and bias reduction techniques. In P. Biemer *et al.* (Eds), *Measurement Errors in Surveys* (pp. 127–144). New York: John Wiley & Sons, Inc.

Goodman, L. (1973). The analysis of a multidimensional contingency table when some variables are posterior to the others. *Biometrika*, 60, 179–192.

Goodman, L. (1974). Exploratory latent structure analysis using both identifiable and unidentifiable models. *Biometrika*, 61, 215–231.

Haberman, S. J. (1979). *Analysis of Qualitative Data*, 2 vols. New York: Academic Press.

Hagenaars, J. A. (1990). *Categorical Longitudinal Data: Log-Linear Panel, Trend and Cohort Analysis*. Newbury Park: Sage.

Hagenaars, J. A. (1998). Categorical causal modeling: Latent class analysis and directed log-linear models with latent variables. *Sociological Methods and Research*, 26, 436–487.

Jabine, T., Loftus, E., Straf, M., Tanur, J. and Tourangeau R. (1984). *Cognitive Aspects of Survey Methodology: Building a Bridge between Disciplines*. Washington, DC: National Academy Press.

Kuha, J. and Skinner, C. (1997). Categorical data and misclassification. In L. Lyberg *et al.* (Eds), *Survey Measurement and Process Quality* (pp. 633–670). New York: John Wiley & Sons, Inc.

Langeheine, R., Pannekoek, J. and van de Pol, F. (1996). Bootstrapping goodness-of-fit measures in categorical data analysis. *Sociological Methods & Research*, 24, 492–516.

Lazarsfeld, P. F. and Henry, N. W. (1968). *Latent Structure Analysis*. New York: Houghton Muffins.

Magnac, T. and Visser, M. (1999). Transition models with measurement errors. *Review of Economics and Statistics*, 81, 466–474.

Martini, A. (1988). Retrospective versus panel data in estimating labor force transition rates: comparing SIPP and CPS. In *Proceedings of the Social Science Section, American Statistical Association* (pp. 109–114). Washington, DC: American Statistical Association.

Poterba, J. and Summers, L. (1986). Reporting errors and labor market dynamics. *Econometrica*, 54, 1319–1338.

Singh, A. C. and Rao, J. N. K. (1995). On the adjustment of gross flows estimates for classification errors with application to data from the Canadian Labour Force Survey. *Journal of the American Statistical Association*, 90, 1–11.

Skinner, C. and Torelli, N. (1993). Measurement error and the estimation of gross flows from longitudinal economic data. *Statistica*, 3, 391–405.

Sudman, S. and Bradburn, N. M. (1973). Effects of time and memory factors on response in surveys. *Journal of the American Statistical Association*, 68, 805–815.

Tourangeau R., Rips, L. J. and Rasinski, K. A. (2000). *The Psychology of Survey Response*. Cambridge: Cambridge University Press.

van de Pol, F. and Langeheine, R. (1997). Separating change and measurement error in panel surveys with an application to labour market data. In L. Lyberg *et al.* (Eds), *Survey Measurement and Process Quality* (pp. 671–688). New York: Wiley.

Vermunt, J. K. (1996). *Log-Linear Event History Analysis: A General Approach with Missing Data, Latent Variables and Unobserved Heterogeneity*. Tilburg: Tilburg University Press.

Wiggins, L. M. (1973). *Panel Analysis: Latent Probability Models for Attitude and Behavior Change*. Amsterdam: Elsevier.

CHAPTER 22

A Comparison of Graphical Models and Structural Equation Models for the Analysis of Longitudinal Survey Data

Peter W. F. Smith, Ann Berrington and Patrick Sturgis

University of Southampton, UK

22.1 INTRODUCTION

Graphical chain modelling (GCM) and structural equation modelling (SEM) are two approaches to modelling longitudinal data. Both approaches have their origins in path analysis and provide pictorial representations of the association between variables, which are usually ordered temporally. Both methods also aim to identify the direct and indirect effects of one variable on another. While the SEM approach specifies a single model for the complete system of variables being studied, the GCM approach permits a model for the complete system to be built up by fitting a sequence of submodels.

In this chapter we briefly discuss the similarities and differences of the GCM and SEM approaches to modelling univariate recursive systems and their application to complex survey data. We identify the strengths and limitations of each approach. By using an example of the relationship between changes in labour force status following entry into parenthood, and changes in gender role attitude, we illustrate their use with survey data. A sample of 632 women, childless and aged between 16 and 39 in 1991, taken from the British Household Panel Survey (BHPS), is used. This survey is a particularly useful resource for this type of analysis since it is nationally representative, continuous since 1991 and collects a wide range of socio-demographic and attitudinal information (Taylor *et al.*, 2007). Information on employment and parenthood status is obtained annually and attitude statements concerning gender roles are collected biennially in 1991, 1993, 1995 and 1997.

Methodology of Longitudinal Surveys P. Lynn

22.2 CONCEPTUAL FRAMEWORK

Recent interest has focused on the role that values and attitudes play in life-course decisions. For example, in the UK there has been considerable debate on the extent to which women's attitudes towards work and family care are exogenous (Crompton and Harris, 1998; Hakim, 1998). Whilst numerous studies based on cross-sectional data have demonstrated significant differences in attitudes according to current family formation status, panel data are required to identify the potential causal direction of the relationship. As noted by Lesthaeghe and Moors (2002), in most longitudinal analyses researchers have either focused on selection effects of attitudes (attitudes as predictors of later events) or on adaptation effects of attitudes (events as determinants of change in attitudes). For the analyses in this paper, we focus on the reciprocal relationship between changes in labour force status (following entry into parenthood) and changes in gender role attitude, and how we can best model it using longitudinal survey data.

The conceptual framework, presented in Figure 22.1, conjectures how the variables are related to each other. (See Berrington *et al.* (2008) for a fuller description.) Here the set of variables is partitioned into blocks. The first block contains our background control variables (age, highest level of education and whether or not the respondent's mother worked when the respondent was 14 years old). The rest of the blocks are temporally ordered. Hypothesized associations are represented by arrows. For example, gender role attitude in 1991 is hypothesized to explain variations in the risk of becoming a parent between 1991 and 1993, which is a selection effect. Becoming a parent between 1991 and 1993 is hypothesized to be associated with changes in gender role attitude during the same period, which is an adaptation effect. Women are thought likely to adjust their working hours upon entry into motherhood. Labour force participation may also be predicted directly by gender role attitudes. The background variables may predict any of the later outcomes, including initial attitude, attitude change and the likelihood of becoming a parent or of leaving the labour force. For clarity, these hypothesized relationships are indicated by the large black arrow in Figure 22.1.

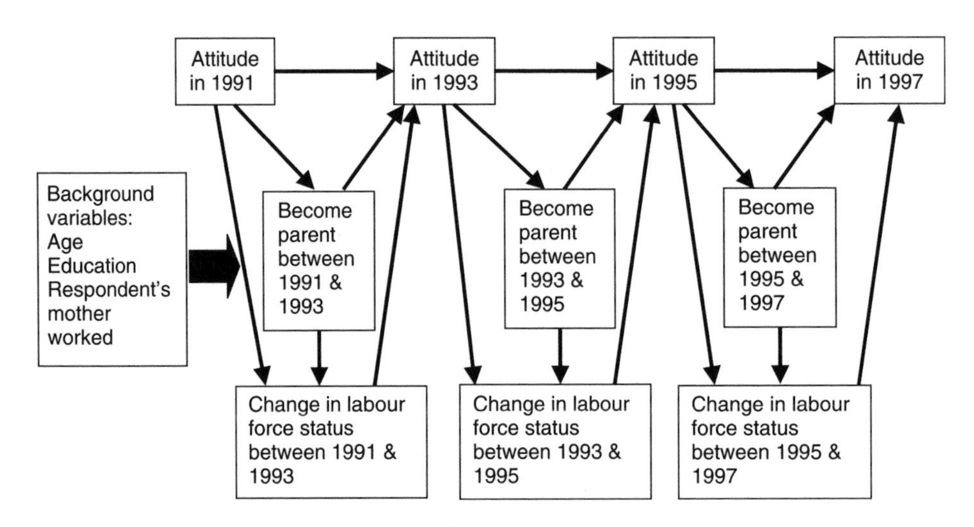

Figure 22.1 Conceptual framework.

Gender role attitude is derived from six items, each measured on a five-point scale from strongly agree to strongly disagree, asked biennially in the BHPS from 1991. The six questions are: (1) a preschool child is likely to suffer if his or her mother works; (2) the whole family suffers when the woman has a full-time job; (3) a woman and her family would all be happier if she goes out to work; (4) both the husband and wife should contribute to the household income; (5) having a full-time job is the best way for a woman to be an independent person; and (6) a husband's job is to earn money, a wife's job is to look after the home and family. Becoming a parent is defined as having a first live birth and hence refers solely to biological parenthood. Change in labour force status is based on changes in the number of hours worked. Movement from being a full-time student into paid work is included as an increase in work hours and movement in the other direction as a decrease. Becoming unemployed or economically inactive due to illness is defined as a reduction in work hours, whilst leaving the paid labour force to take on family responsibilities is referred to as leaving for family care. In summary, change in labour force status has four categories: (1) no change in hours of paid work, (2) increase in hours of paid work, (3) decrease in hours of paid work, and (4) begin family care.

22.3 GRAPHICAL CHAIN MODELLING APPROACH

The graphical chain model (see, for example, Whittaker, 1990; Cox and Wermuth, 1996) is a stochastic model specified via a mathematical graph that is a set of nodes and edges. Nodes represent variables and undirected edges (lines) represent significant associations between pairs of variables. Asymmetric relationships between variables, i.e. one anticipating the other, are represented through directed edges (arrows). Chain graphs allow us to specify directions of causality. Variables are entered into the model in a series of blocks. These reflect the temporal ordering of the variables and the assumed causal ordering of relationships. Hence, directed edges can only flow from variables in a preceding block to variables in subsequent block(s). Fundamental to graphical modelling is the concept of conditional independence. In a graphical chain model a missing edge or arrow corresponds to a conditional independence between the two variables given all the variables in the current and preceding blocks. Graphical chain models are ideally suited to longitudinal data since the temporal ordering of variables can be used to help specify the *a priori* causal ordering of variables. For some applications of graphical chain models to survey data, see Mohamed *et al.* (1998), Magadi *et al.* (2004) and Evans and Anderson (2006).

A graphical chain model can be fitted to our data by fitting a sequence of regression models. First, gender role attitude in 1991 is regressed on the background variables. Second, becoming a parent between 1991 and 1993 is regressed on the background variables and attitude in 1991. Next, change in labour force status between 1991 and 1993 is regressed on the background variables, becoming a parent between 1991 and 1993 and attitude in 1991. The sequence continues by regressing each of the 'attitude', 'becoming a parent' and 'change in labour force status' variables in turn on the background variables and the hypothesized determinants as indicated in the conceptual framework in Figure 22.1.

To permit the use of standard regression techniques when fitting the graphical chain model, we restrict our attention to modelling the observed variables. Hence, we first created a gender role attitude score by coding the responses to each of the six items, in such a way that a higher score indicates a more liberal gender role attitude, and then summing them. The score ranges from 6 to 30, with a mean in 1991 of 21.2 and a standard deviation of 3.7.

One advantage of graphical chain models is their ability to handle response variables of different types: continuous, binary and multicategory. When the response is continuous, as is the case for the attitude score, we use linear regression models. When the response is binary, as is the case for whether or not the respondent becomes a parent, or multicategory, as is the case for the respondent's change in labour force status, we use binary and multinomial logistic regression models, respectively. Furthermore, sample weights, clustered samples and item nonresponse can be handled using the standard techniques developed for the component models. A disadvantage in the GCM approach considered here is that it does not use the information in the repeated measures or model the reciprocal relationship simultaneously. Hence, it is not easy to assess the overall goodness of fit of the model to the data. Also, it is difficult to explicitly adjust for measurement error without introducing a separate stage of latent variable modelling.

22.4 STRUCTURAL EQUATION MODELLING APPROACH

SEM is a framework for statistical analysis that brings together confirmatory factor analysis, path analysis and regression (see, for example, Bollen, 1989; Finkel, 1995). SEM can usefully be conceived of as having two components, or stages. The first is the measurement model in which the relationships between a set of observed (manifest) variables and a set of unobserved (latent) variables are specified. The second is the structural model, which specifies causal and associational relationships between measured constructs, whether latent or observed.

We specify a measurement model for our data in which gender role attitude is a latent variable measured by the six indicator items. The latent variable represents a common factor that influences the observed items. The square of the standardized factor loading of an item reflects the proportion of the item variance that is explained by the common factor. By definition, one minus the square of the factor loading is the proportion of variance in the item that is measurement error. Relative to summing scores over the individual items to form a composite scale, as is common practice, the latent variable approach enables us to make a correction for differential measurement error across items, which is reflected by variation in the magnitude of the factor loadings. Measurement error in independent variables is known to result in attenuated effect sizes, so this error correction should reduce the probability of Type II errors in our anlayses (Bollen, 1989). Because we have repeated measures of the same items on the same individuals over time, we estimate covariances between the error terms for the same indicator at each time point, as the unmeasured causes of responses to each item are likely to remain stable over time. Failing to account for covariances between error terms in panel studies can bias estimates of both stability and cross-lag parameters (Williams and Podsakoff, 1989).

In summary, SEM includes the measurement models for attitude in the four waves, all of the structural paths tested for in GCM and correlations between the errors in the repeated attitude items over time. All of the regression parameters and error correlations are estimated simultaneously. As well as reducing measurement error, another advantage of the SEM approach is that it uses the information in the repeated measures and models the reciprocal relationship simultaneously. Hence, it is possible to measure the overall goodness of fit of the model to the data and it is straightforward to test for equality of effects across time. A disadvantage in the SEM approach is that it cannot easily handle unit nonresponse, attrition, sample weights and clustering. Also, categorical response variables, particularly those with more than two categories, can cause problems for maximum likelihood estimators. However, recent developments in these areas have substantially improved the flexibility of SEM models for dealing with complex survey data, nonresponse and categorical endogenous variables (Muthén, 2001). Mplus (Muthén and Muthén, 1998–2007) and LISREL (Scientific Software International, 2006) are two statistical package that are more advanced in these areas (Stapleton, 2006).

22.5 MODEL FITTING

Initial modelling revealed a very strong association between becoming a parent and change in labour force status. The vast majority of women taking on family care between one time period and the next were new mothers. Preliminary modelling revealed that becoming a parent was not, of itself, significantly associated with attitude change but that it was the change in labour force status, particularly the change to family care, that was significant in the graphical chain model. Therefore, only a binary variable contrasting 'begin family care' with 'did not begin family care' was included in the final graphical chain and structural equation models. Note that the binary variable, whether or not the respondent begins family care, is estimated using a probit regression in the SEM framework and a logistic regression in the GCM framework.

The BHPS is a stratified and clustered sample with weights to adjust for unequal selection probabilities and panel attrition (see Taylor *et al.*, 2007, for details). In the first instance, to permit comparison of the GCM and SEM approaches, these complexities are ignored. We refer to this as the 'simple' approach. To identify whether taking into account the complex survey design and unit response would make a difference to our substantive conclusions we repeat the GCM analyses taking account of the stratification (using region as the stratifying variable), clustering and weighting. We refer to this as the 'complex' approach. The sample is restricted to women who gave a full interview in every year from 1991 to 1997. Since our sample considers only childless women of childbearing age, only a small percentage belong to a household containing at least one other sample member. Hence our cluster variable refers to the primary sampling unit – i.e. postcode sector. The wave 7 longitudinal respondent weights take account of initial unequal selection probabilities into the BHPS sample and subsequent differential nonresponse at waves 1–7. Linearization-based variance estimators (Binder, 1983; Skinner, 1989) are used to produce robust estimates that are reported alongside the naïve estimates in Tables 22.1 and 22.2. (Note that we also repeated the SEM analyses taking account of complex survey design and weights, and found that the impact on the point estimates and standard errors is very similar to that with GCM.)

The GCM analyses were undertaken using the Stata software with the svy commands (Stata Corp, 2007) when taking account of the complex survey design and weights. The SEM analyses, including those incorporating adjustments for stratification, clustering and weights, were undertaken using the Mplus software (Muthén and Muthén, 1998–2007). The full SEM results are available from the authors on request.

22.6 RESULTS

Figure 22.2 depicts the selected graphical chain model using the estimates from the 'simple' approach. Here the full lines indicate significant positive effects (i.e. attitude becomes more liberal) and the dashed lines indicate significant negative effects (attitude becomes less liberal), both at the 5 % level. The corresponding diagram for the structural equation model is the same except that attitude in 1995 is no longer a significant predictor of beginning family care between 1995 and 1997. A test for the equality of the effects of lagged attitude on attitude is significant (meaning that the effect of lagged attitude changes over time), whereas there is no evidence against equality of the effects of beginning family care on attitude, that is, after controlling for the other variables the adaptation effect is of the same magnitude over time.

Table 22.1 presents the parameter estimates for the four linear regressions used to construct the graphical chain model. For each year the first column present the estimates from the 'simple' approach and the second column contains those from the 'complex' approach. Note that the standard error appears in parentheses beneath each estimate. Here we find that all three background variables are associated with gender role attitude in 1991, with younger, more educated women whose mothers worked when they were aged 14 years tending to have more liberal attitudes. For 1993, 1995 and 1997 we are modelling change in attitude, since attitude two years previous is included in the model. Here we find that, in general, background variables are not associated with changes in attitude, whereas women who begin family care tend to become more traditional. Hence, there is a significant adaptation effect. Table 22.2 presents the parameter estimates for the logistic regression of beginning family care. The evidence for a selection effect is weaker. If there is one, then, as expected, more traditional women are more likely to give up work to undertake family care.

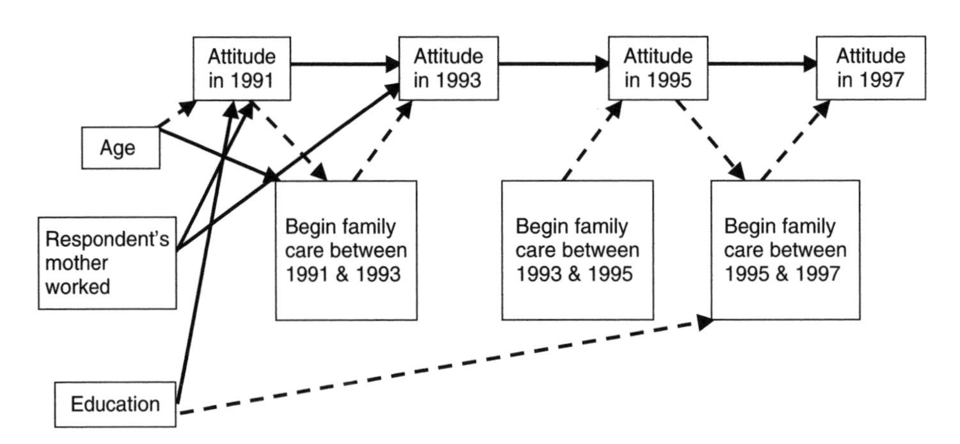

Figure 22.2 Significant associations found in the GCM using the 'simple' approach. Full lines indicate significant positive effects and dashed lines indicate significant negative effects, both at the 5 % level.

Table 22.1 Parameter estimates for the linear regression models for attitude: 'simple' and 'complex' approaches compared.

	Attitude in 1991		Attitude in 1993		Attitude in 1995		Attitude in 1997	
	Unweighted	Weighted	Unweighted	Weighted	Unweighted	Weighted	Unweighted	Weighted
Constant	20.72**	20.87**	9.70**	10.01**	6.47**	6.33**	6.76**	6.43**
	(0.33)[a]	(0.38)	(0.69)	(0.93)	(0.69)	(0.69)	(0.64)	(0.75)
Age (Ref: 16–21 years)								
22–29	– 0.98**	– 1.14**	– 0.06	– 0.13	– 0.49*†	– 0.56*†	– 0.25	– 0.18
	(0.33)	(0.33)	(0.26)	(0.26)	(0.25)	(0.27)	(0.24)	(0.28)
30–39	– **1.12****	– **1.11***	– 0.43	– 0.43	– 0.03	– 0.12	– 0.44	– 0.42
	(0.39)	**(0.43)**	(0.30)	(0.32)	(0.29)	(0.32)	(0.28)	(0.30)
Education in 1991 (Ref: Less than A level)								
A level or above	1.03**	1.01**	0.06	0.32	0.27	0.18	0.16	0.26
	(0.29)	(0.30)	(0.22)	(0.22)	(0.22)	(0.23)	(0.21)	(0.23)
Mum worked (Ref: No)								
Yes	1.03**	1.00**	0.68**	0.67**	0.24	0.30	0.33	0.37
	(0.27)	(0.30)	(0.23)	(0.25)	(0.23)	(0.24)	(0.22)	(0.21)
Attitude two years previous			0.52**	0.50**	0.68**	0.68**	0.65**	0.66**
			(0.03)	(0.04)	(0.03)	(0.03)	(0.03)	(0.03)
Began family care in the preceding two years (Ref: No)								
Yes			– 1.58**	– 1.54**	– 1.91**	– 1.97**	– **1.34****	– **1.15***
			(0.41)	(0.46)	(0.41)	(0.55)	**(0.42)**	**(0.49)**

*Significant at 5 % level; **significant at 1 % level; †overall age is not significant at the 5 % level. Figures in bold denote a change in statistical significance level between the 'simple' and 'complex' approach.

[a] Standard errors are naïve for the unweighted parameter estimates and robust for the weighted estimates.

Table 22.2 Parameter estimates for the logistic regression models for beginning family care: 'simple' and 'complex' approaches compared.

	Begin family care between 1991 and 1993		Begin family care between 1993 and 1995		Begin family care between 1995 and 1997	
	Unweighted	Weighted	Unweighted	Weighted	Unweighted	Weighted
Constant	– 0.32	0.27	– 1.53	– 1.37	– 0.30	0.23
	(0.89)	(1.05)	(0.97)	(0.99)	(0.95)	(0.81)
Age (Ref: 16–21 years)						
22–29	**1.10****	**0.94***	0.27	0.17	0.02	– 0.20
	(0.40)	**(0.44)**	(0.37)	(0.39)	(0.36)	(0.36)
30–39	0.63	0.58	0.20	0.04	– **0.81**	– **1.09***†
	(0.47)	(0.55)	(0.43)	(0.43)	**(0.53)**	**(0.52)**
Education in 1991 (Ref: Less than A level)						
A level or above	– **0.37**	– **0.58***	0.19	0.07	– 0.78*	– 0.82*
	(0.31)	**(0.29)**	(0.32)	(0.30)	(0.35)	(0.33)
Mum worked (Ref: No)						
Yes	0.29	0.25	– 0.24	– 0.04	0.09	– 0.23
	(0.32)	(0.35)	(0.32)	(0.29)	(0.34)	(0.39)
Attitude at beginning of period	– 0.14**	– 0.16**	– 0.06	– 0.06	– **0.10***	– **0.11****
	(0.04)	(0.04)	(0.05)	(0.05)	**(0.05)**	**(0.04)**

*Significant at 5 % level; **significant at 1 % level; †overall age is not significant at the 5 % level. Figures in bold denote a change in statistical significance between the 'simple' and 'complex' approach.

Adjustments for stratification and clustering generally impact on the standard errors but not on the point estimates, whereas including weights affects the magnitude of both the point estimates and the standard errors (see Skinner *et al.*, 1989). We find that the weighted parameter estimates are generally similar to the unweighted estimates and that the robust standard errors are generally larger than those assuming a simple random sample, but this is not always the case. In the linear regression models for attitude there are two parameters where the estimates become less significant, moving from the 1 % to 5 % level (indicated by bold figures in 22.1). In the logistic regression models for beginning family care, there is one parameter where the estimate becomes less significant and three parameters where the estimates become more significant. However, only in one case do we find contradictory evidence for the presence of an edge after accounting for the complex sample design and weights: the association between education and beginning family care between 1991 and 1993 becomes significant at the 5 % level (with a slightly more negative parameter estimate and slightly smaller standard error). Note that adjusting for the complex sample design and weights within the SEM framework had a similar effect, that is, changes were of a similar magnitude and in the same direction.

Standardized parameter estimates from the SEM are presented in Tables 22.3 and 22.4. They have been standardized only with respect to the latent attitude variables. Hence, in Table 22.3, they are the estimated average amount (in standard deviations) by which the latent attitude variables change for a one-unit change in the predictor variable. For attitude two years previous, this corresponds to a change of one standard deviation, since all the latent attitude variables have been standardized, whilst for the categorical variables this refers to the difference between the category of interest and the reference category. For example, for whether or not the respondent began family care, it is the difference between "yes" and "no". For comparison, standardized parameter estimates for the (naïve) linear and logistic regression models used to construct the GCM are also presented in Tables 22.3 and 22.4. Here the attitude variables constructed by summing

Table 22.3 Standardized parameter estimates for the regressions of attitude (using the 'simple' approach).

	Attitude in 1991		Attitude in 1993		Attitude in 1995		Attitude in 1997	
	Linear regression	SEM	Linear regression	SEM	Linear regression	SEM	Linear regression	SEM
Age 22–29	−0.269**	−0.272**	−0.017	0.104	−0.137*†	−0.048	−0.071	−0.015
Age 30–39	−0.305**	−0.319**	−0.124	−0.106	−0.009	0.099	−0.123	−0.143
Education in 1991								
A level or above	0.282**	0.396**	0.017	−0.009	0.074	0.050	0.046	−0.042
Mum worked	0.282**	0.340**	0.198**	0.229**	0.066	−0.022	0.093	0.056
Attitude two years previous			0.552**	0.717**	0.642**	0.839**	0.664**	0.829**
Began family care			−0.458**	−0.169**	−0.530**	−0.308**	−0.377**	−0.187*

*Significant at 5 % level; **significant at 1 % level;
†overall age is not significant at the 5 % level.

Table 22.4 Standardized parameter estimates for the regression of beginning family care (using the 'simple' approach).

	Begin family care between 1991 and 1993		Begin family care between 1993 and 1995		Begin family care between 1995 and 1997	
	Logistic regression	SEM	Logistic regression	SEM	Logistic regression	SEM
Age 22–29	0.609**	0.517**	0.149	0.123	0.012	0.007
Age 30–39	0.346	0.280	0.108	0.079	−0.445	−0.378
Education in 1991						
A level or above	−0.202	−0.131	0.103	0.098	−0.431*	−0.342*
Mum worked	0.158	0.177	−0.130	−0.098	0.049	0.068
Attitude at beginning of period	−0.279**	−0.292**	−0.105	−0.093	−0.193*	−0.176

*Significant at 5 % level; **significant at 1 % level.

the responses to the six items have been divided by their sample standard deviations. Since the SEM used probit regressions for the binary responses, the parameter estimates for the logistic regression modes are standardized by dividing them by the standard deviation of the standard logistic distribution: $\pi/\sqrt{3}$ (see Agresti, 2002, p. 125 for more details).

The conclusions from the significant parameter estimates for the SEM are very similar to those from the linear and logistic regression models. As noted when discussing Figure 22.2, the main difference is the lack of a significant selection effect of attitude in 1995 on beginning family care between 1995 and 1997. When the magnitudes of the significant standardized parameter estimates are compared for GCM and SEM, they are surprisingly similar given the difference in the approaches. The largest discrepancies are for the effects of beginning family care on family role attitude, where the estimates from the SEM are smaller in magnitude. To the extent that these differences are systematic, this could be a result of controlling for correlated measurement error in the responses to the six items used to measure the attitude.

22.7 CONCLUSIONS

Individual attitudes of women in our sample change over time. They appear to be adjusting their attitudes in response to life-course events, such as beginning family care, rather than the 'decisions' being influenced by their attitudes. In other words, in this particular sample, the adaptation effects are stronger than the selection effects. Our graphical chain model and our structural equation model gave very similar conclusions. However, we would not wish to overemphasize the generality of this point; with different data and different models, agreement between SEM and GCM should not always be anticipated. We have already noted a number of advantages and disadvantages with each approach. Graphical chain models allow the researcher to fit relatively simple models, often already familiar to the social science researcher, for various types of response (continuous, binary, ordinal, etc.). By using the Markov properties of chain graphs (e.g. Whittaker, 1990; Cox and Wermuth, 1996), we can draw conclusions about conditional independence and dependence structures of subsets of variables, and hence identify direct and indirect pathways through which explanatory variables are related to the outcome of interest. The SEM approach has the advantage of being able to decompose associations between variables into direct and indirect effects and also permits the inclusion of multiple indicator latent variables, which are useful for identifying and adjusting for measurement error. However, not all SEM software packages allow for complex survey designs and the inclusion of weights, though this is an area currently experiencing rapid development.

Of particular importance in the analysis of survey data are the ability to correct for measurement error (more readily achieved within the SEM approach) and the ability to correct for complex sample designs and to incorporate weights (more readily achieved within the GCM approach). Our comparison of parameter estimates from the GCM and SEM approaches suggests that failure to correct for measurement error in the gender role attitude may result in an overstatement of the selection effect of attitudes on labour force behaviour. At the same time our comparison of the estimates from the 'simple' approach and 'complex' approach, which takes account of stratification, clustering and weights, suggests that failure to take account of clustering will lead to an overestimation of precision, whilst the inclusion of weights affects both the magnitude of the point estimates and their estimated variance. In this example the impact of weighting on our substantive conclusions is fairly minimal, perhaps because of the relative size of the weights. For this subgroup, the weights had a median of 0.85, a lower quartile of 0.68 and an upper quartile of 1.17. They range from 0.18 to 1.78 and hence do not reach the trimming level of 2.5 used in BHPS (Taylor *et al.*, 2007). More consideration of the need to adjust for weights, and their likely impact on the results, is required when they have a larger variance.

ACKNOWLEDGEMENTS

We thank the Editor and two referees for their helpful comments. Data from the British Household Panel Survey were collected by the Institute for Social and Economic Research, University of Essex and are made available by the UK Data Archive. This work arises from the Modelling Attitude Stability and Change using Repeated Measures Data project funded by the UK Economic and Social Research Council Research Methods Programme, grant number H333250026.

REFERENCES

Agresti, A. (2002). *Categorical Data Analysis* (2nd edn.). Hoboken, NJ: John Wiley & Sons, Inc.

Berrington, A., Hu, Y., Smith, P. W. F. and Sturgis, P. (2008). A graphical chain model for reciprocal relationships between women's gender role attitudes and labour force participation. *Journal of the Royal Statistical Society, Series A*, 171, 89–108.

Binder, D. A. (1983). On the variances of asymptotically normal estimators from complex surveys. *International Statistical Review*, 51, 279–292.

Bollen, K.(1989). *Structural Equations with Latent Variables*. New York: John Wiley & Sons, Inc.

Cox, D. R. and Wermuth, N. (1996). *Multivariate Dependencies*. London: Chapman & Hall.

Crompton, R. and Harris, F. (1998). Explaining women's employment patterns: 'orientations to work' revisited. *British Journal of Sociology*, 49, 118–136.

Evans, G. and Anderson, R. (2006). The political conditioning of economic perceptions. *Journal of Politics*, 68, 194–207.

Finkel, S. (1995). *Causal Analysis with Panel Data*. London: Sage.

Hakim, C. (1998). Developing a sociology for the twenty-first century: preference theory. *British Journal of Sociology*, 49, 137–143.

Lesthaeghe, R. and Moors, G. (2002). Life course transitions and value orientations: selection and adaptation. In R. Lesthaeghe (Ed.), *Meaning and Choice: Value orientations and life course transitions* (pp. 1–44). The Hague: NIDI-CBGS.

Magadi, M. Diamond, I., Madise, N. and Smith, P. W. F. (2004). Pathways of the determinants of unfavourable birth outcomes in Kenya. *Journal of Biosocial Science*, 36, 153–176.

Mohamed, W. N., Diamond, I. and Smith, P. W. F. (1998). The determinants of infant mortality in Malaysia: a graphical chain modeling approach. *Journal of the Royal Statistical Society, Series A*, 161, 349–366.

Muthén, B. O. (2001). Second-generation structural equation modeling with a combination of categorical and continuous latent variables: New opportunities for latent class/latent growth modeling. In L. M. Collins and A. Sayer (Eds.), *New Methods for the Analysis of Change* (pp. 289–322) Washington, DC: American Psychological Association.

Muthén, L. K. and Muthén, B. O. (1998–2007). *Mplus User's Guide* (4th edn.). Los Angeles, CA: Muthén & Muthén.

Scientific Software International (2006). *LISREL, Version 8.8*. http://www.ssicentral.com/lisrel/index.html.

Skinner, C. J. (1989). Domain means, regression and multivariate analysis. In C. J. Skinner, D. Holt, and T. M. F. Smith, (Eds), *Analysis of Complex Surveys* (pp. 59–87). Chichester: John Wiley & Sons, Ltd.

Skinner, C. J., Holt, D. and Smith, T. M. F. (1989). *Analysis of Complex Surveys*. Chichester: John Wiley & Sons, Ltd.

Stapleton, L. M. (2006). An assessment of practical solutions for structural equation modeling with complex sample data. *Structural Equation Modeling*, 13, 28–58.

Stata Corp (2007). *Survey Data Reference Manual, Version 10*. College Station, TX: Stata Corporation.

Taylor, M. F., Brice, J., Buck, N. and Prentice-Lane, E. (2007). *British Household Panel Survey User Manual Volume A: Introduction, Technical Report and Appendices*. Colchester: University of Essex.

Whittaker, J. (1990). *Graphical Models in Applied Multivariate Statistics*. Chichester: John Wiley & Sons, Ltd.

Williams, L. J. and Podsakoff, P. M. (1989). Longitudinal field methods for studying reciprocal relationships in organizational behavior research: toward improved causal analysis. In B. Staw, and L. Cummings (Eds.), *Research in Organizational Behavior* (pp. 247–292). Greenwich: JAI Press Inc.

Index

Methodology of Longitudinal Surveys P. Lynn

WILEY SERIES IN SURVEY METHODOLOGY

Established in Part by WALTER A. SHEWHART AND SAMUEL S. WILKS

Editors: *Robert M. Groves, Graham Kalton, J. N. K. Rao, Norbert Schwarz,* Christopher Skinner

The ***Wiley Series in Survey Methodology*** covers topics of current research and practical interests in survey methodology and sampling. While the emphasis is on application, theoretical discussion is encouraged when it supports a broader understanding of the subject matter.

The authors are leading academics and researchers in survey methodology and sampling. The readership includes professionals in, and students of, the fields of applied statistics, biostatistics, public policy, and government and corporate enterprises.

*BIEMER, GROVES, LYBERG, MATHIOWETZ, and SUDMAN · Measurement Errors in Surveys
BIEMER and LYBERG · Introduction to Survey Quality
BRADBURN, SUDMAN, and WANSINK · Asking Questions: The Definitive Guide to Questionnaire Design—For Market Research, Political Polls, and Social Health Questionnaires, *Revised Edition*
BRAVERMAN and SLATER · Advances in Survey Research: New Directions for Evaluation, No. 70
CHAMBERS and SKINNER (editors · Analysis of Survey Data
COCHRAN · Sampling Techniques, *Third Edition*
COUPER, BAKER, BETHLEHEM, CLARK, MARTIN, NICHOLLS, and O'REILLY (editors) · Computer Assisted Survey Information Collection
COX, BINDER, CHINNAPPA, CHRISTIANSON, COLLEDGE, and KOTT (editors) · Business Survey Methods
*DEMING · Sample Design in Business Research
DILLMAN · Mail and Internet Surveys: The Tailored Design Method
GROVES and COUPER · Nonresponse in Household Interview Surveys
GROVES · Survey Errors and Survey Costs
GROVES, DILLMAN, ELTINGE, and LITTLE · Survey Nonresponse
GROVES, BIEMER, LYBERG, MASSEY, NICHOLLS, and WAKSBERG · Telephone Survey Methodology
GROVES, FOWLER, COUPER, LEPKOWSKI, SINGER, and TOURANGEAU · Survey Methodology
*HANSEN, HURWITZ, and MADOW · Sample Survey Methods and Theory, Volume 1: Methods and Applications
*HANSEN, HURWITZ, and MADOW · Sample Survey Methods and Theory, Volume II: Theory
HARKNESS, VAN DE VIJVER, and MOHLER · Cross-Cultural Survey Methods
KALTON and HEERINGA · Leslie Kish Selected Papers
KISH · Statistical Design for Research
*KISH · Survey Sampling
KORN and GRAUBARD · Analysis of Health Surveys
LESSLER and KALSBEEK · Nonsampling Error in Surveys
LEVY and LEMESHOW · Sampling of Populations: Methods and Applications, *Third Edition*
LYBERG, BIEMER, COLLINS, de LEEUW, DIPPO, SCHWARZ, TREWIN (editors) · Survey Measurement and Process Quality
MAYNARD, HOUTKOOP-STEENSTRA, SCHAEFFER, VAN DER ZOUWEN · Standardization and Tacit Knowledge: Interaction and Practice in the Survey Interview

*Now available in a lower priced paperback edition in the Wiley Classics Library.

PORTER (editor) · Overcoming Survey Research Problems: New Directions for Institutional Research, No.121

PRESSER, ROTHGEB, COUPER, LESSLER, MARTIN, MARTIN, and SINGER (editors) · Methods for Testing and Evaluating Survey Questionnaires

RAO · Small Area Estimation

REA and PARKER · Designing and Conducting Survey Research: A Comprehensive Guide, *Third Edition*

SÄRNDAL and LUNDSTRÖM · Estimation in Surveys with Nonresponse

SCHWARZ and SUDMAN (editors) · Answering Questions: Methodology for Determining Cognitive and Communicative Processes in Survey Research

SIRKEN, HERRMANN, SCHECHTER, SCHWARZ, TANUR, and TOURANGEAU (editors) · Cognition and Survey Research

SUDMAN, BRADBURN, and SCHWARZ · Thinking about Answers: The Application of Cognitive Processes to Survey Methodology

UMBACH (editor) · Survey Research Emerging Issues: New Directions for Institutional Research No. 127

VALLIANT, DORFMAN, and ROYALL · Finite Population Sampling and Inference: A Prediction Approach